Nuclear Science and Technology

This interdisciplinary book series publishes monographs and edited volumes in all areas of nuclear science and technology, covering developments in theoretical research, experiments, technology and applications. The series attempts to bring together experts and professionals from all over the world, to provide readers with a library on a wide range of topics in physics, engineering, medicine and environmental sciences. Topics include but are not limited to:

Nuclear physics and interdisciplinary research;
Nuclear energy science and engineering;
Nuclear fusion and high temperature plasmas;
Synchrotron radiation applications, beamline technology;
Accelerator, ray technology and applications;
Nuclear chemistry, radiochemistry, radiopharmaceuticals, nuclear medicine; and
Nuclear electronics and instrumentation.

The books are intended for researchers, professionals, and students in the field, as well as those interested in staying current with the latest advances.

Yoshiaki Oka

Nuclear Power and Society

Issues on Economy, Safety, Environment and Law

Springer

Yoshiaki Oka
Emeritus Professor, Graduate School
of Engineering
The University of Tokyo
Bunkyo, Tokyo, Japan

ISSN 2948-1856 ISSN 2948-1864 (electronic)
Nuclear Science and Technology
ISBN 978-981-97-8950-4 ISBN 978-981-97-8951-1 (eBook)
https://doi.org/10.1007/978-981-97-8951-1

The original submitted manuscript has been translated into English. The translation was done using artificial intelligence. A subsequent revision was performed by the author(s) to further refine the work and to ensure that the translation is appropriate concerning content and scientific correctness. It may, however, read stylistically different from a conventional translation.

Translation from the Japanese language edition: "Genshiryoku Hatsuden to Shakai: Keizai, Anzen, Haikibutsu, Houseido no Kadai" by Yoshiaki Oka, © Yoshiaki Oka 2024. Published by Amazon. All Rights Reserved.

This Springer imprint is published by the registered company Springer Nature Singapore Pte Ltd.
The registered company address is: 152 Beach Road, #21-01/04 Gateway East, Singapore 189721, Singapore

Preface

Will economic liberalization lead to the supply of cheap electricity?

Why has nuclear power come to be considered the most dangerous form of power generation?

Is radioactive waste an unsolvable problem?

The use of technology carries risks, but how safe is safe enough, and how should we make that judgment?

How did the international agency's view that "radiation exposure should be kept as low as reasonably achievable" affect the risks associated with the evacuation from the TEPCO Fukushima accident?

What happened as a result of the nuclear proponents explaining safety to the public?

What form of power generation was built in large numbers due to economic liberalization? What happened to the cost of power generation as a result?

How did the U.S. nuclear power industry achieve both safety and economic improvements after the accident at Three Mile Island?

Is the "safety technology" explained by nuclear engineers the same as the "safety" understood by the public?

Nuclear power has been in use for over half a century, but there are many issues at its intersection with society. This book is the world's first to comprehensively address and examine these issues.

While nuclear power often causes a stir in society, most of these issues are related to its intersection with society. The information about nuclear power that is circulating in society, including that intended for the general public, is often fragmented and biased. The nuclear power sector in Japan is collectivist, so the information may not always be robust.

Unlike many general books on nuclear power, this book does not simply compile the author's arguments. Instead, it gathers information based on facts and reliable evidence, such as reports from Western government agencies, and presents the results of inductive and logical considerations based on these. It does not deduce from a proposition, such as the proposition that nuclear power should be used. The issues addressed in this book cannot be deductively considered based on laws derived from the truths of nature, as is the case with scientific disciplines.

As a specialized book, it is the first in the world to quantitatively show the risks other than radiation exposure during a nuclear accident (risks associated with evacuation). As a policy, it recommends not applying the ALARA principle of the International Commission on Radiological Protection to the public in low-dose areas during a nuclear accident, and the need for a cost-benefit approach to Japan's nuclear safety, regulation, and disaster prevention, in order to learn from the lesson of increased disaster-related deaths due to the delayed lifting of evacuation orders in the TEPCO Fukushima accident. It introduces statistics on the number of public deaths associated with accidents in various power generation methods and research results in risk psychology, warning that objective safety and subjective safety will never logically intersect. If they are made to intersect, it will convey an image of danger.

Although it is a large book, by reading the conclusion of Chap. 5 and the summary and consideration sections at the end of each chapter, one can understand the overview. Busy readers can start reading from these sections. The intended readership includes not only the general public but also politicians, administrative officials, media personnel, those interested in issues related to nuclear power, energy, environment, waste, safety, communication, economics, legal systems, etc., nuclear power-related personnel, university faculty, students, and graduate students.

Chapter 1 describes the history and current status of nuclear power use. For example, it is not well known, but in the United States, economic liberalization from the 1980s had a significant impact on new power source investments. The actual economic performance of hydropower and nuclear power is different from the economic performance of investments, so as a result of liberalization, American consumers ended up having to buy expensive electricity. This is explained by showing data on actual power costs.

Chapter 2 discusses safety and communication. This chapter summarizes the overview, progress, and countermeasures of the TEPCO Fukushima accident, as well as the lessons learned after ten years since the accident, enabling a comprehensive understanding of the accident. For the first time in the world, it analyzes the data of disaster-related deaths from the TEPCO Fukushima accident from the perspective of evacuation risks and benefits. Based on these results, it states that the internationally used concept of radiation protection, ALARA which is "as low as reasonably achievable of avoiding radiation exposure", has prolonged evacuation and increased the number of disaster-related deaths.

Regarding communication, it introduces the results of risk perception studies in the United States in the 1980s, stating that the subjective safety understood by the public and the objective safety understood by experts (safety technically achieved) are logically different. Humans are instinctively sensitive to risk information, and safety (risk) information leaves a three times stronger impression. It warns that without realizing this, the more experts talk about the safety of nuclear power to the public, the more they implant the image that nuclear power is dangerous.

Chapter 3 discusses the decommissioning of nuclear facilities and radioactive waste. It states that waste treatment and disposal is an activity of a circular society, and radioactive waste is part of industrial waste. It introduces the decommissioning and waste disposal of nuclear facilities in Western countries. It states that the treatment and disposal of radioactive waste are a series of continuous activities. It also introduces examples and lessons from Finland, Sweden, France, the United States, and Japan, regarding geological disposal.

Chapter 4 is about the legal framework for the use of nuclear power, discussing non-proliferation and safeguards, nuclear damage compensation system, structure and thinking of the Atomic Energy Act, laws and organizations related to safety regulations, and safety review of nuclear power plants in Japan. Not limited to nuclear power generation, technology is used because it has benefits. The benefits of use should outweigh the risks of use, which is the basic idea of the Atomic Energy Act. This concept has been incorporated into US nuclear regulations and emergency response for a long time, but it is not in Japan's nuclear regulations, so it points out that improvement is necessary.

Chapter 5 is the conclusion, summarizing each chapter and the author's observations. At the end, it states that Japan does not have a legal system where an independent institution with investigative authority investigates the accountability of administrative agencies (explanation of policies by documents and responsibility for the results) in all fields, not limited to nuclear power, and publishes the report. It states that this is a self-evaluation of each ministry and agency in Japan. It points out that this may be the root cause of Japan's long-term stagnation, as it encourages dependence on the country's system and budget by corporations and others.

Bunkyo, Japan Yoshiaki Oka

Introduction

There are many issues at the intersection of nuclear power generation and society, but so far, they have been discussed only in a fragmented and one-sided manner. This book is the world's first to comprehensively address and examine these issues.

In this book, while presenting information based on the history of nuclear power utilization and related information from Western government agencies, etc., these have been organized. Chapter 1 deals with the intersection of nuclear power and economics, Chap. 2 with the intersection of nuclear power and humanities, Chap. 3 with the intersection of nuclear power and environment, and Chap. 4 with the intersection of nuclear power and law, albeit modestly. Since it is written without using nuclear-specific terms, it should be readable for anyone interested in nuclear power generation.

This book is written not only for the general public but also for administrators, politicians, and media personnel. In Chap. 5, the author expressed his opinion on nuclear power generation in Japan as a conclusion. Each chapter has a summary and consideration section at the end. There is also a way to start reading from these.

This book should be a reference for professionals already involved in nuclear power utilization. The themes covered in this book are likely of interest to many experts, but they probably have not investigated and deeply considered them themselves. Understanding should deepen by reading the cited references and thinking for oneself, rather than swallowing the author's findings whole. The exercise problems should serve as a trigger for problem recognition.

In nuclear engineering education at universities, the theory and technology of nuclear reactors are taught, but the issues at the intersection with society are not taught. This is a drawback of compartmentalized education, and this book can be used as a textbook, exercise book, and supplementary reader. Since the description of each theme is concise, it would be good to use reference materials in conjunction with this book when using it as teaching material.

The author has been interested in the issues at the intersection of nuclear power generation and society since he was a university professor, but at that time, his job was research and education in nuclear engineering, so he could not fully consider it. After the accident at Tokyo Electric Power Company's Fukushima Daiichi Nuclear Power

Plant in 2011, the author was appointed chairman of the reorganized Japan Atomic Energy Commission of the Cabinet Office in 2014, searched in English, understood the situation in the West through business trips, etc., and thought about this issue. Even when the author verbally told the results to those involved in nuclear power, they were not always understood. There is a lot of information about nuclear power generation circulating in society, including fragmented and biased information. The author thought it necessary to compile it into a book. He has had the opportunity to encounter the transition of nuclear power utilization, mainly in the West, over about 40 years. Based on that experience, the author wanted to convey the results of his research, along with his considerations, to the next generation, which was also a motivation for writing this book.

Each chapter is based on an independent theme, and at the end of each chapter, a section titled "Summary and Discussion" is provided, summarizing the chapter and presenting the author's thoughts. The sections prior to this mainly consist of organized descriptions of the collected information, although there are parts where the author's opinion is added. The author considered whether it would be better to separate his opinion from the research results, but decided against it as separating them would make the issues and discussions become general and abstract, resulting in a lack of focus. The references in this book are public documents. It would be appreciated if the reader does not simply accept the author's discussions as they are, but thinks about them based on the references. Even if the reader disagrees with the author after reading the references, it would be appreciated if they search in English and look for evidence to form their own thoughts.

Each chapter deals with a broad theme that could be a book in itself and is independent. There are scattered repetitions, but they are not organized. In social issues, one fact can be related to multiple fields. At the end of Chap. 5, the author discusses the issue of accountability in Japanese administration, which is not a problem unique to nuclear power. This issue was noticed while reading reports from Western accounting inspection agencies and is perceived as the root cause of Japan's prolonged stagnation. What do the readers think? At the end of each chapter, exercise questions are provided. It would be appreciated if they are used for discussion. The author researched and discussed the history of "Nuclear power reactor development" and prepared a manuscript, but decided to make it a separate book.

Contents

About the Author

Yoshiaki Oka Emeritus Professor at the University of Tokyo. Former Chairman of the Japan Atomic Energy Commission (2014–2020). Since taking charge of the Commission, he has been examining the issues at the intersection of nuclear power usage and society.

He graduated from the Department of Nuclear Engineering, Faculty of Engineering, University of Tokyo in 1969, completed master's and doctoral programs at the University of Tokyo, and received Ph.D. In 1974, he became an assistant at Nuclear Engineering Research Laboratory of the University of Tokyo, located in Tokai-mura, Ibaraki-prefecture. Then he was promoted as a lecturer, and associate professor, and in 1989, a professor. He was engaged in research and education in nuclear engineering at the University of Tokyo for 36 years. He was also in charge of the operation and joint use of the University of Tokyo's nuclear reactor facility. His specialty is conceptual nuclear reactor design. The supercritical-pressure light water-cooled reactor (SCWR) which he developed the concept was selected as a Generation IV nuclear reactor and is being researched worldwide. In 2005, he led the establishment of the Professional Nuclear School and Department of Nuclear Engineering and Management of the University of Tokyo. He led the creation of the *Nuclear Engineering Textbook Series* in Japanese (Ohm-sha) and *An Advanced Course in Nuclear Engineering* in English (Springer) and also edited and authored them by himself. In 2007, he was the leader of the Ministry of Education, Culture, Sports, Science and Technology's Global COE Program in the field of nuclear energy. In 2010, he retired from the University of Tokyo and was engaged in the establishment and operation of the Joint Department of Nuclear Engineering at the Waseda University. From 2014 to 2020, he served as the chairman of Japan Atomic Energy Commission. He served as the President of the Atomic Energy Society of Japan in 2008, a member of Board of Directors of American Nuclear Society between 2001 and 2004. His books include *Nuclear Reactor Design* (Springer 2014), *Super Light Water Reactors and Super Fast Reactors* (Springer, 2010), and several others in both Japanese and English. He has published about 230 peer-reviewed research papers in English. ScholarGPS™ ranks him the second as the scholar in "nuclear reactors" and 16th in "nuclear power".

List of Figures

Chapter 1
Nuclear Utilization

In this chapter, we first discuss nuclear science, which is the foundation of nuclear utilization. Then, we discuss the use of radiation, another field of nuclear utilization alongside energy use. We look back at the history of human energy use and discuss energy self-sufficiency and energy security. We discuss the economics of various power generation methods and examine the relationship between economic liberalization, power liberalization, investment in new power generation facilities, and stable power supply. Furthermore, we introduce the history and current status of nuclear energy use in the world, the development of power reactors, the continuous safety improvement activities in the power industry, the improvement of nuclear regulations in the United States, the new construction of nuclear power plants, nuclear power technology from the history of innovation, and computer-aided design, construction, and maintenance technology.

1.1 Nuclear Science and the Nobel Prize

The universe was born from a nuclear reaction called the Big Bang about 13.8 billion years ago. Stars, the sun, and the Earth exist as a result of this. The light from stars and the sun is generated by nuclear reactions. The energy brought to Earth by sunlight is the source of Earth's biological activity. The interior of the Earth is heated by the heat generated when unstable atomic nuclei emit radiation and change into stable atomic nuclei. This energy is released as volcanic activity and hot springs.

Since the Big Bang occurred a long time ago, most of the elements on Earth are already stable atomic nuclei, but there are still a few unstable atomic nuclei that emit radiation. For example, uranium and potassium-40. Uranium is an unstable atomic nucleus that emits radiation and repeatedly changes into other atomic nuclei, eventually becoming a stable atomic nucleus, lead. Radium and radon are unstable atomic nuclei along the way. Uranium is contained in small amounts in rocks and

Y. Oka, *Nuclear Power and Society*, Nuclear Science and Technology,
https://doi.org/10.1007/978-981-97-8951-1_1

sand, so radon is generated from the ground. Since radon is a gas, it is contained in the air and is taken into the human body by breathing, and radon is the main cause of radiation exposure from naturally occurring radiation. Potassium-40 is also a naturally occurring unstable atomic nucleus. Since potassium is a constituent element of human muscles, radiation is also slightly emitted from our bodies.

Furthermore, radiation pouring down from the universe reacts with the stable atomic nuclei of the Earth's atmosphere and water vapor, and new unstable atomic nuclei (for example, carbon-14 and tritium) are generated, which emit radiation.

Carbon-14 is generated from nitrogen-14, a constituent element of the atmosphere. Carbon-14 is contained in carbon dioxide, and animals and plants take it in during their lifetime, so by measuring its remaining proportion, we can know the age of fossils. Tritium is an isotope of hydrogen and is used as a tracer to investigate the movement of underground water. The field of nuclear power utilization is shown in Table 1.1, and the types of radiation are shown in Fig. 1.1.

Humans have been using the energy that the sun brings to Earth since ancient times. It was not until the twentieth century that we were able to artificially generate nuclear reactions on the ground and utilize their energy and radiation. Research on nuclear science has been conducted since the end of the nineteenth century, and its mechanism has been gradually elucidated, and we have been able to utilize nuclear reactions and the resulting radiation.

The Nobel Prize began to be awarded in 1901, and the first recipient of the Physics Prize was Roentgen, who discovered X-rays. The Curies won the Physics Prize in 1903 for their research on radioactivity. Their daughter, Irène Joliot-Curie and her husband Frédéric Joliot-Curie, won the Chemistry Prize in 1935 for their research on artificial radioactive isotopes. Japanese people such as Hideki Yukawa and Masatoshi Koshiba have contributed to the development of nuclear science through theoretical physics and experiments and have received awards.

Table 1.1 Field of nuclear utilization

Radiation utilization: diagnosis, cancer treatment, sterilization, analysis, material development, processing, non-destructive testing, exploration, precision measurement, dating, breed improvement, pest control, food preservation, scientific research etc.
Energy utilization: nuclear power generation, ship propulsion, district heating, power sources for deep space exploration, etc.

Fig. 1.1 Types of radiation

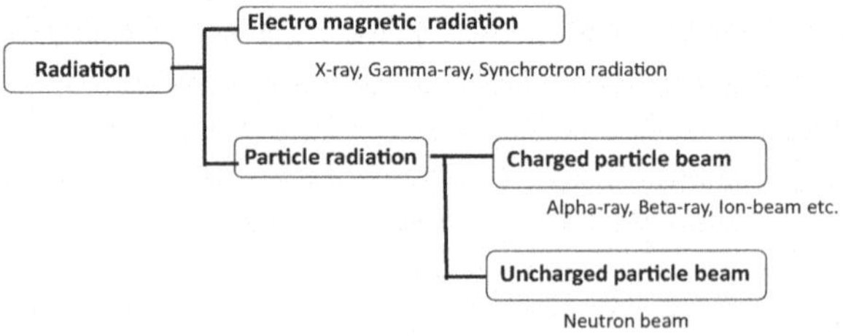

Through research in radiation, radioactive materials, and theoretical physics, nuclear science has developed, enabling mankind to utilize nuclear energy and radiation. Madame Curie used natural radioactive materials in her research, but with the invention of accelerators and nuclear reactors, it became possible to artificially generate radioactive materials and produce radiation. Nobel laureates in these fields account for about 25% of the physics prize and about 15% of the three prizes in natural sciences (physics, chemistry, medicine/physiology), indicating that research in the nuclear field has had a significant impact on the development of science and technology (JAEC 2017, p. 98).

1.2 Radiation Usage: Radiation is the New Light Acquired by Mankind

Light (visible light) separates into seven colors when it passes through a prism. The spectrum is what light is divided into by its wavelength. The seven colors of light each have different wavelengths, and since they are divided by wavelength, they are a spectrum. The spectrum obtained varies depending on the material of the prism, so by measuring the spectrum, we can learn about the characteristics (properties) of the prism as a material. The method of studying the properties of a substance by shining light on it and measuring the spectrum of its reflection or transmission is called spectroscopy.

Since it was discovered in the nineteenth century that light (visible light) is a type of electromagnetic wave, it has become possible to investigate the properties of matter using electromagnetic waves of various wavelengths (frequencies). Furthermore, in the twentieth century, accelerators were invented, making it possible to also use high-speed particles. Radiation is a general term for high-energy (short wavelength) electromagnetic waves and high speed particles. There are various types of particles, including electrons, positrons, protons, neutrons, muons, K mesons, ions, etc., and their interactions with matter are also diverse and unique to each particle. By irradiating matter with not only high-energy electromagnetic waves but also high-speed particles and measuring their reflection or absorption, it has become possible to understand the properties of matter from various perspectives. The light used has expanded from visible light to radiation, dramatically expanding the scope and capabilities of spectroscopy. *Radiation can be considered as a new light.*

Various types of accelerators have been developed that can not only accelerate particles with charges, such as electrons, protons, and ions but can also generate high-energy electromagnetic waves using accelerated electrons. Neutrons do not carry a charge, but they can be generated by causing nuclear reactions by colliding accelerated protons or ions with target materials. Neutrons can also be extracted from research reactors as neutron beams. These high-energy electromagnetic waves and particle beams can be irradiated onto materials, and their properties can be investigated by measuring scattering and absorption. Among high-energy electromagnetic waves, the electromagnetic waves radiated in the tangential direction when electrons or positrons close to the speed of light are bent in a magnetic field are called synchrotron radiation. Laser light is a powerful light (electromagnetic wave) of almost a single wavelength with excellent directivity and convergence, which can be generated using a laser. The properties of a material can also be investigated by irradiating it with synchrotron radiation or laser light.

The flow of high-speed particles or high-energy electromagnetic waves in the form of a beam in the same direction is called a "quantum beam". There are more than 20 major quantum beam facilities in Japan, which are used for measuring the properties of various materials and for material improvement. Large facilities are available for joint use and are widely used from product development by private companies to basic research. Figure 1.2 shows the large synchrotron radiation facility SPring-8 located in the Harima Science Park City in Hyogo Prefecture. Electrons are generated and accelerated in a linear accelerator, further accelerated to energies close to the speed of light in an accelerator called a synchrotron and then injected into a storage ring and stored by circulating. When high-speed electrons are bent by the magnetic field of the magnets in the storage ring, high-energy electromagnetic waves with directivity, called synchrotron radiation, are radiated in the direction of electron travel (tangential direction of the storage ring). This synchrotron radiation is used to analyze various materials, and a wide range of research is being conducted from nanotechnology and biotechnology to industrial applications. Synchrotron radiation, which was considered a nuisance accompanying large accelerator experiments when it was discovered, has now become a wonderful method that can analyze and evaluate the characteristics of various materials by generating synchrotron radiation using large accelerators. SPring-8 is a joint-use facility open to researchers from industry, academia, and government both in Japan and abroad, and anyone can use it if they apply and are selected. It is operated by the RIKEN, a national research and development corporation. In addition to this, there are synchrotron radiation facilities in Aichi, Saga, Hiroshima, Shiga, and Ibaraki prefectures. Furthermore, in response to the increasing demand for synchrotron radiation, a next-generation synchrotron radiation facility is being built in Sendai as a project by a public-private regional partnership, aiming to start operation in 2024. National Institutes for Quantum Science and Technology (QST) is in charge of its development and operation.

The high-intensity proton accelerator facility, J-Park, is located in Tokai Village, Ibaraki Prefecture, and produces secondary particle beams such as neutrons, muons, neutrinos, and K mesons from high-intensity proton beams accelerated to nearly speed of light. It is a facility that conducts research to explore the origins of matter and life from the observation of atomic and molecular structures, and to solve the mysteries of the universe from the study of elementary particles and atomic nuclei. J-Park consists of a linear proton accelerator, multiple synchrotrons, which are circular accelerators, and a group of experimental facilities, and is available for joint use as well as for researchers at J-Park to conduct their own research. J-Park is a joint project of the Japan Atomic Energy Agency and QST. Neutron beams can also be

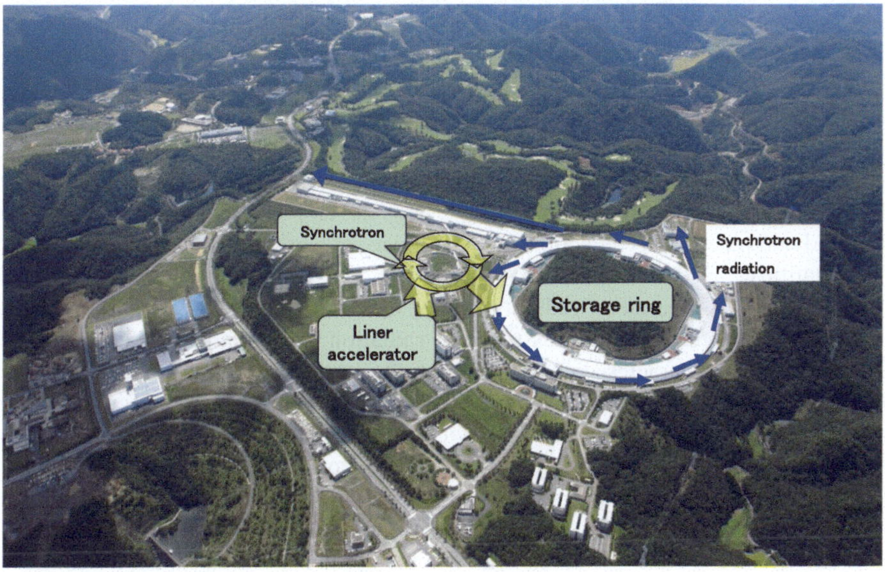

Fig. 1.2 Spring-8: synchrotron radiation facility. *Source* SPring 8

generated in research reactors, and are used in material research at research reactors of the Japan Atomic Energy Agency and Kyoto University.

Accelerators are used not only for material research but also for basic scientific research to explore the origins of the universe and matter, and particle physics research is conducted at the High Energy Accelerator Research Organization (KEK) in Tsukuba City. Research and applications investigating chemical reactions by irradiating materials with radiation, known as radiation chemistry, are conducted at Takasaki Institute for Advanced Quantum Science of QST in Takasaki City, Gunma Prefecture. Cancer treatment using accelerated ions is conducted at QST hospital n in Chiba City. Accelerators are advancing in acceleration methods, and it is believed that smaller ones will soon be used for treatment in city hospitals. Research and application fields that require large accelerators are also expected to have new developments and various applications due to the miniaturization of accelerators.

Quantum beams can not only observe the structure of matter at the atomic and molecular level but also use the energy of quantum beams to process matter at the atomic and molecular level. Furthermore, they can be used for treatments at the cellular level, such as cancer treatment. Because radiation interacts with matter, it is widely used in fields such as medicine, industry, agriculture, environmental conservation, and academia. Its economic scale is larger than that of nuclear energy use (Okada 2017; JAEC 2019).

In the medical field, X-rays are used for diagnosing fractures and dental issues through radiography. Subsequently, devices that analyze and display cross-sectional images of the body using a computer (computer tomography, CT) were developed and have become widely used worldwide. Japan has the highest share of CT devices in the world. Furthermore, various methods such as Single Photon Emission Computed Tomography (SPECT), which detects radiation emitted from a drug containing a radioactive isotope administered into the body and concentrates in specific organs, and converts its distribution into cross-sectional images, Positron Emission Tomography (PET), Radioimmunoassay (RIA), and even Nuclear Magnetic Resonance (NMR), which does not use radiation, have been developed. Each method is utilized to its advantage for the detection and diagnosis of various cancers and physical abnormalities.

In the field of treatment, radiation therapy, which irradiates cancer with radiation, is performed using various types of radiation. The treatment methods vary depending on the characteristics of the radiation and the type of cancer. The method of implanting a small radiation source into the cancerous area is used for the treatment of prostate and breast cancer. A device called a Gamma Knife arranges a large number of radioactive isotopes that emit gamma rays around the head so that the direction of gamma ray emission from each source concentrates on the area to be treated, thereby treating brain tumors. Ion beams, unlike gamma rays, can concentrate their energy on the cancerous area even from a single source. In Japan, a treatment method using accelerated heavy ions has been developed at the Radiation Medical Science Center (now Institute for Quantum Medical Science of QST in Chiba) and has become covered by insurance for the treatment of prostate cancer. Furthermore, research and development are being conducted on treatment methods that accumulate radioactive substances that emit alpha rays in cancer and perform diagnosis and treatment simultaneously.

Sterilization of medical devices is performed by radiation. Since sterilization can be done after packaging the instruments, it is cheaper and safer than methods using heat sterilization or chemical substances. In the development of pharmaceuticals, isotopes that emit radiation are used to investigate which organs or affected areas the pharmaceuticals accumulate in. In university research, radioactive isotopes are widely used in medical and biological fields.

Radiation allows us to observe substances. By observing, we can clarify mechanisms and create materials with new functions. For example, radiation is used in the development of high-temperature superconducting materials, magnetic materials, interface functional materials, energy-saving materials, hydrogen storage materials, lithium-ion batteries, space microprocessors, and genetic reagents.

In the industrial field, radiation is used not only for observation and creation, but also for processing. Radiation generators are used for material modification and creation, precision measurement, and non-destructive testing. In material modification and creation, it is used in the manufacture of semiconductors, tires with excellent wear resistance, button batteries and fuel cell membranes, ultra-heat-resistant silicon carbide, wound covering materials, and the development of useful and harmful metal capture materials. Examples of precision measurement include measuring the thickness of metal plates and thin films and the level of water, while examples of non-destructive testing include diagnosing the deterioration of structures, flaw detection, oil exploration, and airport baggage inspection. Furthermore, in the field of environmental conservation, it is useful for the decomposition and removal of nitrogen oxides and sulfur oxides, and the decomposition of organic compounds such as dioxins.

In the field of science and technology, the structural analysis of the atomic arrangement of materials and imaging using radioactive isotopes are contributing to the development of academia and technology. The dating of fossils can be done by measuring the remaining ratio of radioactive carbon-14 to the stable isotope carbon-12. Carbon-14 is generated by the reaction with cosmic rays and taken in by plants and animals while they are alive through the atmosphere and water. The long-term movement of groundwater in strata can also be measured by the ratio of carbon-14.

In the field of agriculture, radiation is used for variety improvement using mutations, pest control, and food irradiation. Variety improvements such as increased yields of rice, wheat, cotton, and fruit are being made by radiation breeding. In pest control, the melon fly in Okinawa has been eradicated by the sterile insect release method (a method of raising males sterilized by radiation and releasing them into the wild), making it possible to sell its fruit in mainland Japan. The sterile insect release method has also been successful in Mexico, enabling the export of fruit to the United States.

Food irradiation with radiation is only used for preventing sprouting of potatoes in Japan, but it has been widely used in the United States for a long time. Research began following President Eisenhower's "Atoms for Peace" speech in 1953, and in the 1960s, the U.S. Food and Drug Administration (FDA) approved its use for pest control in wheat and flour. In the 1980s, it was applied to fruits, vegetables, spices, pork, and poultry, and in the 1990s, it was applied to ground meat and beef and used to remove pests and pathogens. In 1992, the World Health Organization (WHO) recommended food irradiation as an "ideal food preservation technology". Sterilization by radiation is superior to methods using refrigeration or preservatives, and its wider use is being called for in Japan.

While radiation is used in various ways in modern society, this book deals with the use of nuclear energy, so I will not discuss it further. Those interested should refer to Chapter 7 of the White Paper on Nuclear Energy posted on the Atomic Energy Commission's website, or the book subtitled "In Search of Marie Curie's Dream" (Waltar 2004), furthermore, I would like you to refer to the websites of the institutes of QST Kansai Photon Science Institute, Large Synchrotron Radiation Facility (Spring-8), Japan Atomic Energy Agency's High Intensity Proton Accelerator Facility (J-PARK), and the Quantum Science and Technology Committee and Quantum Beam Utilization Promotion Subcommittee of the Ministry of Education, Culture, Sports, Science and Technology, which is the responsible ministry.

In order to further develop the use of radiation, it is necessary to understand both the bottleneck of radiation use and the experience of using radiation. The bottleneck of radiation use is a negative cycle where "due to excessive fear of radiation", "even if radiation is used, it is not said to be used", and therefore, "the benefits of radiation are not conveyed". For example, in Japan, sterilization and pest control by radiation irradiation of food is extremely limited compared to the United States. Not only what is already being done in the United States, but also from the perspective of environmental protection, radiation irradiation should be useful for reducing agricultural, aquatic, livestock products and food loss in the future.

Radiation exists in the natural environment. Radiation from space is largely attenuated in the atmosphere, but it reaches the ground. The radiation dose received by astronauts staying at the space station is much larger than on the ground. There are natural radioactive materials that emit radiation in the ground and on the surface. Worrying about the health effects of trace amounts of radiation is more harmful than beneficial. The knowledge about the health effects of radiation is abundant due to the experience of radiation and nuclear power use over a century from the early twentieth century. Since the TEPCO Fukushima accident, unified basic materials on the health effects of radiation, etc. have been created and published by the Ministry of the Environment, and general information has been enriched in Japan. Regulations on the use of radiation and nuclear power are based on scientific findings. The existence of regulatory agencies as independent agencies from the promotion side is also a characteristic of the nuclear and radiation field. It may be difficult for the general public to understand the scientific basis of the health effects of radiation. In that case, it might be good to trust the independent Nuclear Regulation Authority. We hope that these will eliminate the negative cycle of radiation use in Japan.

In other fields as well, as knowledge and technology advance, they are replaced by simpler methods, but the use of radiation and radioisotopes has repeatedly gone through a history of being replaced by other methods. The field of radiation is strictly regulated, so if there are other methods, they are easily replaced. For example, residential fire alarms used to use a radioactive source that emits alpha rays, and when the alpha rays could not reach the detector due to smoke, an alarm was triggered. Now, they use light-emitting diodes, and when the light is blocked by smoke, a signal is emitted. Radioisotopes have been widely used for disease elucidation and treatment by injecting them into the human body or experimental animals and measuring the accumulation in organs or lesions with a radiation detector, but due to the hassle of managing radioisotopes due to regulations, they are being replaced by fluorescence methods where applicable. Radioisotopes are still overwhelmingly used in the medical and biological fields, and in the radiation and nuclear power fields.

Methods using radiation have been the first opportunity to create new substances, but it is said that they have repeatedly gone through a history of the same substances being made by other methods. For example, fullerenes (made up of only carbon atoms connected in a chain to form a closed shell) were first made in very small amounts using a laser method. This discovery won the Nobel Prize, but later it was mass-produced by the arc discharge method. The blue light-emitting diode, for which a Japanese person won the Nobel Prize, was first made by Akasaki and Amano using a low-speed electron irradiation method, but the light was dim. Later, Nakamura achieved bright light emission using a high-temperature annealing method. This does not mean that radiation is not useful. Rather, it is necessary to pay attention to the fact that radiation and particle beams have played an indispensable role as a means of creating new substances (Okada 2017, p. 31). Not limited to this, new discoveries, new products, and performance improvements have been made using accelerators and radiation. For example, various achievements obtained using SPring-8 and J-Park should be referred to. *Radiation is a new light*, and its use is expected to continue to evolve without stopping. Radiation has a large negative image associated with exposure in the public eye, but *radiation is a "new light" that mankind has handled*, and it brings blessings in all fields of our lives, industries, and research and development. The use of radiation, taking advantage of its non-contact, non-destructive characteristics, is expected to be used and developed more and more in the future.

1.3 Energy Usage: Nuclear Energy is the New Fire Acquired by Mankind

Through the use of fire and tools, mankind has prospered. By these means, mankind reigns at the top of the food chain. Figure 1.3 shows the relationship between mankind and energy.

Primitive humans, born in East Africa a million years ago, survived mainly by gathering food from plants. Later, in Eastern Europe 100,000 years ago, there were hunter-gatherers who obtained food through hunting. They used fire to burn wood for heating and cooking. Mankind began to use fire and tools, and by doing so, has overwhelmed other creatures and reigned at the top of the food chain to this day. Around 5000 BC, agriculture began in the region from Mesopotamia, known as the Fertile Crescent, to Egypt. They cultivated grains and used the energy of

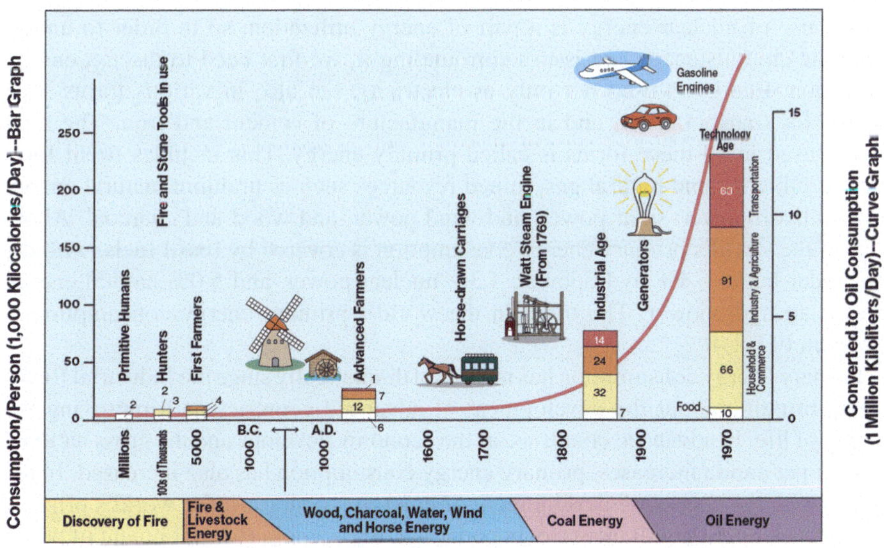

Fig. 1.3 Relationships between humans and energy. *Source* JAERO

livestock for this purpose. They can be called early agricultural people. Around 1400, in Northwestern Europe, there were people known as advanced agricultural people. They used coal for heating, water wheels and windmills for agriculture, and livestock for transportation. In the late eighteenth century, Watt invented the steam engine in Britain, and coal began to be used as energy. This led to the Industrial Revolution. By using fossil fuels as energy, tools could be used on a large scale. In the late nineteenth century, the invention of generators and alternating current transmission, and electric motors driven by electricity, made electricity a new convenient form of energy use. By using these energies, mankind has been able to enjoy various conveniences, health, and culture. Mankind has not only stood at the top of the food chain, but also gained health, culture, and convenience. In the twentieth century, oil replaced coal as an energy source. Natural gas is also used for power generation. Coal, oil, and natural gas are called fossil fuels because they are the energy of the sun stored in plants and animal plankton and fossilized. Fossil fuels are the main players in modern energy sources. Energy consumption has increased significantly since the use of fossil fuels. Nuclear energy has been used for nuclear power generation since the second half of the twentieth century. Hydroelectric power generation began shortly after the invention of the generator, in the late nineteenth century. The large-scale use of solar and wind power as electricity began in the 1980s.

Fossil fuels such as coal, oil, and natural gas, as well as natural energy such as sunlight, wind power, and hydropower, all originate from the energy of the sun, so all energy, not limited to nuclear power, is energy originating from nuclear reactions. The direct use of energy brought about by nuclear reactions, such as nuclear power generation, has only begun in the last half century. *Nuclear energy is the new fire that mankind has acquired*. There is potential for further development through future exploration.

The use of nuclear energy is a part of energy utilization, so in order to understand the circumstances and issues surrounding it, we first need to discuss energy utilization. Energy is used not only as electricity, but also in various forms such as fuel for transportation, and in the manufacture of cement and iron. The total energy used in all these forms is called primary energy. This includes fossil fuels such as oil, coal, and natural gas, mined resources such as uranium, natural energy such as hydropower, solar power, and wind power, and wood and charcoal. About 84% of the world's primary energy consumption is covered by fossil fuels, with the remainder being 6.4% hydropower, 4.3% nuclear power, and 5.0% natural energy other than hydropower. The trend in the world's primary energy consumption is shown in Fig. 1.4.

Primary energy consumption has increased dramatically since the Industrial Revolution, bringing about the development of advanced countries and improving the quality of life. In advanced countries, as the economy develops and the gross national product per capita increases, primary energy consumption has also increased. In the future, with the economic development of emerging countries, the world's primary energy consumption will increase. In particular, the economic development of populous countries such as China and India requires a large amount of energy. China is

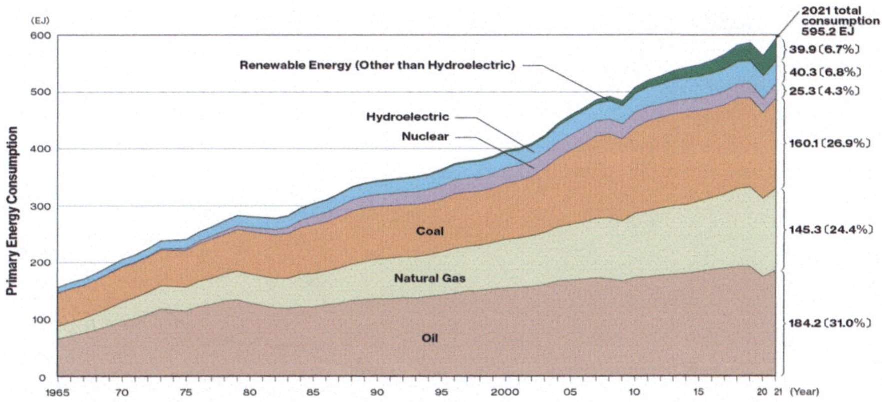

Fig. 1.4 Primary energy consumption in the world. *Source* JAERO

already consuming 22% of the world's primary energy due to its economic development. Most of the primary energy in China and India (86% in China, 92% in India, 2019) is supplied by fossil fuels such as coal. The United Nations predicts that Africa's population will increase in the second half of this century. Not only in China and India, but also for the development of African countries, a huge amount of primary energy will be needed in the future.

1.4 Energy Self-sufficiency and Energy Security

Over 80% of the world's primary energy is covered by fossil fuels. The issue of energy security became globally apparent due to Russia's invasion of Ukraine in 2022. In Europe, countries such as Germany and Italy were heavily dependent on natural gas supplies from Russia. Italy had abolished nuclear power plants after the Chernobyl accident and electricity rates were extremely high, but the cost of electricity was kept down because Russian natural gas was cheap. Germany had laid a submarine pipeline in the North Sea to import natural gas from Russia, providing cheap fossil fuel resources for industrial use and combining it with natural gas-fired power generation to compensate for the natural variation in renewable energy generation, aiming to stabilize power supply and expand the use of renewable energy. However, Russia invaded Ukraine, considering the increase in former Soviet Union countries joining NATO as a security threat. Economic sanctions against Russia by Western countries led to a global surge in prices of natural gas and other fossil fuels. In Japan, many nuclear power plants have been shut down since the TEPCO Fukushima accident, and most of the power supply depends on thermal power generation, so the effect of the weak yen and the rise in natural gas and coal prices have been a double punch, causing electricity rates to rise significantly in 2022.

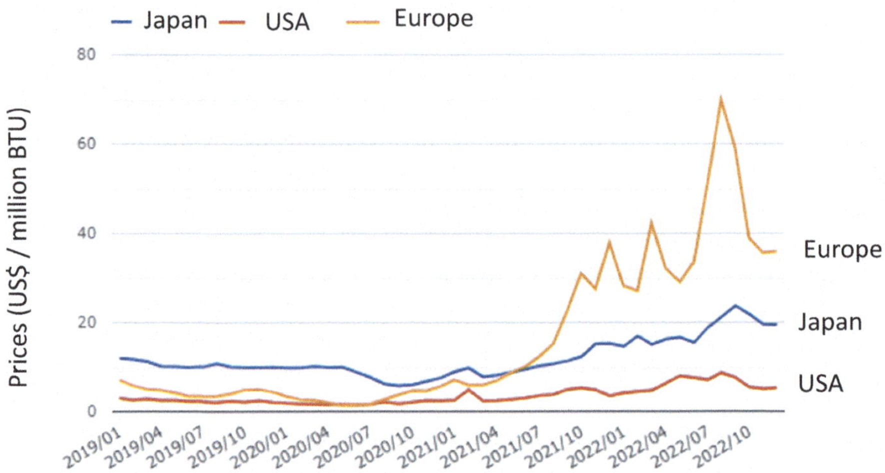

Fig. 1.5 Natural gas prices (2019–2022). *Source* https://ecodb.net/commodity/group_ngas.html. World Bank-Commodity Market (worldbank.org)

Figure 1.5 shows the natural gas prices (monthly changes) from 2021 to 2024. In 2022, European prices multiplied, and Japan's import prices also more than doubled. This is shown in US dollar terms, but the yen depreciated significantly in 2022, so the price in Japan in yen terms in 2022 is even higher. Figure 1.6 shows the crude oil prices (annual changes) from 1980 to 2022. Crude oil prices have been rising since around 2005, fell once in 2015, but are high again in 2022. Natural gas prices change similarly to crude oil prices, so their annual changes are similar to the annual changes in crude oil prices. Figure 1.7 shows the coal prices (annual changes) from 1980 to 2022. Coal prices also rose significantly in 2022.

Russia's main exports are natural gas, crude oil, and coal. After the Soviet Union collapsed in 1991, Russia went through a period of chaos for several years, with President Yeltsin transitioning to a market economy. The people were struggling with inflation, high prices, and a shortage of goods (METI 1993). After the resolution of the financial crisis in 1998, Russia maintained positive growth for ten years due to the rise in energy prices such as crude oil, achieving remarkable growth. The crude oil prices, which fell sharply in the second half of 2008 due to the impact of the global financial crisis, turned to rise again from the first quarter of 2009, and the Russian economy regained its growth trend (METI 2012). Since 2022, Russia has been under economic sanctions by Western countries due to its invasion of Ukraine. Putin served as president from May 2000 for eight years, and from 2012 to the present (2023). During his first term as president, the Russian economy grew for eight consecutive years, unemployment and poverty were halved, real wages tripled, and the satisfaction of the Russian people with their lives significantly increased. In Russia, it is said that President Putin has a high approval rating, especially among the elderly who experienced the economic crisis after the collapse of the Soviet Union. It is no exaggeration to say that Russia was able to escape from the crisis after

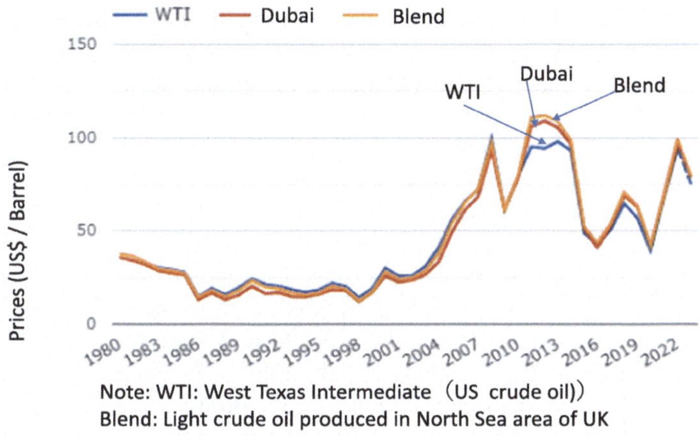

Fig. 1.6 Crude oil prices (1980–2023). *Source* https://ecodb.net/commodity/group_ngas.html. World Bank-Commodity Market (worldbank.org). *Note* WTI: West Texas Intermediate (US crude oil)blend: light crude oil produced in North Sea area of UK

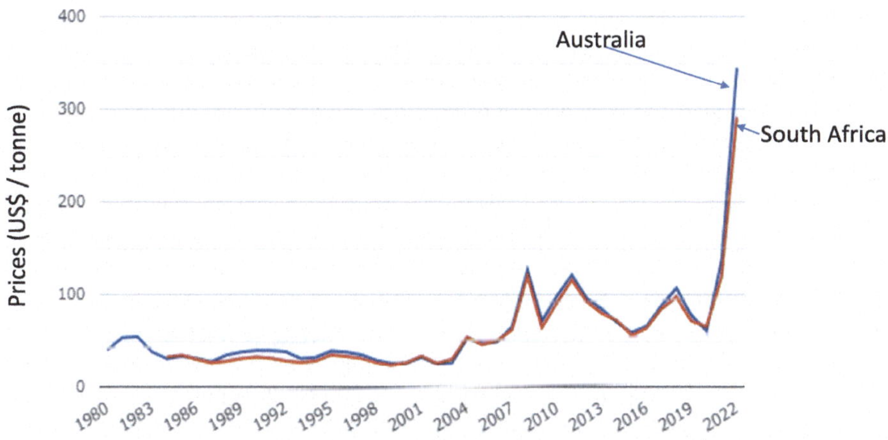

Fig. 1.7 Coal prices (1980–2022). *Source* https://ecodb.net/commodity/group_ngas.html. World Bank-Commodity Market (worldbank.org)

the collapse of the Soviet Union thanks to its own energy resources. From a security perspective, the former Soviet Union exerted influence on neighboring countries, and the Soviet army invaded Eastern European countries in events such as the Hungarian Uprising in 1956 and the Prague Spring in 1968. Even after becoming Russia, it has invaded Georgia, Ossetia, Tajikistan, Chechnya, North Caucasus, and Ukraine (from 2014). There are many countries in Asia and Africa that have friendly or neutral relations with Russia.

Energy is necessary not only for industry but also for all life activities. Foreign currency is needed to import energy resources such as coal and natural gas. In Japan, money that could be used for domestic economic activities is paid overseas as import payments, which has a negative impact on domestic economic activities such as employment. Supply shortages or disruptions of oil and natural gas raise energy prices such as electricity and gasoline, affecting industries and life activities through prices. The exchange rate also affects the price because it is imported.

If a country heavily depends on other countries for energy resources, there is a risk that its political independence will be threatened. For example, the United States has long maintained a basic policy of trying to keep the oil import dependency rate low for security reasons. President Eisenhower implemented a quota policy in 1959 to limit oil imports to 9% of U.S. consumption. Since then, the U.S. oil import dependency has been increasing. In 1973, the Middle East War and the oil crisis occurred, and President Nixon announced the Energy Independence Project. The United States had banned the export of domestically produced crude oil until the end of 2015 (Council on Foreign Relations 2023). Japan entered World War II against the United States in 1941, directly triggered by the U.S. ban on oil exports to Japan. At that time, the United States supplied 60% of the world's crude oil. Russia has also exerted political pressure on neighboring countries through the supply of natural gas and oil. For every country, securing energy resources is closely related to national security.

The energy import dependency of major countries is shown in Fig. 1.8. The energy shown in this figure is also called primary energy, which is used not only for power generation, but also for transportation, heating, industrial use, etc. Nuclear fuel, which is the fuel for nuclear power generation, can be used for about four years once loaded into a reactor. Since nuclear fuel generates about a million times the energy of fossil fuels, it is easy to stockpile. Uranium resources are abundant in developed countries such as Australia and Canada, so there is no fear of supply disruption. Nuclear power generation can be treated as quasi-domestic energy, so the degree of import dependence can be considered excluding nuclear power generation. The figure also shows the degree of import dependence excluding nuclear power.

Since the late 2000s, the United States has increased production of shale gas and shale oil, so it is now able to meet its natural gas and oil demand domestically and has become an exporter of primary energy. The UK was an exporter of primary energy for about 20 years from around 1980 due to the North Sea oil field. Even now, the degree of import dependence is low. France has a high proportion of nuclear power generation in its power supply composition, but it imports almost all of its fossil fuels, and its import dependence is 45%. Germany produces coal domestically and contributes to renewable energy generation such as wind power, so its import dependence is 65%. Japan depends on imports for its fossil fuels. Even excluding nuclear power, Japan's primary energy import dependence is 89%. In other words, the self-sufficiency rate of primary energy is 11%, which is extremely low compared to other OECD countries. South Korea also depends on imports for its fossil fuel resources, and its import dependence is high. The self-sufficiency rate is 19%.

Japan's import dependence was about 80% in fiscal 2010 before the Great East Japan Earthquake, due to the contribution of nuclear power generation, but it was still about one-third to half of the self-sufficiency rate of the UK, France, and Germany.

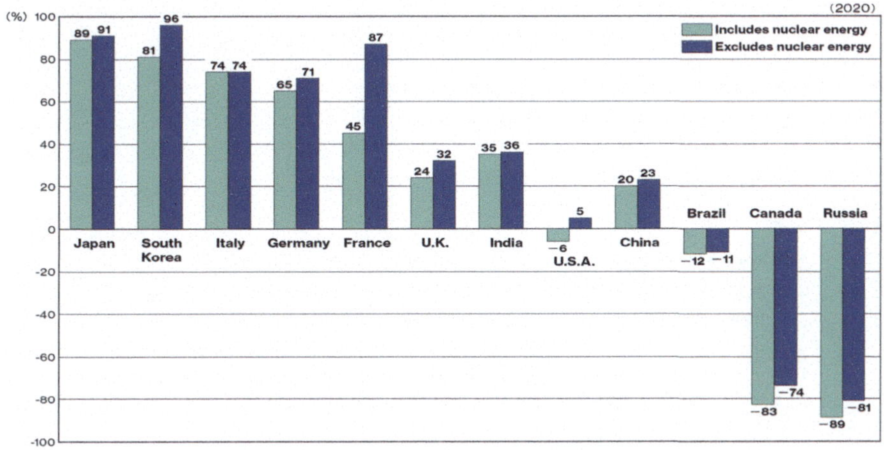

Fig. 1.8 Dependence on imported energy sources in major countries. *Source* JAERO

If the supply of natural gas and oil in Japan is insufficient or interrupted for a long period of time, not only will prices rise, but all social activities such as the economy, medical care, and culture will be greatly affected. The Japanese archipelago is not connected to the continent by natural gas pipelines or power lines. Even if it were connected, there is no guarantee of stable supply due to political situations, etc. Japan is in a unique environment in the world, isolated in terms of energy.

China produces a large amount of coal domestically, so its import dependence is low at 20%. Incidentally, the proportion of coal in primary energy consumption is 59% in China and 56% in India. In these countries, coal is the most widely used for power generation. Coal is also widely used in many emerging countries. Coal-fired power generation in China accounts for 68% of power supply, and in South Korea, it accounts for 45%. The proportion of nuclear power generation is 3% and 23%, respectively. Coal is widely used because it is less unevenly distributed worldwide compared to oil and natural gas, and because it is a solid, it is easy to transport using regular transportation infrastructure such as railways and ships. Oil and natural gas are unevenly distributed in the Middle East and need to be transported by pipeline or dedicated ships. Natural gas is the fuel for gas turbine power generation, but in countries like Japan that are not connected to the production area by pipeline, it needs to be liquefied and transported by dedicated ships, and it is not easy to stockpile.

The natural fluctuation of renewable energy power generation output is compensated by thermal power generation using natural gas. This is because natural gas thermal power generation uses aircraft jet engine technology, and like aircraft during takeoff and landing, it excels in the ability to respond to sudden starts and stops and load (power generation output) changes. Renewable energy is a self-sufficient energy resource, but since the power generation output varies depending on the weather, it is necessary to compensate for the fluctuation with other power generation methods for stable supply. France, which has a large proportion of nuclear power generation,

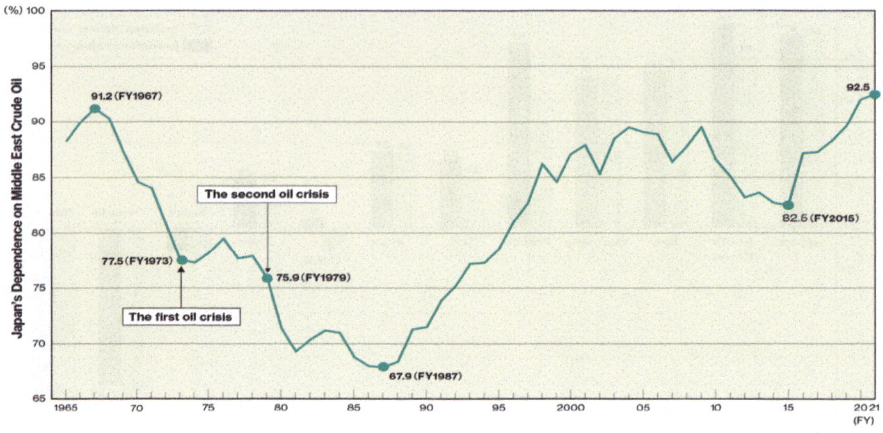

Fig. 1.9 Japan's dependence on Middle East crude oil imports. *Source* JAERO

adjusts the power output of nuclear power plants to cope. Power demand fluctuates even within a day, so adjusting the power output of nuclear power plants has been done in France for some time.

Japan depends on the Middle East for more than 90% of its crude oil imports. Figure 1.9 shows the transition of dependence on the Middle East for crude oil imports in Japan.

Japan has needed a large amount of energy resources along with economic development since the 1960s. The dependence on the Middle East once declined due to the oil crisis of the 1970s, but it has been rising since the 1990s, and in fiscal 2020, it has exceeded 90% again. The reason is that there are a large amount of cheap oil resources in the Middle East. The Middle East is politically unstable, and if the safety of navigation in the Strait of Hormuz is threatened, world energy prices will soar. Since oil can be stored in tanks, Japan has a stockpile of oil for about 200 days or more, combining national and private reserves. The United States and Denmark have more than 700 days, the UK has 270 days, and Germany and France have about 120 days.

Primary energy resources each have their own characteristics. In Japan, in order to supply electricity stably, it is necessary to use a balanced combination of power generation methods such as thermal, hydro, nuclear, and renewable energy. This is called the energy best mix.

Energy is deeply related to international politics, economics, and social activities. Securing energy resources and stable, inexpensive supply is a critical issue that will determine our future and that of Japan.

1.5 Economic Rationality (Japan's Electricity Rates)

The change in the average unit price of electricity in Japan is shown in Fig. 1.10. Electricity rates change with fluctuations in crude oil prices and greatly affect economic activity. Japan's electricity rates have traditionally been higher than other countries, but they have risen since the Great East Japan Earthquake. This is due to the increase in thermal power generation using fossil fuels due to the shutdown of nuclear power generation. Nuclear power generation, which was 25.1% in 2010 before the earthquake, has dropped to a few percent, and thermal power generation has increased from 65 to 77%. The burden of introducing renewable energy has also been added to electricity rates. Renewable energy increased from 2.2 to 9.2%, but since it was introduced by a fixed price purchase system, the purchase cost in 2020 reached 3.8 trillion yen, and the burden on electricity rates will continue. While promoting the introduction of renewable energy, it is necessary to suppress the burden on the people. The CIF price of crude oil in the figure refers to the price of crude oil upon arrival in Japan, which includes insurance premiums and freight charges. Japan's electricity rates are determined considering fluctuations in the prices of crude oil and natural gas.

The slogan representing Japan's energy policy is 3E + S. The 3E stands for Energy security, Economy, and Environment, and S stands for Safety. The goal is to achieve self-sufficiency, economic efficiency, and environmental compatibility while prioritizing safety. The goal for energy security is to achieve about 25%, which is higher than the 20% before the Great East Japan Earthquake, by 2030. The goal for economy is to reduce electricity costs from the current level. The goal for the environment is to achieve a greenhouse gas reduction target comparable to that of

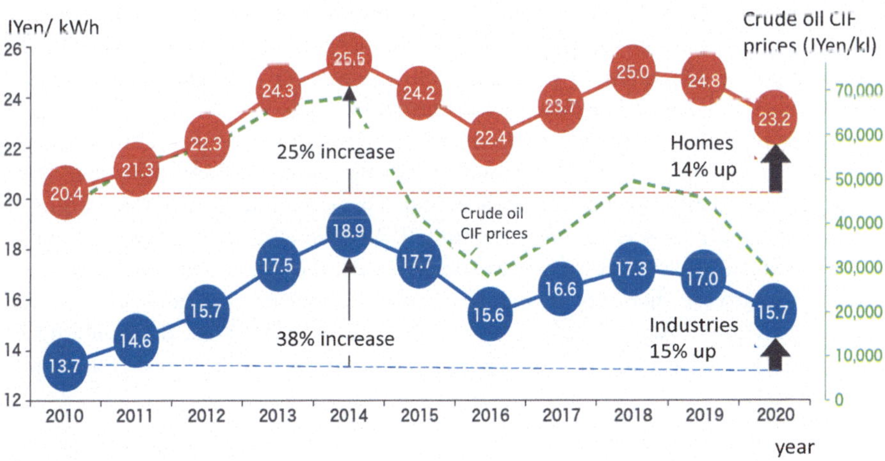

Fig. 1.10 Changes in average electric power rates in Japan. *Source* METI, Japan's Energy, Japan_energy_2021.pdf (meti.go.jp)

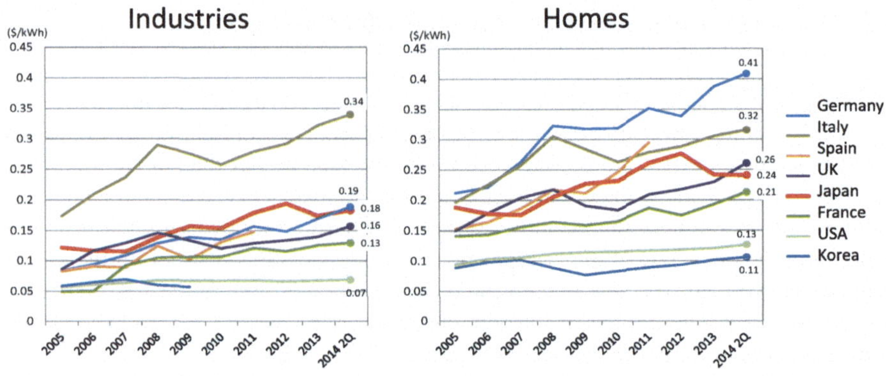

Fig. 1.11 Comparison of electricity prices. *Source* IEA energy prices and taxes

Europe and the United States. Based on this policy, the ideal state of primary energy supply and power source composition is shown.

A comparison of electricity rates with other countries is shown in Fig. 1.11. Japan's electricity rates have been high among major countries, for both industrial and domestic use, even before the Great East Japan Earthquake. Japan's industrial electricity rates are 2.6 times those of the United States and 1.4 times those of France. South Korea, like Japan, has no energy resources, but its electricity rates are less than half of Japan's. South Korea's economic development since the 1990s has been supported by cheap electricity rates. It is said that nuclear power, which was introduced in large quantities in South Korea since the 1990s, has contributed to South Korea's development. In 2019, about 40% of South Korea's electricity was supplied by coal-fired power, about 26% by nuclear power, and about 26% by natural gas-fired power. Korea Electric Power Corporation (KEPCO) is a private company owned by the Korean government and government-affiliated banks. It is believed that cheap power generation by coal and nuclear power and efficient management of KEPCO have made cheap electricity rates possible.

Italy has high electricity rates for both industrial and domestic use. This is thought to be due to the lack of energy resources and the complete abolition of nuclear power generation after the Chernobyl accident. The economic growth rate of OECD countries in the 2000s was the lowest in Italy (0.4%), and the second lowest in Japan (0.7%). In Italy, after the surge in electricity prices, there is a negative correlation between the growth disparity in industry and the share of electricity costs. Even though electricity costs are only a few percent of the total cost, the doubling of electricity prices has certainly reduced net production, according to analysis results (Nomura 2015).

Economic rationality is the principle of competition in a global modern society, and neglecting it will mislead Japan's future. Instead of relying on other countries or people, it is necessary for individuals and each organization to incorporate domestic

and foreign history and facts, strive and make efforts based on economic rationality, and prepare the environment for it.

1.6 Power Generation Cost: Investment Economic Feasibility and Actual Economics Feasibility

The indicators that represent the cost of power generation include the levelized cost of electricity (LCOE) and the production cost. The former is an average power generation unit price over the entire lifecycle of building, operating, and decommissioning a power plant, calculated assuming the operating years, equipment utilization rate (operating rate), and discount rate (the rate used to convert future value into present value). It is a method used for comparing the power generation costs of various power generation methods and for making investment decisions when building a new power plant. The latter is calculated from the expenses (fuel costs, maintenance costs, personnel costs, etc.) required for the annual operation of the power plant, excluding construction costs. The former is a calculated value, and the latter is calculated based on the actual cost incurred for power generation each year. In Japan, there seems to be no public value for the latter, but in the United States, it is compiled by the EIA (Energy Information Administration), a federal government statistical agency.

In Japan, an example of calculating the levelized cost of electricity (LCOE) was published by the Power Generation Cost Verification Working Group of the Agency for Natural Resources and Energy in 2015. The results are shown in Fig. 1.12. The operating years of nuclear power generation, coal-fired power generation, and natural gas-fired power generation are 40 years, the equipment utilization rate (operating rate) is 70%, onshore wind power has an operating year of 20 years, equipment utilization rate of 20%, and solar power (residential) has an operating year of 20–30 years, and the equipment utilization rate is 12%. The cost includes capital costs, operating and maintenance costs, fuel costs, CO_2 countermeasure costs, policy costs, and in the case of nuclear power generation, additional safety measure costs and accident risk response costs are considered (Matsuo 2015). Furthermore, taking into account subsequent developments, the estimated results of the power generation costs by power source for 2020 and 2030 were announced in July 2021.

Figure 1.12 is shown per kWh, which may be difficult to understand, but the power generation capacity of each power generation method varies greatly. The power generation capacity of a single nuclear power plant (set at 1200 MW) is much larger than that of a single renewable energy power plant (set at 4 kW for residential use, 2 MW for mega solar). The cost of accident measures for the TEPCO accident is enormous, but the additional safety measure costs and accident risk response costs per kWh shown in Fig. 1.12 are small because the amount of electricity generated by a nuclear power plant over 40 years is enormous, and if you divide the additional costs, the impact on power generation costs becomes limited.

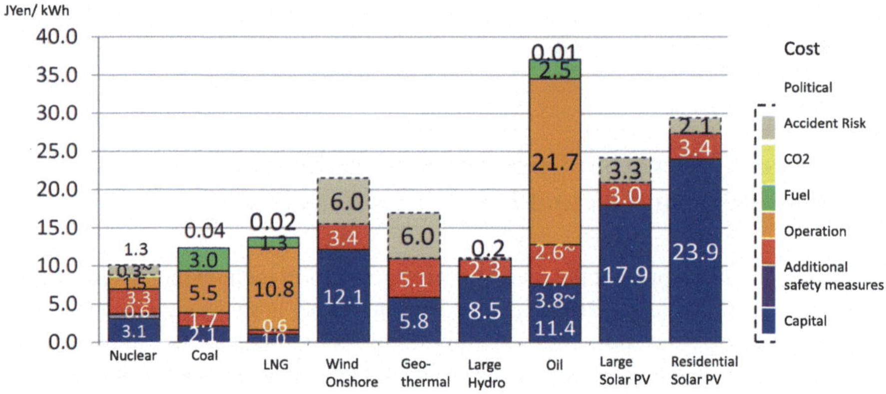

Fig. 1.12 Levelized cost of electricity (Japan). *Source* Matsuo Y. IEEJ November 2015

As mentioned in the section on liberalization, the fact that the power generation capacity of a single unit is small, like renewable energy, and the cost required for construction (the total amount of capital costs) is small, is an advantage for construction due to the small investment risk in a liberalized economic environment. Natural gas-fired power generation has smaller capital costs and lower investment risk compared to coal-fired and nuclear power generation, so construction investment has progressed under a liberalized economic environment. It has already been introduced on a large scale worldwide. The total investment amount for the construction of renewable energy is also small, and the investment risk in a liberalized economic environment is small.

LCOE is often used as an indicator of power generation costs. New power plant construction investments are also considered in this way. However, upon closer consideration, it is one-sided.

LCOE cannot be determined without determining the period of operation of a newly built power plant, i.e., the investment recovery period. LCOE is a value calculated in this way. However, the investment recovery period is not necessarily the actual usage period of the power plant (the period until permanent shutdown), so it may not be the actual power generation cost. This is typical for hydropower, but also for nuclear power. Natural gas-fired power plants, solar power plants, wind power plants, etc., remove old facilities and rebuild them as they age, so the actual usage period and the investment recovery period are roughly the same. Hydropower plants and nuclear power plants continue to use by updating old facilities such as turbine generators. They have no lifespan.

Large hydropower plants require a huge investment for new construction, but they continue to operate even after the investment recovery period has passed. For example, the hydropower plants in the Tennessee Valley (TVA) built by the United States during the Great Depression under the New Deal policy, the Aswan High Dam on the Nile in Egypt, and many hydropower plants in Japan. According to the EIA data, the power production cost (Production cost) of hydroelectric power plants in the U.S. is extremely low at about 1.1 cents/kWh. Many hydroelectric power plants were built more than 50 years ago, so the recovery of construction investment is complete, and the cost of electricity production can be considered close to the actual generation cost.

There are about 100 nuclear power plants in the United States, most of which started operation in the 1970s and 1980s. The period for recovering investments has already ended for many of them. The cost of electricity production includes the cost of equipment repair and renovation, so the cost of electricity production (the cost incurred annually for generation) can be considered close to the actual generation cost. In the United States, some nuclear power plants have obtained permission to extend their operating period to 80 years, contributing to low electricity rates. According to the EIA, the cost of electricity production from nuclear power in the United States is about 2.4 cents per kWh. The breakdown is about 1.1 cents for operating costs, about 0.6 cents for repair costs, and about 0.7 cents for fuel costs per kWh. Incidentally, renewable energy (solar and wind power) compensates for output fluctuations in gas turbine combined cycle power generation, so the average generation cost, which includes these, is shown to be 2.8 cents/kWh. It is rational to compare with others in terms of average generation cost. This average generation cost has dropped by about half in ten years from 5.8 cents in 2009. The reason is the reduction in fuel costs for gas turbine power generation due to the decrease in natural gas prices caused by the shale gas revolution. The changes in the generation cost (cost of electricity production) of natural gas power in the United States since the 2000s and the changes in natural gas prices can be understood by drawing them on one graph (EIA 2019).

Although shale gas exploration is being conducted in many countries, few countries have started mining on a commercial scale. Shale gas is mined by injecting water into the shale where it exists and fracturing it. A large amount of water is required for mining. Most of China's shale gas resources are located inland, but due to the complexity of the strata and the lack of water resources available for mining inland, economic mining has not yet been achieved. In Canada, shale gas has started to be mined. It is said that the reason why shale gas can be used in large quantities in the United States is because there were natural gas pipelines. The United States also needs to build pipelines to use a large amount of natural gas from Canada. Japan needs to import it by liquefying it and using dedicated tankers. Stockpiling is also difficult. Natural gas is not as easy to transport as solid uranium or coal, but it is a suitable power generation method to compensate for the natural fluctuations of renewable energy. The recoverable reserves of conventional natural gas are greater than those of shale gas. The price of natural gas fluctuates greatly due to the relationship between demand and supply, as already mentioned. In Japan, the impact of exchange rate fluctuations is added to this.

It is good to use nuclear power plants for a long time: there is no lifespan if the equipment is replaced, and they can supply power cheaply and stably over a long period.

Hydropower plants require a large amount of money for construction, but the operating and maintenance costs are small. Even after the construction investment is recovered, they generate electricity cheaply for a long time. For example, the group of power plants at the Tennessee Valley Dam, built in the United States in the 1930s as a countermeasure against the Great Depression, is still supplying cheap electricity. The Aswan High Dam in Egypt was built on the Nile River in the 1960s with the help of the Soviet Union, and after 50 years, major repairs such as replacing generators were carried out. The Kurobe River No. 4 Power Station was built in the 1960s in Japan and continues to generate electricity after major repairs such as replacing generators. Hydropower plants require a large amount of money to build dams, but there is no lifespan if generators and other equipment are replaced. Nuclear power plants are similar to hydropower plants, with high construction costs, but there is no lifespan if equipment such as generators that age over time are replaced, and they can supply electricity cheaply and stably over a long period of time, like hydropower plants. In the United States, nuclear power plants that have obtained permission to operate for 80 years from the start of operation by applying for operating permit renewal every 20 years are emerging. In France, too, Electricité de France (EDF) is carrying out major repairs of its nuclear power plants, called "Grand Carénage", from 2014 to 2025.

The Abolition of Nuclear Power Plants in Japan is a National Loss

After the TEPCO Fukushima accident, new regulatory standards were established by the Nuclear Regulation Authority, and nuclear power plants cannot operate until the construction work that meets these standards is completed. Furthermore, in Japan, operating permit renewals are limited to once every 20 years. There are quite a few nuclear power plants that power companies have abolished because it is impossible to recover the cost of construction work for new regulatory measures within the 20-year extended operation period, especially those with small output. They could have supplied cheap electricity if used for a long time, which is a national loss.

After the TEPCO Fukushima accident, other countries conducted stress tests on nuclear power plants, that is, resilience evaluations to see if nuclear power plants can withstand even considering the lessons of the accident, and took measures where they could not withstand and where the lessons were reflected. However, in Japan, operating nuclear power plants were stopped and responses to the lessons of the TEPCO Fukushima accident were made. No other country has done this. More than 10 years after the accident, many plants are still stopped. Since there has been no major tsunami since 2011, the same accident did not occur in these 13 years even if nuclear power plants were operated in Japan and safety inspections and construction work in response to new regulatory standards were carried out systematically. There was no need to stop nuclear power plants and respond to new regulatory standards.

The concept of balancing risk and benefit is missing in Japan's safety, regulation, and disaster prevention. This is behind this problem. The need for regulation of a certain technology is because there are benefits to using that technology. This is stated at the beginning of the Handbook of Nuclear Law, which the International Atomic Energy Agency made public in 2003 and is introduced as reference material in Chap. 4. *Consideration of the balance between risk (cost) and benefit is necessary for Japan's use of nuclear power.*

"There is no zero risk", is a phrase sometimes used by those involved in Japan's nuclear power industry. However, if they use such a phrase, they should realize that the concept of cost-benefit is absent in Japan's regulations and safety measures. The extent to which safety is considered safe is a responsibility of nuclear engineers. If they think it is enough to agree on safety goals with the public, they divert their attention from the essence of the problem and neglect insight into other issues. Isn't this a lesson learned from the failure to prevent the TEPCO Fukushima accident? *Safety goals should be kept as performance goals as in the U.S.*

In the future, to improve self-sufficiency and reduce electricity rates, the proportion of nuclear power generation in Japan should be increased. Reducing the proportion of nuclear power generation not only raises electricity rates but also loses money that could have been used domestically due to the import of natural gas, worsening Japan's economy and employment. This is a suicidal act for Japan.

1.7 Economic Liberalization, Power Liberalization, and Investment in New Sources

In the United States, in response to the economic stagnation of the 1970s, President Reagan promoted market-oriented economic management. This included a smaller government, privatization of public enterprises, and promotion of competition through deregulation. In the UK, Prime Minister Thatcher also promoted the privatization and deregulation of state-owned water, electricity, gas, telecommunications, railways, and airlines.

Liberalization of the electricity market involves easing entry regulations into the electricity market and introducing market principles. Note that power liberalization is different from the privatization of power companies. Even after power liberalization in Europe, the governments of France and Sweden own more than half of the shares of power companies. Japanese power companies have always been private companies. The Electric Power Development Corporation (EPDC) in Japan, which was a national policy company, was privatized in the early 2000s.

In the United States, power liberalization was promoted in the 1990s to encourage free competition in the power market. The opening of the transmission network and the establishment of independent system operators to independently operate it were recommended. The liberalization of the retail market was left to the discretion of each state. From the mid-1990s, 24 states decided to liberalize retail, but the movement stopped around 2000 when a power crisis (major blackouts, price fluctuations, unstable supply, etc.) occurred in California. Currently, about one-third of all states in the United States are classified as liberalized states. Note that electricity rates are higher in liberalized states than in non-liberalized states.

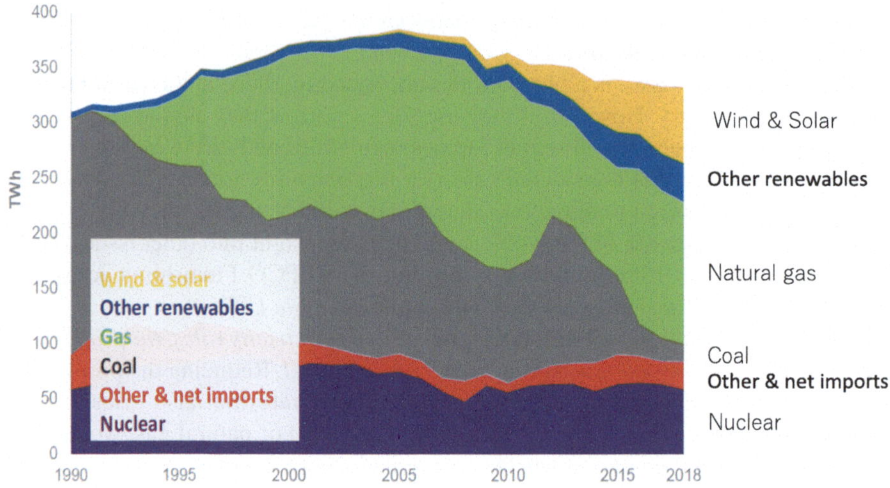

Fig. 1.13 Changes in UK power generation. *Source* Department for Business, Energy & Industrial Strategy, UK ENERGY IN BRIEF 2019, p. 28 www.gov.uk/government/statistics/uk-energy-in-brief-2019

The transition of the UK's power generation from 1990 to 2018 is shown in Fig. 1.13. As shown, the introduction of natural gas power generation has been rapidly increasing since around 1990. This is thought to be due to the supply of natural gas from the North Sea oil fields and the effects of economic liberalization and power liberalization. Wind power and solar power have been increasing since around 2005.

A comparison of the power generation capacity and power generation amount in the United States in 1990, 2000, and 2019 is shown in Fig. 1.14. The capacity of natural gas power generation facilities has been increasing since 1990. Since 2000, the capacity of renewable energy generation facilities has been increasing. The amount of power generated by natural gas has been significantly increasing since 2000. The amount of power generated by renewable energy is not increasing as much as the capacity of power generation facilities. It is thought that this is because the operating rate is low due to it being natural energy, and it is supplemented by natural gas power generation. Nuclear power generation has not changed in capacity since 1990, but the amount of power generated has increased. The proportion of nuclear power in the power generation capacity is not large, but the amount of power generated is large, and the amount of power generated in 2019 is more than that of renewable energy.

The transition of power generation costs (power production costs) in the United States is shown in Fig. 1.15. The power generation cost of natural gas in 2000 was nearly three times that of nuclear and coal. However, as shown in Fig. 1.14, the capacity of natural gas power generation facilities has been increasing since 1990. This period coincides with the period of economic liberalization in the United States. The reason is thought to be that natural gas power generation is compatible with the investment environment of economic liberalization. Even if the power generation

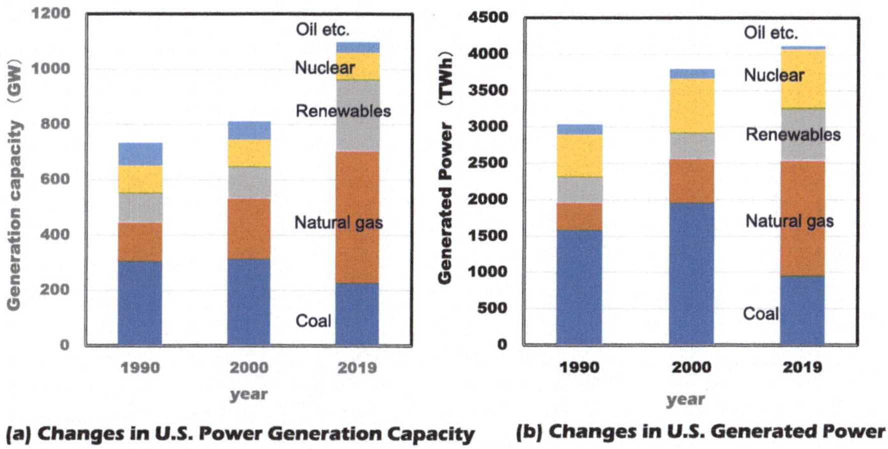

Fig. 1.14 Changes in U.S. power generation capacity and generated power. *Source* EIA, Electricity explained, "Electricity generation, capacity and sales in United States". Last updated March 19, 2020

cost is high, natural gas power generation, which has a small investment amount per unit, has a short investment recovery period and a small investment risk, so new construction has progressed. Even if the power generation cost is high, electricity is essential for modern life, so it can be sold. Even with liberalization, there are restrictions on transmission and distribution, and electricity cannot be bought freely from anywhere. As a result, consumers in the United States ended up buying expensive electricity. Not only natural gas power generation, but also renewable energy has a very small investment amount per unit compared to hydroelectric power plants, nuclear power plants, or coal-fired power plants. There are also ones that can be introduced at home. Therefore, it is thought that the introduction has progressed.

What I want to point out is that natural gas power generation was not introduced with subsidies. *In the investment environment of economic liberalization, power generation methods with low investment risk were preferred even if the power generation cost was high.* LCOE cannot be determined unless the investment recovery period (operating period) is determined. However, *hydroelectric power and nuclear power require a large amount of investment to build, but they can be used beyond the investment recovery period, so they can supply cheap electricity in the long term. This feature does not fit in the environment of private investment in economic liberalization. As a result, consumers have no choice but to buy expensive electricity.* This is believed to be the reason why electricity rates are higher in deregulated states than in non-deregulated states in the United States.

The UK is a country that is dealing with the investment risk issue of new nuclear power plant construction in deregulated electricity countries. The UK has created a system called the RAB model, which improves the fixed price purchase system to reduce the investment risk of new nuclear power plant construction, and is trying to support the construction of new nuclear power plants. The UK is supporting the

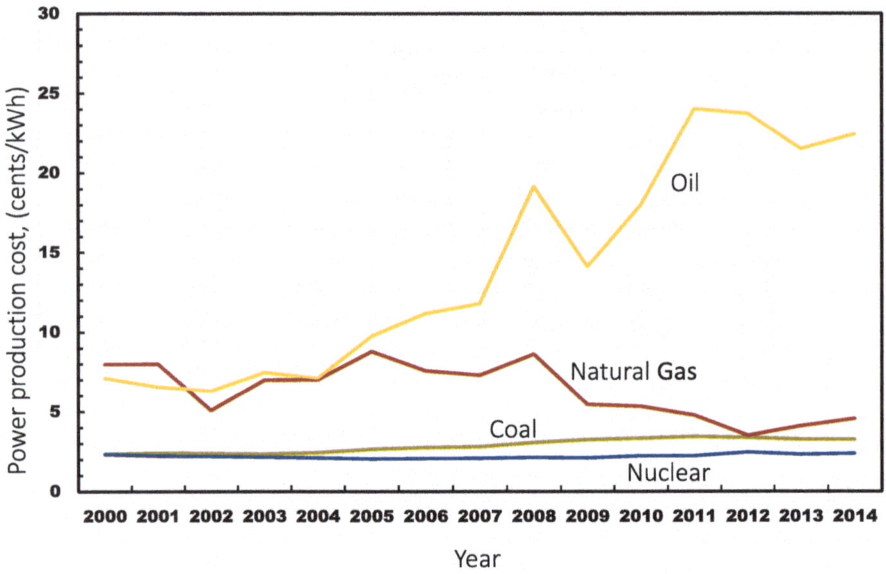

Fig. 1.15 Changes in U.S. power production cost. *Source* EIA

introduction of renewable energy and nuclear power generation in response to global environmental issues through a fixed price purchase system. Globally, the investment environment for power generation facilities and government involvement in new plans vary by country, and there are countries where the investment risk issue of nuclear power plants is not manifest.

In the United States, venture companies are proposing modular nuclear power plants that line up several small output reactors. This is one approach to the investment risk issue. The economic improvement of light water reactors has been achieved by increasing output, so if the output is reduced to just one unit, the cost per power generation output will increase. The concept of a Small Modular Reactor (SMR) is to use multiple small output reactor modules to take advantage of the learning effect of design and manufacturing. Natural gas power generation is a modular power plant. It will be interesting to see whether large output nuclear power plants or small modular nuclear power plants will be built more in the future. In the UK, plans for the construction of large output nuclear power plants are underway. Because the construction of large output nuclear power plants preceded in Europe and the United States, problems such as construction delays have become apparent, but care is needed because there are many common parts with large reactors in the construction of small modular reactors. Support is also being provided for the spread of renewable energy, but in the case of renewable energy, the investment amount is much smaller than that of nuclear power plants, and the investment recovery period is short, so the problem of construction investment risk is not large. The spread of renewable energy has issues separate from construction investment, such as access to transmission lines and energy storage. The construction of long-distance transmission lines involves large costs and a long period.

Notes: Natural Gas Power Generation: Gas turbines have evolved as jet engines for aircraft. Large-scale gas turbines for power generation have been developed, and power plants using them have been built since around 1990. Natural gas is burned to rotate the gas turbine. The emitted gas is high temperature, so it is led to a steam boiler to create high-temperature and high-pressure steam to rotate a steam turbine. This method of power generation is adopted, which is called gas turbine and steam turbine combined cycle power generation, and it is the principle of natural gas power generation. The power generation capacity of one gas turbine is about 150,000 kilowatts, which is smaller than the 1 million kilowatt class of nuclear power generation, but the same power generation modules are lined up and built, and the total power generation capacity of the power plant is also 1 million kilowatt class. Natural gas power plants are built in Japan in places like Tokyo Bay, Mikawa Bay, and Osaka Bay.

Renewable energy such as solar and wind power fluctuates with changes in sunlight and wind speed, and power output fluctuates. Natural gas power generation is used to compensate for this output fluctuation. Since gas turbines have evolved as aircraft engines, they can rapidly increase or decrease output in a short time during takeoff and landing. Therefore, it is suitable for compensating for the fluctuation of power output of renewable energy. Power demand changes between daytime and nighttime. This fluctuation (daily load fluctuation) and response (following power demand changes in power output) are carried out by natural gas power and pumped-storage hydroelectric power. Nuclear power generation can also follow daily load, and it is done in France and other countries.

Nuclear Power Generation is Cheap, Stable, and Safe

The actual power generation cost of nuclear power, as seen in the example of the United States in Fig. 1.15, or in France where the proportion of nuclear power generation exceeds 70% as shown in Fig. 1.11, is cheap. Nuclear power plants can operate for several years even if the supply of uranium fuel is interrupted. Nuclear fuel is significantly smaller in volume than coal or natural gas, making it easy to stockpile. Uranium resources are abundant in developed countries and concerns about stable supply are small. Uranium enrichment facilities are in the Western countries. Japan can manufacture fuel assemblies for use in nuclear power plants domestically. On the other hand, natural gas and crude oil are large in volume, the amount that can be stockpiled is limited, and resources are concentrated in politically unstable countries. Natural gas and crude oil have a large price fluctuation range. The price fluctuation of natural gas from 2005 to 2012 is shown in Fig. 1.16. In the case of imports, it is also affected by exchange rate fluctuations. Natural gas and crude oil prices were on an upward trend from around 2021, but soared in 2022 due to Russia's invasion of Ukraine. Stable energy supply is essential for national security and the economy and health of the people.

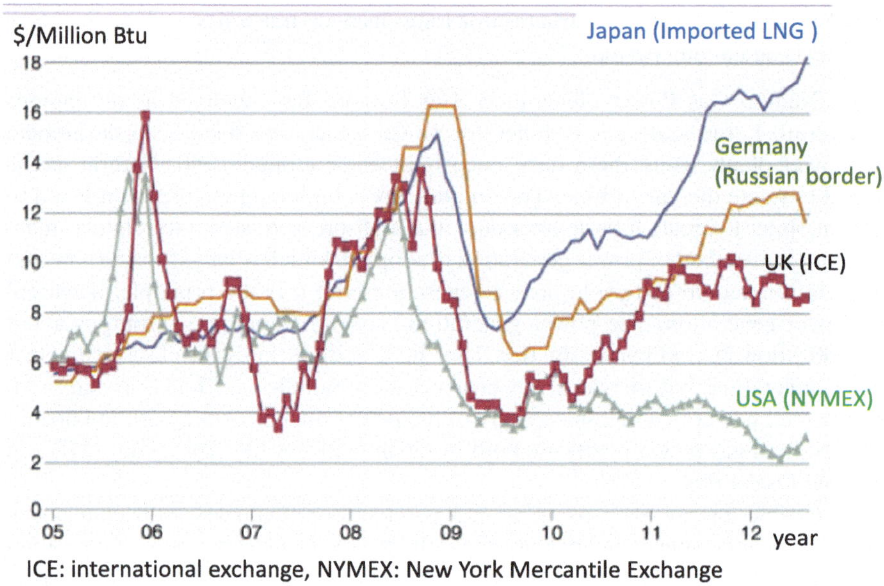

Fig. 1.16 Natural gas price volatility. *Source* JOGMEC. http://www.jogmec.go.jp/library/contents8_05.html

In the 2011 TEPCO Fukushima accident, Japan's nuclear power generation was halted, and was replaced by thermal power generation. Although a large amount of foreign currency was lost due to the import of fossil fuels, fortunately, at that time the yen was strong, and subsequently the prices of crude oil and natural gas fell, so the electricity rates were not greatly affected. However, there had been warnings that if the prices of crude oil and natural gas soared and the yen depreciated, Japan would be hit with a double punch. This became a reality due to Russia's invasion of Ukraine in 2022. Electricity rates rose by 30%. We are being made aware of the harsh aspects of the energy issue. It would have been better if we could operate nuclear power plants while conducting inspections and construction under new regulatory standards.

When discussing safety, as shown in Table 2.9 of Chap. 2, the only public deaths due to radiation from nuclear power plant accidents are the eight children who died of thyroid cancer at the time of the Chernobyl accident. There were none in the TEPCO Fukushima accident. When compared per unit of power generation, nuclear power has a lower mortality rate than coal, oil, natural gas, bio power, hydro, solar, and wind energy (Brook et al. 2014).

Isn't nuclear power being talked about too much, and talked about too much, in terms of safety (radiation risk)? The harm caused by rumors, mental instability, and increased disaster-related deaths due to delays in evacuation, may be far greater than the health damage caused by radiation. "We want nuclear power to become 'normal'," this is the opinion of the expert in the UK that we will introduce in Chap. 2.

In Japan, after World War II, a system where nine power companies each supplied power exclusively in their respective regions continued for a long time. The electricity rates were determined by a method called the cost-plus method, which added a certain profit to the costs necessary to supply power (fuel costs, equipment costs, operating costs, personnel costs, transmission and distribution costs, etc.). As a result, power plants were built and power was stably supplied, but there was no competition between power companies, and Japan's electricity rates remained high. For example, compared to South Korea, which, like Japan, has scarce domestic energy resources and relies on imports for most of its fossil fuels, Japan's electricity rates are about twice as high. In Japan, electricity liberalization has been promoted since around 1995, but full-scale liberalization such as full retail liberalization and separation of transmission and distribution was carried out after the Great East Japan Earthquake and is still in progress. Even if we liberalize and introduce competition, rates will not necessarily go down unless the people and politics monitor it. For example, in Japan, communication used to be a monopoly business of Nippon Telegraph and Telephone Public Corporation. The corporation was privatized in 1985, and multiple companies entered the market afterward. However, even after communication was liberalized, mobile phone charges were high. Improvements have been made under the leadership of Prime Minister Suga in 2021, and mobile phone communication charges have been significantly reduced.

In the economic liberalization and power liberalization environments of the United States and the UK, there were no investments made in the power transmission network or new nuclear power facilities. The reason why new nuclear power plant construction did not take place in the United States for 20 years after the 1980s was not because of safety concerns, but because in the environment of economic liberalization and power liberalization. Huge initial investments were required and investments in nuclear power plants, which take a long time to recover. Furthermore, even at existing nuclear power plants, in the power liberalized states of the United States, the power generation cost is high for small, single-unit plants, and some plants have stopped operating. The reason for this is that in favorable weather conditions, a large amount of electricity from renewable energy can be supplied, even at zero cost, making nuclear power and coal-fired power generation disadvantaged in the liberalized wholesale electricity market. However, as weather conditions vary, power liberalization ultimately leads to issues with stable supply. There is a need to create a way to meet both the challenge of preparing an investment environment for nuclear power plants and transmission lines, which require huge investments, and the challenge of providing cheap and stable power supply.

1.8 Spread of Renewable Energy and Stable Power Supply

In Japan, a large-scale introduction of renewable energy such as solar power and wind power is being promoted. Except for biomass power generation that uses imported biomass as fuel, renewable energy is self-sufficient energy, and its introduction is desirable for Japan, which has an extremely low self-sufficiency rate. On the other hand, because the power generation cost of renewable energy is high, in the 2010s, its introduction was promoted at the expense of the public by adding it to electricity charges through a fixed price purchase system. Controlling this burden is a challenge for energy policy. In this fixed price purchase system, because a business environment that eliminates investment risk systemically is guaranteed for renewable energy power producers, there has been a lot of investment with an asset management sense, and most of it is not considered sustainable power source investment. However, the introduction is progressing. The power generation amount of natural energy fluctuates. In the future, it will be necessary to closely watch how renewable energy power generation will contribute to Japan's stable and inexpensive power supply.

The stable supply of electricity is another major issue that needs to be addressed with the large-scale introduction of renewable energy. Electricity is used in every aspect of our lives and industries, and its interruption can have a significant impact. The stable supply of electricity mentioned here does not refer to the improvement of energy self-sufficiency by reducing the import of fossil fuels, but to the prevention of power outages. The electricity generated by solar power and wind power is "direct current", and it does not have the ability to withstand fluctuations in power demand, so if measures are not taken, its large-scale introduction could cause power outages. The electricity we use is "alternating current". Thermal power generation, hydroelectric power generation, and nuclear power generation all generate alternating current by rotating a turbine generator in a magnetic field. The motors that power homes and factories rotate in sync with the generators at the turbine power plants. When the power consumption of homes and factories increases, the load on the turbine generators at the power plants increases, slowing down the rotation. Power companies detect this and increase the power output of their power plants to cover the increased load, maintaining a constant rotation speed and ensuring a stable power supply.

Wind power generation generates electricity by rotating windmills, but since the strength of the wind is not constant, transmitting it as is would negatively affect the quality of the electricity. Therefore, it is converted to direct current once using electronic devices, and then converted back to stable alternating current for transmission. The power from solar power generation is also converted from direct current to alternating current by electronic devices for transmission. These electronic devices are vulnerable to sudden changes in the flow of electricity, so if an accident occurs somewhere on the transmission line, it is necessary to stop power generation once for protection. Therefore, solar and wind power generation have a weaker "resilience" when an accident occurs on the transmission line, and are more likely to stop in a chain reaction.

One possible measure for the stable supply of electricity is to balance supply and demand by borrowing power from neighboring regions or increasing the output of thermal power generation. However, if the balance of a large amount of power supply is suddenly lost, it will affect the power supply in neighboring regions, causing the mutual supply to be cut off, and even if you try to start up a thermal power plant that was stopped, it will be impossible to do so immediately, leading to a power outage. For example, in South Australia in 2016, where 50% of the power was supplied by wind power, a statewide blackout occurred when multiple lightning strikes damaged transmission lines, causing all wind power plants to stop at once and cutting off power from neighboring regions.

In another example, on September 6, 2018 (Thursday), a major power outage (blackout) occurred throughout Hokkaido, triggered by the Hokkaido Eastern Iburi earthquake. The earthquake damaged equipment at thermal power plants near the epicenter, causing two thermal power plants to stop. Wind farms stopped due to the impact of the frequency drop caused by this, and then hydroelectric power plants stopped due to broken transmission lines. Furthermore, these effects also caused another thermal power plant to stop, resulting in a blackout 17 min after the earthquake. At that time, there was a fear of another major power outage due to a lack of supply during the daytime of the following weekday, so a 20% power saving was requested for all sectors of households, businesses, and industries throughout Hokkaido.

In technical terms, maintaining the quality of electricity demand and supply is called maintaining the stability of the power system. Japan has different frequencies in Kanto and west of it (west of the Fujikawa in Shizuoka Prefecture and Itoigawa in Niigata Prefecture), and frequency conversion is necessary for power interchange, limiting the amount of power that can be interchanged. The power system interconnection capacity between Hokkaido and Honshu (the capacity of transmission lines for interchanging power) is also not large. European countries have a grid-like transmission network, allowing for power interchange, and the stability of the power system is high. In Japan, for the large-scale introduction of renewable energy, technologies that are resistant to changes in the quality and flow of electricity and inexpensive batteries that can reduce the impact of weather are required in the future, but these cannot solve all problems. First, in relation to power liberalization, it is necessary to recognize the current issues, aim for stable power supply, and avoid power outages.

For the stabilization of the power system, it is necessary to secure supply capacity, power quantity, frequency adjustment capacity, and inertial force synchronization capacity (Komiyama 2022, p. 14; Japan Business Federation 2019, p. 12). Securing supply capacity (securing kW, securing the ability to generate power) is about securing power generation facility capacity, which means having many power generation facilities. By having power plants on standby as a reserve, it is possible to secure supply reserve capacity that can cope with the maximum power demand throughout the year. In Japan, high-cost oil-fired power plants were maintained in a state where they could be used at any time for this purpose. Securing power quantity (securing kWh, securing the power actually generated) is about securing the amount of power that can respond to short-term power demand, such as changes in daily power demand. For example, in response to changes in the power generation amount of renewable energy due to weather, natural gas-fired power plants are started or stopped in a short time. By generating power by releasing water from the upstream reservoir to the downstream reservoir at pumped-storage hydroelectric power plants, the power generation amount according to demand is secured. Securing frequency adjustment capacity (short-term supply-demand adjustment ability), the ability to respond to kW changes is the ability to handle sudden changes in the amount of electricity flowing through the power system due to the shutdown or damage of power plants, substations, transmission and distribution facilities, and transmission systems caused by lightning strikes or earthquakes. This includes the ability to draw power from unaffected neighboring regions. Turbine generators, which involve heavy objects called steam turbines rotating at high speeds, have a high ability to continue rotating and possess inertia and synchronizing forces. Power plants with turbine generators, namely thermal, nuclear, and hydroelectric power plants, can ensure inertia and synchronizing forces by using their power generation to a certain extent.

By ensuring these four requirements, the ability to avoid power system shutdowns (supply capacity, or "adequacy" in technical terms) and the ability to suppress the spread of system failure impacts (security) can be enhanced. Nuclear power generation meets all of supply capacity, power quantity, frequency adjustment ability, and inertia and synchronization forces, and can contribute to stable power supply and stabilization of the power system (Komiyama 2022).

In Japan, due to power system reforms (electricity liberalization), the traditional power supply system by power companies (such as TEPCO and KEPCO) that have consistently handled generation, transmission, and sales for each region is transitioning to a system realized through the collaboration of generation, transmission and distribution, and retail. In the power generation field, simply forming a free wholesale power (kWh) market will reveal a medium to long-term supply capacity shortage. Since electricity basically cannot be stocked, it is necessary to always secure a supply capacity (power source) that exceeds peak demand. Once a power source is installed, it is economically rational in the short term to operate it as much as possible and earn income as long as the sales income exceeds variable costs such as fuel costs. Under these conditions, liberalizing (marketizing) the wholesale trade of electricity will cause the contract price in the wholesale market to converge to the variable cost of the most expensive power source among the contracted ones (Japan Business Federation 2019).

In the United States, economic liberalization has led to the construction of many new natural gas power plants, which require a small investment and carry a low investment risk, followed by a large number of renewable energy power plants such as solar and wind power. Renewable energy power plants have extremely low variable costs because they do not require fuel. In the liberalized wholesale electricity market, when the weather conditions allow for a large amount of renewable energy generation, the wholesale electricity price drops to zero. As a result, natural gas power plants, coal-fired nuclear power plants, and even single-unit nuclear power plants, which require fuel and have a small power output, have been forced to shut down in states with liberalized electricity due to their relatively high power generation costs. Since weather conditions vary, the shutdown of these power plants, which can supply cheap electricity in the long term, has threatened the stable supply of electricity and led to higher electricity rates. This is why electricity rates are high in states with liberalized electricity in the United States. Some states in the U.S. recognize the value of nuclear power plants in terms of climate change measures and are trying to keep them operational.

In Japan as well, electricity liberalization has been implemented, and a large amount of renewable energy has been introduced with policy support in the form of surcharges on electricity bills. As a result, when the weather conditions are good, the oversupply of electricity drives down market prices even further. For example, in May 2021, when solar power generation increased in Kyushu, output control of solar power, output control of thermal power, implementation of pumped-storage power generation, and transmission of electricity from Kyushu to Honshu were carried out, but the wholesale electricity price was zero during the day (Komiyama 2022, p. 11). If this situation continues, the outlook for recovering the fixed costs of power generation investments in thermal power, nuclear power, or hydroelectric power will deteriorate, and power generation investments will not be made in the medium to long term, leading to concerns about a shortage of supply capacity. This could expose the lives of the Japanese people and business activities, which rely on a stable supply of electricity, to uncertainty.

Electricity is also a public good. It is not possible to leave its supply to the free market, and it is necessary to address both competition activation and public interest issues. The Japanese government also recognizes this as a problem. As a measure, it is trying to build a system that breaks down the value associated with power generation and trades each value individually (Japan Business Federation 2019). In addition to the four indicators mentioned above regarding the stable supply of electricity, there is also environmental value. An example of environmental value is the value of generating electricity from non-fossil sources. Nuclear power generation has the advantage of not only environmental value in terms of preventing global warming, but also requiring a small land area for plant installation and having a small impact on ecosystems. It also employs a large number of people and plays a role in regional revitalization.

Europe has countries that produce natural gas and coal, but Japan imports almost all of its fossil fuels. While European countries are connected by a power grid, Japan's power grid is not connected to other countries, and even if it were, relying on it would pose a security threat. In Japan, it is necessary to consider the burden on the public and promote the introduction of solar power and wind power, but in order to do so, it is necessary to demonstrate in practice that a method of storing a large amount of electricity can be realized at a low cost. Currently, in Japan, renewable energy is used in combination with thermal power generation using natural gas and pumped-storage hydroelectric power generation. Nuclear power is an essential power source for Japan's stable power supply. Like France, there is also a method of introducing a large amount of nuclear power, operating in load-following mode, and combining it with renewable energy. This method has already been put into practical use and has been proven. *Nuclear power can take advantage of its characteristics when used stably over a long period of time.*

1.9 World's Nuclear Power Usage and Development of Power Reactors

Humanity has progressed socially and culturally by using fire. By using fire for cooking, it became easier to take in nutrients, it became possible to act even on cold nights, and it became possible to live in cold regions. Fire is obtained by burning wood, animal and plant oils, and fossil fuels (things that became coal or oil underground due to the carbonization of ancient plants and animal plankton). The energy of fire is generated by a reaction in which the carbon contained in these reacts with oxygen in the air to become carbon dioxide.

Nuclear energy is energy obtained by nuclear reactions. Per unit mass, about a million times the energy of chemical reactions like fire is generated by nuclear reactions. Nuclear energy is generated when a heavy (large atomic number) nucleus like uranium splits into two or more nuclei, or when two light nuclei like deuterium and tritium fuse to form one nucleus. The former is called a nuclear fission reaction, and the latter is called a nuclear fusion reaction.

Nuclear energy is the "new fire" that humanity has acquired. We will describe how nuclear energy has historically become usable.

1.9.1 The Beginning of Nuclear Energy Use

The nuclear fission reaction was discovered in Berlin in 1938 by Hahn and Strassmann. Furthermore, the Italian genius physicist Fermi successfully initiated a sustained nuclear fission reaction (nuclear fission chain reaction: a phenomenon where neutrons generated by the fission of uranium cause other uranium in the vicinity to undergo fission, and fission continuously occurs mediated by neutrons) for the first time artificially at the University of Chicago in 1942. This made it possible to generate a large amount of energy from nuclear fission sustainably. Nuclear fusion reactions have been discovered by physicists who have been engaged in particle accelerator research since the 1920s, such as Cockcroft, and are still used as one of the methods to generate neutrons in accelerators, but they have not yet been able to generate nuclear fusion energy sustainably and continuously, more than the energy input to cause the nuclear fusion reaction.

The scientists who fled from Europe to the United States realized the potential to create a bomb far more powerful than using gunpowder through nuclear fission reactions. Einstein sent a letter to President Roosevelt in 1939, stating that an atomic bomb has the ability to destroy a harbor and its surroundings with just one, and to counter Germany, research should be accelerated. Under the leadership of the military, the United States succeeded in developing the atomic bomb, and in 1945, a bomb using enriched uranium was used in Hiroshima, and a bomb using the artificial element plutonium was used in Nagasaki.

The hydrogen bomb utilizes nuclear fusion reactions, and the United States successfully conducted an explosion test in 1952. It should be noted that nuclear fission reactions are used to detonate hydrogen bombs. The crew of a fishing boat, in the South pacific ocean was exposed in 1954 to radioactive substances (white radioactive dust that accumulated on the ship) that were radioactivated by elements of coral reefs and seawater due to atmospheric nuclear tests of hydrogen bombs. It should be noted that the hydrogen bomb is a nuclear fusion reaction, and no fission products (fallout or death ashes) are generated, so it is not exactly correct to say that they were exposed to fallout. The Partial Test Ban Treaty, which prohibits nuclear tests other than underground, was concluded in 1963 by the United States, the UK, the Soviet Union, and others, so since then, these countries have not conducted atmospheric nuclear tests (*note*).

Note: France and China did not initially participate in this treaty. The experiment conducted by China in 1980 was the last atmospheric nuclear test in the world. At that time, the radioactive dust released into the atmosphere was carried by the westerlies, and when it rained, the environmental radiation level in Japan increased. The half-life (the time it takes for the radioactivity to decay by half) of cesium-137 and strontium-90, which are fission products, is about 30 years. These substances released in the atmospheric nuclear tests of nuclear-weapon states have only decayed to about half of their original levels and remain on the world's land surface and in seawater. Their total amount is far greater than that derived from nuclear accidents. After the TEPCO Fukushima accident, reports were made of radioactive cesium being detected in various countries in the Indian Ocean and the South Pacific, but this cesium was not derived from the TEPCO Fukushima accident, but from atmospheric nuclear tests. The fission products derived from China's atmospheric nuclear tests are attached to particles of Chinese soil and sand and have been carried over with yellow sand due to the influence of the westerlies and have been detected for a long period of time at Japan's nuclear facilities' environmental monitors (radiation detectors).

The U.S. Navy, noting the advantage of nuclear-powered submarines being able to submerge for long periods without fuel supply, began research on nuclear power for submarines in 1946 under the command of Admiral Rickover, later known as the "father of the nuclear navy". In 1948, they built a nuclear reactor for submarines and launched the first nuclear submarine, Nautilus, in 1954, which navigated the Arctic Ocean in 1958.

Although the military use of nuclear energy preceded, its peaceful use began with a speech by U.S. President Eisenhower at the United Nations General Assembly in 1953. The establishment of the International Atomic Energy Agency (IAEA), including the USSR, was also proposed and was launched in 1957.

1.9.2 Development of Power Reactors

The United States, the UK, France, the USSR, Canada, and others developed various power reactors. The U.S. Atomic Energy Commission began developing five types of power reactors in 1954. This included the pressurized water light water reactor, which Admiral Rickover developed for use in nuclear submarines and is now the most widely used power reactor in the world (light water is natural water, so called to distinguish it from heavy water, which consists of deuterium and oxygen). In a pressurized water light water reactor, the fuel rods containing the nuclear fuel that emits energy as heat from the nuclear fission reaction are cooled with water. The heated high-temperature, high-pressure water is sent to a heat exchanger called a steam generator to create steam, which turns the rotor of a generator connected to a steam turbine to generate electricity.

The difference between nuclear power generation and thermal power generation is shown in Fig. 1.17. As shown, the equipment that generates steam for power generation differs in that it is a boiler for thermal power generation, while for nuclear power generation it is a nuclear reactor. However, the principle of power generation is the same for both. Thermal power plants generate steam by burning fossil fuels, while nuclear power plants generate steam using the energy produced by the nuclear fission of uranium. In this diagram, a boiling water type light water reactor, which directly generates steam inside the reactor vessel, is shown. The steam generated in the nuclear reactor rotates the turbine, and electricity is generated by the generator. The steam discharged from the turbine is cooled and returned to water in the condenser and is sent to the nuclear reactor by the feedwater pump. The cooling water for the condenser is seawater at power plants along the coast, and river water inland. Large cooling towers that cool the water returned to the river with air are provided at power plants inland. Since all nuclear power plants in Japan are located along the coast, they cool the condenser with seawater. Nuclear power plants were developed using the technology of coal-fired power plants at the time as a reference.

Alternating current power is sent from the power plant to the consumption area, and motors of electrical appliances in factories and homes rotate in sync with the generator. The frequency is 50 Hz in Kanto and 60 Hz in Kansai in Japan. Since the frequency fluctuates depending on the load, it is necessary to suppress the fluctuation, and the power generation amount is controlled by detecting subtle changes in rotation speed. If the balance between power generation and power load is greatly disrupted, a major power outage will occur. As already mentioned, solar power has a different principle of power generation, so equipment for maintaining frequency has been added and used.

There are various types of nuclear reactors for power generation, but all of them generate high-temperature, high-pressure steam in a steam generator using the heat generated by nuclear fission to generate power. This method is the same as thermal power generation. Steam turbines rotate the impeller with steam. The method of guiding steam into a cylinder and reciprocating a piston was used in steam locomotives, but both are steam engines. In thermal power generation, the heating of the

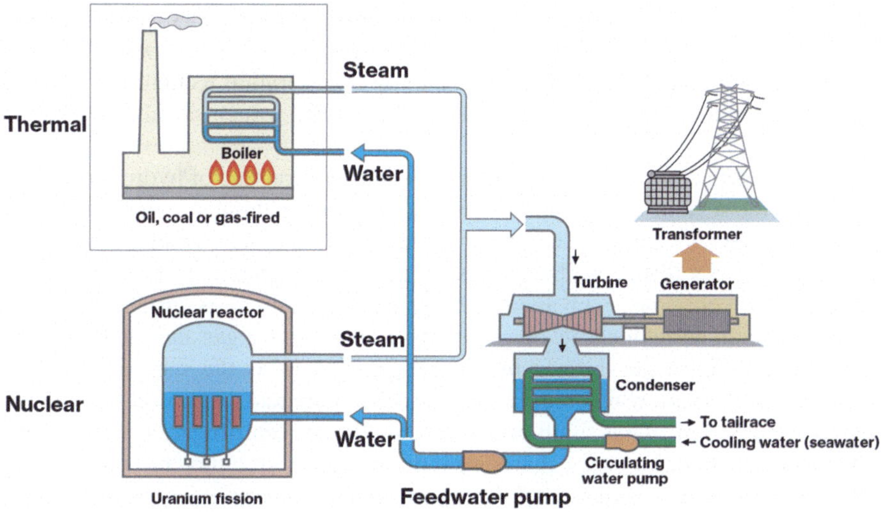

Fig. 1.17 Difference between thermal and nuclear power plants. *Source* JAERO

boiler is performed by burning coal or heavy oil. In nuclear power generation, the nuclear reactor plays the role of the boiler (high-temperature, high-pressure steam generator) of the steam engine. Thermal power generation and nuclear power generation have many technical commonalities, and there is a technical group called the Japan Thermal and Nuclear Power Generation Technology Association in Japan.

The material used in the reactor when Fermi achieved a sustained nuclear fission reaction (also known as a nuclear fission chain reaction) at the University of Chicago was graphite (a crystalline substance of carbon). Neutrons are needed to cause a nuclear fission reaction, but since graphite absorbs a small proportion of neutrons, it was possible to realize a nuclear fission chain reaction even when using natural uranium as a nuclear fuel. Subsequently, reactors using graphite were also used for military purposes to produce artificial (non-naturally occurring) fissionable material, plutonium, from natural uranium.

The United States developed and operated various power reactors from the 1950s to the 1960s. It is not widely known, but in the first decade, over 100 different types of reactors, including test and experimental reactors, were actually built and tested by both the government and private sector (Loftness 1964). Currently, about 100 power reactors are in operation in the United States, all of which are light water reactors (reactors that use naturally occurring water for cooling). The United States exported light water reactors to the world. Germany, France, Japan, and South Korea initially acquired light water reactor design and manufacturing technology through imports and licensed production from the United States, then nurtured their own light water reactor design and manufacturing capabilities, first using them domestically. Light water reactors are also the most common in nuclear power plants worldwide. The reason why light water reactors dominated the nuclear power market in the United

States and the world is that they are compact compared to reactors using graphite, have lower capital costs, and lower power generation costs. Light water reactors also have a high commonality with thermal power generation technology, and the experience of thermal power generation could be utilized, which can also be cited as a reason.

Light water reactors, due to the high neutron absorption rate of hydrogen atomic nuclei, cannot sustain a nuclear fission reaction with natural uranium fuel, and enriched uranium is required. Therefore, commercial uranium enrichment was previously done by gas diffusion, but is now done by centrifugation. The fissionable uranium, which only makes up about 0.7% of natural uranium, is enriched to 3–4% for use. Uranium enrichment involves turning uranium into a gas and separating the fissionable uranium from the non-fissionable uranium using the very slight difference in weight (mass), which requires large-scale equipment and power. The power required for enrichment has significantly decreased since centrifugation replaced gas diffusion. Centrifugation is a technological innovation in uranium enrichment.

The United States is the world's largest nuclear power generating country, accounting for about 30% of the world's total power generation. The proportion of nuclear power generation in the United States' power supply is about 20%. Many of the nuclear power plants in the United States were built in the 1960s and 1970s, and recently, some have obtained operating permits for 80 years. In the United States, nuclear power generates the most electricity among power generation methods that do not emit greenhouse gases.

Light water reactors need to use enriched uranium (uranium with a higher proportion of fissionable uranium than natural uranium) as fuel, and since the UK and France did not have uranium enrichment facilities in the 1950s, they developed power reactors based on reactors using natural uranium and graphite. The cooling of the reactor was air for military use, but carbon dioxide was used for power reactors.

The UK built 26 Magnox reactors, which use graphite, and 15 Advanced Gas-cooled Reactors (AGRs) domestically for nuclear power generation. Magnox reactors were exported to Japan and Italy. The first commercial nuclear power plant in Japan was this type of reactor. The Advanced Gas-cooled Reactor is still in operation in the UK. The UK later built one Pressurized Water Reactor (PWR) in the 1990s. Since the 1980s, the UK had slowed down the construction of new nuclear power plants due to the availability of crude oil from the North Sea oil fields. However, as the North Sea oil fields are depleting and the country is returning to being an energy importing country, and in response to global environmental issues, the UK has recently been promoting the construction of new nuclear power plants. Currently, the reactors under construction and planned for power generation in the UK are light water reactors.

France initially developed gas-cooled reactors using natural uranium and graphite for power generation, and built eight units in the 1960s. However, due to the scarcity of domestic fossil fuel resources and in response to the first oil crisis, France started operating 44 PWRs in the first 10 years from 1978, and a total of 58 PWRs. The proportion of nuclear power in France's electricity supply is extremely high at about 70%, and electricity rates are also cheap. Including not only power generation but also energy consumption for transportation, industry, and domestic use, the self-sufficiency rate of primary energy (energy used not only for electricity but also for industry, domestic use, and transportation) in France exceeds 50%.

Canada developed and used a power reactor called the Canadian Deuterium Uranium Reactor (CANDU), which uses natural uranium fuel and heavy water. As of the end of 2020, 19 units are in operation in Canada. Canada has exported to Pakistan, India, South Korea, Argentina, Romania, and China from the 1970s to the 2000s. India has manufactured, built, and operated 19 Pressurized Heavy Water Reactors (PHWRs) with its own technology based on the reactors imported from Canada. The power output is smaller than that of light water reactors for both graphite moderated carbon dioxide-cooled reactors and pressurized heavy water reactors, with the UK's Advanced Gas-cooled Reactor at about 600,000 kilowatts, Canada's heavy water reactor from 500,000 to about 900,000 kilowatts, and India's heavy water reactor from 200,000 to 600,000 kilowatts.

Russia's power reactors were independently developed during the Soviet era. One is a Russian type of Pressurized Water Reactor (PWR) called VVER, and the other is a type of reactor called RBMK that uses graphite and is cooled with light water. These reactors were built and operated in Russia and Ukraine during the Soviet era and were also exported to Finland and Eastern European countries. The reactor involved in the Chernobyl accident was an RBMK type. Currently, both types of reactors are still in operation in Russia and other countries. Even after becoming Russia, Russian type light water reactors have been exported to China, India, Bangladesh, etc., and are in operation and under construction.

China has pursued the use of nuclear energy through both the independent development of light water reactors and the import of large light water reactors from France. The first large light water reactor imported from France was built in Guangdong Province and successfully supplied electricity to Hong Kong, achieving business success. Subsequently, large light water reactors were built in various coastal areas. Currently, they have managed to build a large light water reactor called Hualong One, which holds intellectual property rights in their own country, and it has been in operation since 2020. The small light water reactor they developed independently started operating a 300,000-kilowatt nuclear power plant domestically in 1991, exported it to Pakistan, and put it into operation in 2000.

South Korea has imported power reactors from France and Canada but has nurtured its own nuclear power technology based on American-style light water reactors and promoted the use of nuclear energy. By 2024, 25 reactors are supplying about a third of South Korea's electricity. Based on the design and experience of nuclear power plants built and operated domestically, South Korea has exported four nuclear power plants to the United Arab Emirates, with the first power plant in operation in 2021. South Korea, like Japan, is in a situation with similar energy resources and geographical conditions, such as the lack of fossil fuel resources, the need to import liquefied natural gas, and the lack of connection to power lines with other countries, but its electricity rates are about half of Japan's. Japan has much to learn from the efficient management of South Korean power companies and the government's energy and nuclear policies. In addition to low electricity rates, administrative reforms caused by the economic crisis in 1997, and the systems, competitive environment, and overseas orientation created by them, may have led to South Korea's subsequent economic development.

These countries not only use the reactors they have developed for nuclear power generation in their own countries, but also export them. Table 1.2 shows the export destinations and the year when the first exported nuclear power plant started commercial operation. Each country started exporting shortly after starting domestic use. Japan's exports have not yet been successful. Japan is almost the only country in the world that has built many nuclear power plants in its own country but has not exported them. It can be seen that Japan should have considered exporting around 1970 when Kansai Electric Power's Mihama Unit 1 and Tokyo Electric Power's Fukushima Daiichi Unit 1 started operation, i.e., when Japan started using nuclear power in earnest. Japan's use of nuclear power may have been focused on domestic perspectives for a long time. This is unique to Japan. Efforts to export have been made since the 1990s.

As of August 2022, the number of commercially operable power plants is 307 pressurized water reactors (PWR), 61 boiling water reactors (BWR), 47 pressurized heavy water reactors (PHWR), 11 RBMKs, 8 advanced gas-cooled reactors (AGR), 2 fast reactors, and 1 high-temperature gas-cooled reactor (HTGR) (WNA 2022).

Various types of nuclear reactors have been developed, but the reason why light water reactors have dominated the market is that their construction costs were lower than other types of reactors. The reason for the lower construction cost is that light

Table 1.2 Export of nuclear power plants and the year of commercial operation of the first plant

USA (PWR): Belgium 1962, Italy 1965, France 1965, Germany 1969, Spain1969, Switzerland 1969, Japan 1970, Sweden 1975, Korea 1978, Slovenia 1983, Taiwan 1984, Brazil 1985, UK 1995, China 2018
USA (BWR): Italy 1964, Holland 1969, India 1969, Japan 1970, Spain 1971, Swizerland1972, Taiwan 1978, Mexico 1990
UK (GCR): Italy 1964, Japan 1966
France: (GCR): Spain 1972, (PWR): Belgium 1975, South Africa 1984, Korea 1988, China 1994
Canada: (CANDU): Pakistan 1972, India 1973, Korea 1983, Argentina 1984, Rumania 1996, China 2002
Germany (PWR): Holland 1973, Switzerland 1979, Spain 1988, Brazil 2001, (PHWR): Argentina 1974
Sweden (BWR): Finland 1979
Soviet Union/Russia (VVER): Bulgaria 1974, Finland 1977, Armenia 1977, Slovakia 1980, Ukraine 1981, Hungary 1983, Lithuania 1985, Czechoslovakia 1985, China 2007, Iran 2013, India 2014, Belarus 2020 (RBMK): Ukraine 1978, Lithuania 1983
China (PWR): Pakistan 2000
Korea (PWR): UAE 2021

Note BWR: Boiling Water Reactor, PWR: Pressurized Water Reactor, GCR: Gas-Cooled Reactor, CANDU: Canadian Deuterium Uranium, PHWR: Pressurized Heavy Water Reactor, RBMK (Reaktory Bolshoi Moshchnosti Kanalynye), VVER (Voda Voda Energo Reactor)

water requires the smallest necessary volume to efficiently cause a nuclear fission chain reaction compared to graphite or heavy water, which reduces the volume of the nuclear reactor. When the volume of the equipment is small, the cost of manufacturing materials and processes is reduced, and the building that houses the equipment also becomes smaller, reducing the construction cost required for the construction of a nuclear power plant. The reason why the volume of the nuclear reactor becomes smaller when using light water can be explained from the theory of nuclear reactors. As it becomes specialized, this explanation is omitted in this chapter. Please refer to the materials posted on the activity introduction of the Cabinet Office Nuclear Energy Commission's website (Oka 2018a) from page 112 onwards.

1.9.3 Nuclear Research and Development and Energy Use in Japan

Research related to nuclear science in Japan began with the dawn of the twentieth century. Table 1.3 summarizes the history of nuclear utilization in Japan.

In 1903, Hantaro Nagaoka proposed an atomic model. The Institute of Physical and Chemical Research (RIKEN) was established in 1917 as the first comprehensive research institute for basic science in Asia and has since produced many scientists. Yoshio Nishina, while studying in Europe in 1928, published the Klein-Nishina formula for inelastic scattering of gamma rays. Nishina built a cyclotron accelerator at RIKEN in 1937 and conducted research on atomic science and its applications, nurturing nuclear scientists. Nobel laureates in physics have come from Nishina's disciples and grand-disciples. Hideki Yukawa received the Nobel Prize in Physics in 1948 for his meson theory, which he published in 1934.

The research and development of nuclear energy in Japan began when the budget for nuclear research and development was submitted to the Diet in 1954 by Yasuhiro Nakasone and others, following President Eisenhower's 1953 United Nations speech. In 1955, the Atomic Energy Basic Law was established, limiting the research, development, and use of nuclear power to peaceful purposes, and stipulating that it should be based on the three principles of "democracy", "autonomy", and "openness". In 1956, Japan Atomic Energy Commission (JAEC) was established, with Matsutaro Shoriki becoming the first chairman. Shoriki became the first Director-General of the Science and Technology Agency and had a significant influence on the introduction of nuclear power to Japan. The Japan Atomic Energy Research Institute (JAERI) was established in Tokai Village (Tokai mura), Ibaraki Prefecture, and the Japan power demonstration reactor (JPDR) conducted Japan's first nuclear power generation in 1963.

In 1957, Japan Atomic Power Company (JAPC), funded by nine electric power companies and Electric Power Development Co. Ltd (EPDC, also called J-Power), was established. They imported from the UK and built the Tokai Power Station in Tokai Village, which started the first commercial nuclear power generation in 1966. In

Table 1.3 History of nuclear utilization in Japan

1903: Hantaro Nagaoka devised the atomic model
1917: Establishment of RIKEN
1928: Yoshio Nishina presents the Klein-Nishina formula
1937: Yoshio Nishina builds a cyclotron accelerator at RIKEN
1948: Hideki Yukawa Wins Nobel Prize for Meson Theory
1954: Nuclear research and development budget submitted to the Die
1955: Atomic Energy Basic Law enacted
1956: Japan Atomic Energy Commission (JAEC) established
1956: Japan Atomic Energy Research Institute (JAERI, now Japan Atomic Energy Agency, JAEA) established
1957: Japan Atomic Power Co., Ltd. (JAPC) established
1957: National Institute of Radiological Sciences (now Institute for Radiological Science of National Institutes for Quantum Science and Technology, QST) established
Late 1950s-early 60s: Universities open nuclear engineering departments
1963: The first power reactor, Japan Power Demonstration Reactor (JPDR) of JAERI started power generation
1966: The first nuclear power plant (JAPC Tokai Power Station) started commercial operation
1967 Power Reactor Nuclear Fuel Development Corporation (PNC) launched, currently Japan Atomic Energy Agency (JAEA) after Japan Nuclear Cycle Development Institute (JNC)
1970v First commercial light water reactor power plant (JAPC Tsuruga Unit 1) started operation
1970s–80s: Electric power companies constructed and started operation of light water nuclear power plants throughout Japan
1976: Established the Nuclear Engineering Testing Center (later renamed the Nuclear Power Technology Center (NUPEC), disbanded in 2008 after transfer of operations to Japan Nuclear Energy Safety Organization (JNES), etc.)
1978: Nuclear Safety Commission (NSC) established
1980: Established Japan Nuclear Fuel Service Co., Ltd., currently Japan Nuclear Fuel Ltd. (JNFL)
1999: JCO accident in Tokai-mura
2001: Established Nuclear and Industrial Safety Agency (NISA) (Abolished with the establishment of the Nuclear Regulation Authority (NRA) in 2012)
2005: Establishment of Japan Atomic Energy Agency(JAEA) (launched by integration of Japan Atomic Energy Research Institute (JAERI) and Nuclear Fuel Cycle Research and Development Agency (JNC))
2011: TEPCO Fukushima Daiichi Nuclear Power Station accident
2012: Nuclear Regulation Authority (NRA) and Nuclear Regulatory Agency established (Nuclear Safety Commission abolished)

1970, the Tsuruga Power Station Unit 1, a light water nuclear power station imported from the United States, and Kansai Electric Power's Mihama Power Station Unit 1 in Fukui Prefecture began generating power, which was also supplied to the Osaka Expo held that year. In Japan, which was in the midst of a period of high economic growth, expectations for various cutting-edge technologies that would support the future were high, and the recognition that "nuclear power is an energy that can be used for power generation" spread throughout Japan.

From the late 1950s to the early 1960s, nuclear engineering departments were established at universities and nuclear engineering education began. Research on the health effects of radiation was conducted at Hiroshima University, Nagasaki University, the University of Tokyo, Kyoto University, and others, and National Institute of Radiological Sciences was established as a national research institution in 1957. From the 1970s to the 1980s, electric power companies built and operated nuclear power plants throughout Japan. As shown in Fig. 1.18, the annual power generation of nuclear power reached 34% in the late 1990s. After that, the operation rate of Japan's nuclear power plants declined due to criticism of trouble response, etc. Since the 2000s, the amount of coal-fired and natural gas-fired power generation has increased, and the proportion of nuclear power generation in 2010 was 25%. Meanwhile, nuclear power plants in Western countries have achieved improved operation rates and increased power output over the approximately 20 years from the 1990s. During this period, in Japan, neither the improvement of the operation rate nor the increase in output progressed much, and it was left behind by the world. Since the TEPCO Fukushima accident in 2011, new regulatory standards have been implemented by stopping nuclear power plants, so there are few nuclear power plants that are restarted, and the proportion of nuclear power generation is low.

In 1976, the Nuclear Engineering Test Center, under the jurisdiction of the Ministry of International Trade and Industry, was established with the cooperation of the electric power industry, the electrical industry, the construction industry, etc., and

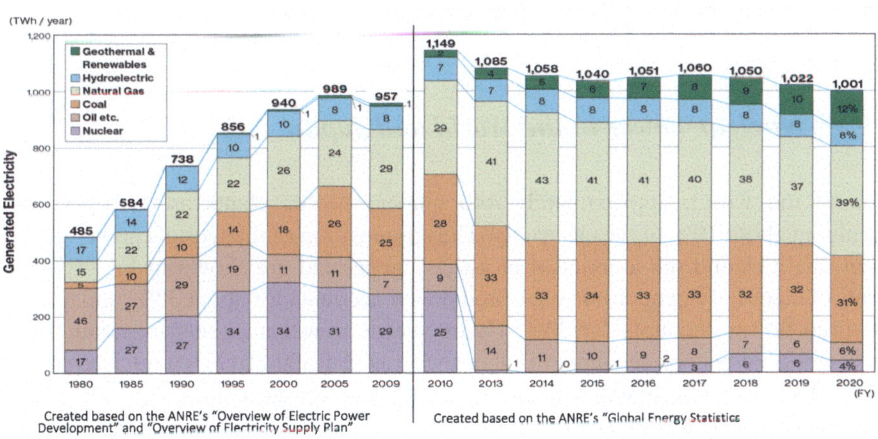

Fig. 1.18 Japan power generation and purchase volume by source. *Source* JAERO

in 1992, it was renamed the Nuclear Power Engineering Center (NUPEC), conducting reliability demonstration tests for nuclear power generation equipment and seismic demonstration tests. In 2003, the regulatory affairs of nuclear power generation were transferred to Japan Nuclear Energy Safety Organization (JNES), the seismic demonstration test was completed in 2005, and the remaining business was transferred to the Institute of Applied Energy (IAE), and the NUPEC was dissolved in 2008. The JNES was integrated into the Nuclear Regulation Authority (NRA) after the Great East Japan Earthquake and was abolished in 2014. The Central Research Institute of Electric Power Company (CRIEPI), operated with funds from electric power companies, was established in 1951 and is active as a joint research institution and think tank for the electric power business. Research related to nuclear technology, socio-economics, civil engineering seismic technology, etc. is also being conducted.

In 1973 and 1978, oil shocks occurred. The urgent need to transition to inexpensive energy sources led to the construction and operation of numerous nuclear power plants throughout Japan. Including those that have already been shut down, 59 reactors were built for power generation. In 1980, Japan Nuclear Fuel Services Co., Ltd. (now Japan Nuclear Fuel Ltd., JNFL), which conducts uranium enrichment, low-level radioactive waste disposal, high-level radioactive waste management (radioactive waste generated from spent fuel reprocessing), reprocessing, mixed oxide fuel manufacturing, and transportation, was established with investments from electric power companies. Uranium enrichment, low-level radioactive waste disposal, and transportation are in operation. As of 2023, work is underway to start the reprocessing and mixed oxide fuel manufacturing operations. Research and development on the disposal of high-level radioactive waste has been conducted since 1976, and activities are underway to locate a disposal site. The Nuclear Safety Commission was established in 1978 but was abolished with the establishment of the Nuclear Regulation Authority in 2012. The Nuclear and Industrial Safety Agency (NISA), established in 2001, was also abolished at this time. In 2005, the Japan Atomic Energy Research Institute (JAERI) and the Japan Nuclear Fuel Cycle Development Institute (JNC) merged to form the Japan Atomic Energy Agency (JAEA).

1.9.4 Nuclear Power in the World and Japan

Around the world, approximately 440 nuclear power plants are in operation, supplying about 10% of the electricity. The composition of the world's power generation facilities and the amount of electricity generated are shown in Fig. 1.19. In terms of electricity generation, coal-fired power accounts for about 38%, and natural gas-fired power accounts for 23%. Hydroelectric power accounts for about 16%, and renewable energy accounts for about 9% (as of fiscal 2017), so nuclear power, which does not emit greenhouse gases, is the second-largest supplier of electricity after hydroelectric power. Nuclear power and coal-fired power have a larger share of electricity generation than their share of power generation facilities. This is because they are used as base load power sources that operate continuously.

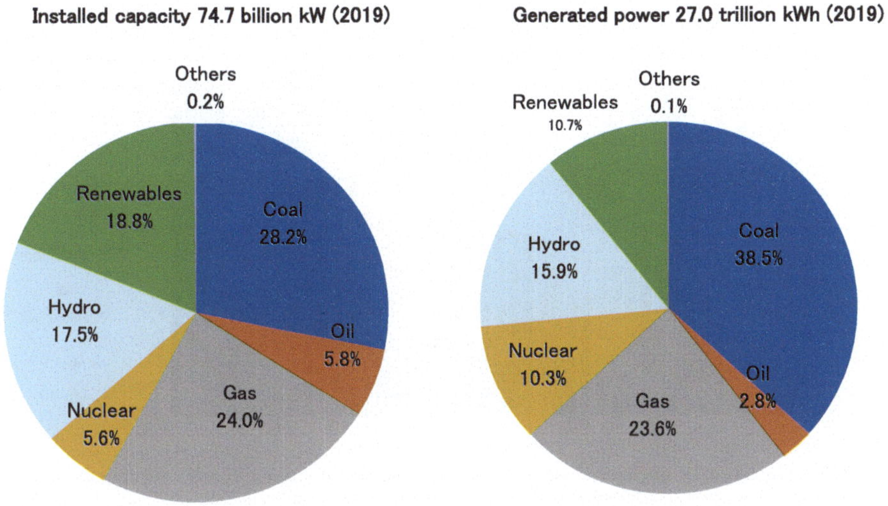

Fig. 1.19 Installed capacity and generated power in the world. *Source* Energy white paper 2022, METI

Since the start of nuclear power generation in the United States, the use of nuclear power has been actively promoted in Japan and other countries around the world, triggered by two oil shocks. Figure 1.20 shows the trend in the world's nuclear power generation capacity (in operation). Since around 1990, the pace of increase in power generation has slowed in North America and Europe. However, in Asia, nuclear power generation has been steadily increasing, in countries such as Japan, South Korea, and China. The power generation capacity of nuclear power in Asia decreased due to the accident at Tokyo Electric Power Company's Fukushima Daiichi Nuclear Power Plant in March 2011, but it has been increasing again since 2014.

Currently, the nuclear power generation capacity of countries around the world is shown in Fig. 1.21 As shown, the United States has the most, followed by France, China, Japan, and Russia, in that order.

As of February 2011, Japan was operating 54 commercial nuclear power plants. Due to the Great East Japan Earthquake, six units of the TEPCO Fukushima Daiichi Nuclear Power Plant, which had an accident, and subsequently, an additional 15 nuclear power plants chose to cease operations. Many of these nuclear power plants were built early on and were small power plants that had been operating for over 40 years after their start, but they also included large power plants. Whether to cease operations was up to each power company. Under the nuclear regulations established in Japan after the accident, extensions of operating periods were only approved once every 20 years, so if the cost of construction for compliance with new regulatory standards could not be recovered in the extended 20 years of operation, it is thought that they had no choice but to cease operations. The reason why many of the power plants that chose to cease operations were small power plants is because they generated less power, and it was judged that the recovery of construction costs

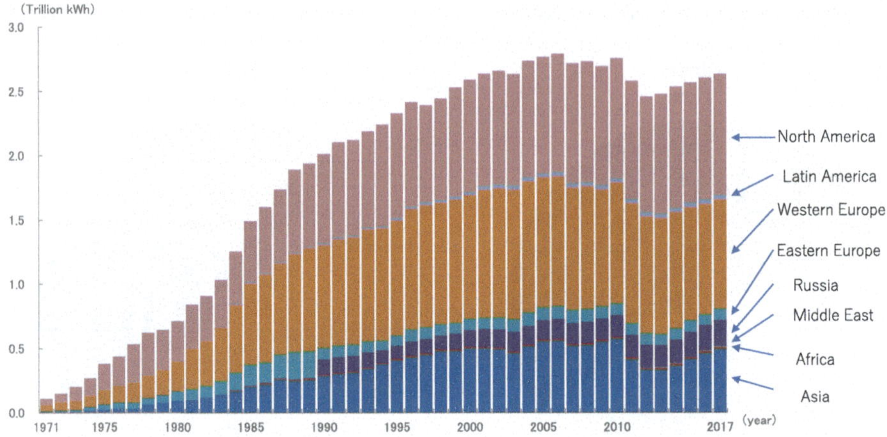

Fig. 1.20 Transition of nuclear power generation in the world. *Source* Energy white paper 2020, METI

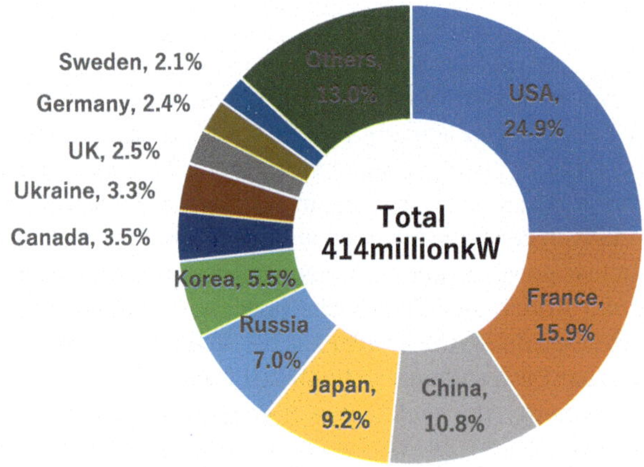

Fig. 1.21 Installed capacity of nuclear power in the world. *Source* Energy white paper 2020, METI

due to the extension of operations would be difficult over 20 years. In Japan, nuclear power plants that have been operating for 30 years from the start are inspected every 10 years and can operate up to 60 calendar years. A revision to the rule that the period of operation stoppage is not included in the calculation of the operating period was made in 2023.

Nuclear power plants in the United States and other countries around the world do not have predetermined operating periods of 40 or 60 years. Whether continued operation is possible is reviewed by the Nuclear Regulatory Commission, and if permission is obtained, it can be extended by 20 years at a time. In the United States, some power plants have obtained permission to operate for up to 80 years. The equipment of nuclear power plants is regularly inspected, and items that have changed over the years due to wear and corrosion can be replaced.

In Japan, the fact that many nuclear power plants, which were supposed to be able to supply cheap electricity if used for a long time, chose to cease operations after the Great East Japan Earthquake, is a major mistake in terms of the national economy and local employment. Like hydroelectric power plants, nuclear power plants have no lifespan and can be used for a long time if equipment is replaced. There is no technical basis for Japan's current decision to allow nuclear power plants to operate for up to 60 years.

1.9.5 Continuous Safety Improvement Activities in the US Nuclear Power Industry: Achieved Both Safety and Economic Improvements

In the United States, a nuclear reactor meltdown occurred at the Three Mile Island nuclear power plant in 1979. The U.S. power industry recognized that there were issues with their operation management, training, experience sharing, and crisis management, and established the Institute of Nuclear Power Operations (INPO) to improve safety. They collected and shared operational and trouble experiences, learned from the mistakes and experiences of other companies, and the INPO reviewed the operation management of each power plant, urging the presidents of power companies that performed poor operations and management to improve. This is referred to as voluntary safety improvement activities or continuous safety improvement activities. As a result, as shown in Fig. 1.22, in the United States, over a decade from 1989, power generation increased by about 50%, and significant events such as accidents decreased to about one-thirtieth. The involvement of the president, focusing on safety, industry support for INPO activities, accountability, and independence are cited as factors for this success (Pate and Ellis 2011).

In the case of nuclear power generation, one company is affected by the accidents of other companies. The key point of INPO's activities, as the slogan "Hostages of each other" suggests, is that the president of power companies with poor safety management are called out at INPO meetings by other presidents, urging them to improve. At the time of INPO's establishment, many power companies across the country were operating nuclear power plants. This mechanism to improve poorly managed companies among them hits the key points of safety management improvement. The information on each company's safety management and troubles provided to INPO is used internally and is not disclosed. Although there have been lawsuits by anti-nuclear groups demanding disclosure, federal courts have dismissed them, citing the many benefits of non-disclosure. In Japan, there is a common impression that everything should be disclosed due to collectivism, but the data obtained from operations is the private property of power companies. The responsibility for operation management lies with the power company. As the example of INPO shows, it is better to use troubles below the regulatory standards not for disclosure but for improving safety by power company management. Safety management is management itself.

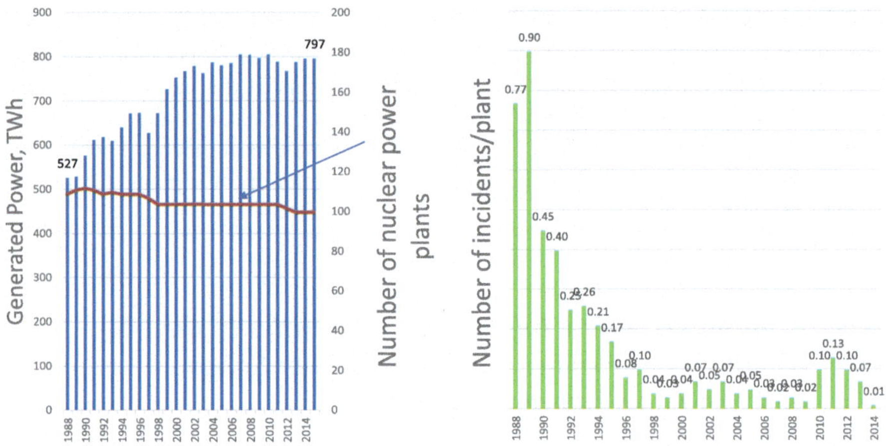

Fig. 1.22 Increased nuclear power generation and improved safety after the TMI accident in USA. *Source* NEI

This is also the experience the author gained from long managing a research reactor facility at the University of Tokyo.

It is noteworthy that the involvement of the president is cited as a factor for the success of INPO's activities. In the United States, there is a saying, "If one power company causes a nuclear accident, it becomes an accident for all power companies". Since accidents and troubles at nuclear power plants attract a lot of public attention, and all companies that operate nuclear power are affected by accidents at other companies, it is important to improve weak organizations and companies in organizational management. It is reasonable to demand the direct involvement of the president, who is the person responsible for company management. Although human factors and safety culture have been issues in ensuring nuclear safety, they can be addressed by organizational management. The method of learning from failures and taking measures is also organizational management.

When discussing economics in Japan, one may hear opinions that safety is neglected, but as shown in Fig. 1.22, it is a fact that safety improvement and economic improvement can coexist. From the success stories of INPO, it can be learned that safety assurance and improvement are important for management and that the software (operation) is more important than the hardware (equipment).

The U.S. Nuclear Regulatory Commission also strives to improve regulations and encourages power companies to improve, rather than cracking down on minor troubles. Slogans such as ROP (Reactor Oversight Process) and Risk-informed/ performance-based regulation are used.

Figure 1.23 shows the U.S. nuclear power utilization system related to continuous safety improvement (also called voluntary safety improvement or self-regulation of safety by power companies in Japan). It can be considered that U.S. nuclear power companies fulfilled their responsibility for causing a stir in the world with the

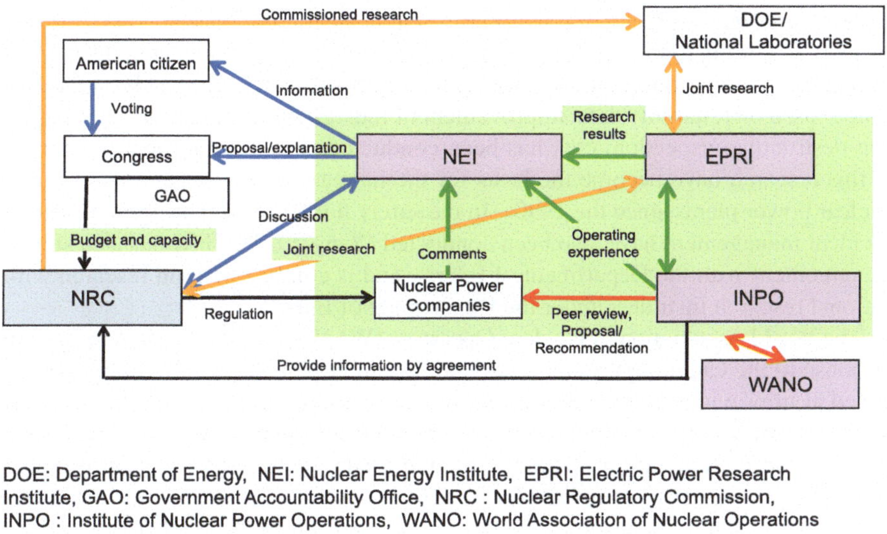

Fig. 1.23 Nuclear power utilization system including continuous safety improvements in USA. *Source* METI, modified by the author

Three Mile Island nuclear accident in 1979 by reducing accident rates and improving operation rates.

The United States is a federal country, so each state has jurisdiction over the power business. The degree of power liberalization is also decided by each state. The United States has a large territory, so there are many power companies, and there is no transmission line covering the whole country. U.S. nuclear power generation was previously conducted by many power companies, but in recent years it has been consolidated into several companies through the acquisition and sale of power plants. Most of them are private companies.

The Nuclear Energy Institute (NEI) is a member organization composed of nuclear power companies and reactor manufacturers and was established in 1994 by integrating several nuclear industry associations. The main role of NEI is to consolidate the U.S. nuclear industry and express opinions and proposals to the Federal Congress and the Nuclear Regulatory Commission on nuclear and energy policies. It also communicates with Congress and the public. In the United States, there is a service called SmartBrief that sends news from each industry to those who wish by email. Information in the nuclear field was originally sent as NEI SmartBrief, but now it is sent by the American Nuclear Society (ANS) under the name of ANS Nuclear Smart Brief.

The Electric Power Research Institute (EPRI) was established in 1972 as an independent, non-profit research institute conducting research on power-related technologies. Several Japanese power companies also participate. In the nuclear power field, before the deregulation of electricity in the United States that began in the 1980s, they conducted research on their own themes, but now they do not conduct basic research and development like national research institutes. Instead, they propose projects according to the needs of participating companies and collect budgets from companies participating in those projects to conduct research. Research on maintenance based on reliability data, improvement of fuel reliability, material degradation, non-destructive inspection, etc., has been conducted, and it is said that the results of this research have become the basis for the increase in power generation at US nuclear power plants since the 1990s. In the safety field, research projects on severe accident management have also been conducted (Yang 2014). EPRI has also signed an agreement with the Department of Energy and is conducting joint research with national research institutes under the Department of Energy.

The World Association of Nuclear Operators (WANO) was established in 1989 in response to the Chernobyl accident as a non-profit organization (a UK corporation) aimed at improving safety, with nuclear power operators from around the world as members. They conduct exchanges of operational information and good practices, analysis and dissemination of events, mutual visits, reviews, etc. Japanese power companies also participate in WANO.

The US Congress determines the budget and staffing of the Nuclear Regulatory Commission. The budgets of the Nuclear Regulatory Commissions in the United States and the UK are mostly covered by fees paid by companies receiving licenses. The Government Accountability Office (GAO) audits and evaluates US government agencies for Congress. In the 1990s, the GAO evaluated the work of the Nuclear Regulatory Commission, which was used to improve nuclear regulation. In the United States, independent administrative agencies such as the Nuclear Regulatory Commission are required to create and publish their work and plans as strategic plans, etc., based on which the GAO evaluates and Congress can encourage improvements.

1.9.6 Continuous Safety Improvement in Japan

In Japan, after the TEPCO Fukushima nuclear accident, the Japan Nuclear Safety Institute (JANSI) was established by power company personnel as an organization for continuous safety improvement (also called voluntary safety improvement), and activities are being conducted with reference to the American INPO. The structure of Japan's use of nuclear power generation, including continuous safety improvement, is shown in Fig. 1.24.

In Japan, ten power companies, such as Tokyo Electric Power Company (TEPCO) and Kansai Electric Power Company (Kansai), were conducting power businesses in each region before the liberalization of electricity. Of these, nine companies excluding Okinawa Electric Power are conducting nuclear power generation. In addition, Japan Atomic Power Co., Ltd. (JAPC), which was established with investment from power companies, is conducting nuclear power generation, and Electric Power Development Co., Ltd. (EPDC) is constructing a nuclear power plant. EDPC was a power company established by the government, but is now privatized, and all these power companies are private companies.

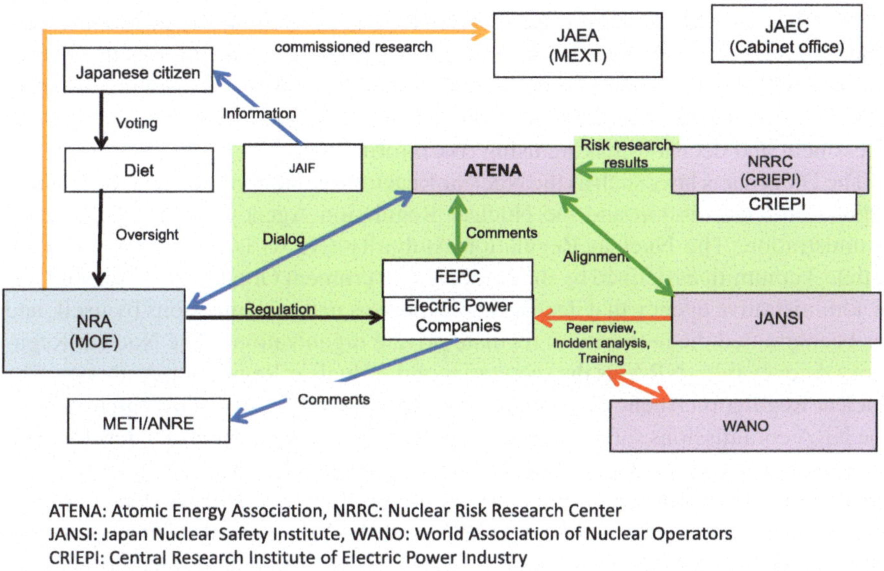

Fig. 1.24 Nuclear power utilization system including continuous safety improvements in Japan. *Source* METI, revised by the author

The Federation of Electric Power Companies of Japan (FEPC) was established in 1952 and is now a private organization with ten power companies, including Okinawa Electric Power, as members. It exchanges opinions among power companies and communicates opinions on the electric power business to the government and the public.

Japan's resource and energy administration is under the jurisdiction of the Agency for Natural Resources and Energy (ANRE), an external bureau of the Ministry of Economy, Trade and Industry (METI), and administration related to resources and energy, such as nuclear power policy and nuclear fuel cycle policy, is being conducted. It compiles the Basic Energy Plan. ANRE has the Advisory Committee for Natural Resources and Energy, and within its Electric Power and Gas Business Subcommittee, there is a Nuclear Power Subcommittee, which is held regularly. Under the Nuclear Power Subcommittee, there are the Voluntary Safety Improvement Working Group and the Radioactive Waste Working Group, etc., and university professors, etc. are appointed as members, held for a certain period as needed, and compile recommendations and roadmaps to report to the Nuclear Power Subcommittee. Electric power operators (power companies) attend the Nuclear Power Subcommittee as expert members.

The Atomic Energy Association (ATENA) is a private organization established in 2018 by power companies and nuclear reactor manufacturing companies, etc., and is an organization that plans safety measures and conducts dialogues with regulators in order to establish continuous safety improvement efforts.

The Nuclear Risk Research Center (NRRC) is a research organization established by power companies at the Central Research Institute of Electric Power Industry (CRIEPI) in 2014 to conduct risk research necessary for voluntary safety improvement. It conducts research on the development and use of probabilistic risk assessment and decision-making using risk information.

The Diet enacts laws such as the Nuclear Reactor Regulation Law, and the Nuclear Regulation Commission and the Nuclear Regulation Agency are in charge of its administration. The Nuclear Regulation Authority (NRA) is one of the so-called Article 3 committees defined by the National Government Organization Act, and it is an administrative agency that decides and expresses national intentions by itself, and has been granted the authority to regulate private organizations. The Nuclear Regulatory Agency (also NRA) is the secretariat of the Nuclear Regulation Authority. The Nuclear Regulation Agency is an external bureau of the Ministry of the Environment. The NRA commissions safety research to national nuclear research and development institutions such as the Japan Atomic Energy Agency (JAEA), which is under the jurisdiction of the Ministry of Education, Culture, Sports, Science and Technology.

The author has already mentioned JANSI and WANO. The enforcement power of INPO's review and WANO's review is different. If INPO's method cannot be directly applied to Japan, The author would like JANSI to not only imitate the form of INPO, but also create a new method and achieve results (Yamasaki 2022). The organizations directly related to Japan's continuous safety improvement and regulatory improvement are as above.

The Central Research Institute of Electric Power Industry (CRIEPI) is a private research institution (general foundation) established in 1951, with hundreds of researchers. It conducts research on not only nuclear technology, but also civil engineering technology, power transmission and distribution technology, and socio-economic technology, mainly with grants from electric power companies. The Socio-Economic Research Institute and the Nuclear Risk Research Center appear to be operated with respect for their independence due to the nature of their research.

The Japan Atomic Industrial Forum (JAIF) was established in 2006 by reorganizing the Japan Atomic Industrial Conference, which was established as a corporation in 1956. Members include electric power companies, reactor manufacturers, civil engineering and construction companies, related companies, local public bodies, and foundations. It has the Kansai Nuclear Power Discussion Group and the Chubu Nuclear Power Discussion Group as branches. It conducts public relations activities, international exchanges, and regional exchanges. It publishes the Nuclear Industry Newspaper and distributes news by email to those who wish.

The Japan Atomic Energy Commission (JAEC) was established in 1956 under the Atomic Energy Basic Law to plan and implement national policies and to ensure democratic administration of nuclear power. It deliberates on cross-ministerial issues and other matters. In addition to expressing opinions on individual issues as JAEC decisions and views, it prepares the "Basic Thoughts on the Use of Nuclear Power" approximately every five years and the Atomic Energy White Paper every fiscal year. The JAEC is a so-called Article 8 committee, which conducts investigations and deliberations by a panel of academic experts, as stipulated in the National Government Organization Act. The Cabinet Office is in charge of its affairs.

1.9.7 Improvement of Nuclear Regulation in the United States

The U.S. Nuclear Regulatory Commission (NRC) was created in 1974 as an offshoot of the Atomic Energy Commission (AEC) and began operations in 1975. In 1979, the Three Mile Island nuclear accident occurred, and the Kemeny Commission report submitted to then-President Carter recommended the need for regulatory reform, along with industry-led reform that led to the activities of the aforementioned INPO (Kemeny 1979).

In the 1980s, there was growing criticism from the industry that inspections and regulations by the NRC involved a lot of paperwork and were not producing substantial effects. The Nuclear Energy Institute (NEI), commissioned a consulting firm to create a report on the actual state of nuclear regulation, which was published in 1994. Initially, the NRC rebutted this report, but the Government Accounting Office (GAO) evaluated the NRC's activities and urged the NRC to respond. The federal congress stepped in, and the Union of Concerned Scientists (UCS) also testified in congress. The fact that INPO's efforts to improve safety, as shown in Fig. 1.22, began to take effect from the 1990s, backed up the industry's claim that regulatory improvements were necessary. The NRC's reform was carried out by Chairperson Shirley Jackson (Jackson 1998). Public hearings were held, and workshops were held based on the model proposed by NEI, and the operation of a new system called the Reactor Oversight Process (ROP) began. The relationship between the NRC and the operators is based on "Trust but verify" (Kondo 2020; Kondo and Yamaguchi 2018).

The U.S. Nuclear Regulatory Commission has many points of reference for Japan's nuclear regulation. The overview is described in Chap. 4.

1.9.8 Probabilistic Risk Assessment

Although it becomes a bit specialized, continuous safety improvement and regulatory improvement are important for Japan's nuclear power generation to supply cheap and stable electricity safely in the future, so I will discuss the probabilistic risk assessment method, a risk assessment method developed in the United States. The probabilistic risk assessment method is used in the United States as a means of rational regulation (improvement of regulation).

In the United States, there has been a question since the 1960s, "How safe is safe enough?" To answer the question, "What is the risk of nuclear power generation?" the then U.S. Atomic Energy Commission conducted the "Reactor Safety Study". The report was published in 1975 by the newly established Nuclear Regulatory Commission. The relationship between the frequency of accidents and the number of deaths at 100 nuclear power plants was determined by the probabilistic risk assessment method and compared with accidents of artificial objects such as aircraft accidents and dam collapses, and natural disasters such as hurricanes and tornadoes (NRC 1975).

Probabilistic risk assessment (PRA) targets all possible accidents, quantitatively evaluates their frequency and impact when they occur, and determines how small the "risk", which is the product of these, is. This is a method to express the degree of safety. Reliability analysis methods called fault trees and accident progression analysis methods called event trees are used. In fault tree analysis, the reliability of nuclear power plant equipment is calculated from the reliability data of the components that make up it. In event tree analysis, accidents are represented by combinations of initial events and the successful operation and failure of related equipment, and the probability (frequency) of accidents is calculated using the reliability of the equipment obtained from fault tree analysis (McCormick 1981). The appendix of the WASH-1400 report contains a vast amount of reliability data used. In Japan, the term Probabilistic Safety Analysis (PSA) has been used, but since it is a method of evaluating risk, it is better to call it Probabilistic Risk Assessment, following the United States.

After "Nuclear Reactor Safety Research", probabilistic risk assessments of several nuclear power plants were conducted and the method was improved. In the late 1980s, the risk of severe accidents at five types of light water nuclear power plants was analyzed using probabilistic risk assessment, and the final report was compiled in 1990 (NRC 1990).

Since then, probabilistic risk assessment has been widely used in U.S. regulations and has been helpful in improving regulations.

The U.S. Nuclear Regulatory Commission (NRC) issued a policy statement in 1995 stating that probabilistic risk assessment should be used "to complement traditional deterministic methods in all areas of regulation, recognizing its current state and limitations". The problem with using the quantitative results of probabilistic risk assessment (the absolute values obtained) is described in NRC staff training materials in relation to safety goals and defense in depth principle (NRC 2002).

Probabilistic risk assessment and safety goals are used for dialogue with regulations, and it can be understood from this material that they are not used for communication with the public. The psychological aspects of communication and risk communication are discussed in Chap. 2. The U.S. Nuclear Regulatory Commission does not conduct risk communication (dialogue) with the public. Moreover, there is no record of risk communication with the public using safety goals. Many explanatory materials on policies and their basis related to U.S. nuclear regulation have been created and made public. Probabilistic risk assessment is conducted in the licensing of nuclear power plants, but the numerical value of the obtained core damage probability itself is not used in the final judgment of licensing.

The concept of cost-benefit (also known as risk-benefit) underlies probabilistic risk assessment methods. The idea of cost-benefit is that the dangers associated with the use of a certain technology should not exceed the benefits of its use. In other words, improvements in safety should be considered in balance with the costs associated with them. In the case of disaster prevention, as written in the U.S. nuclear protection guidelines, the risk avoided by evacuation should not exceed the risk associated with evacuation. As mentioned in Chap. 4, the concept of risk-benefit is stated as a basic principle at the beginning of the International Atomic Energy Agency's handbook on nuclear law. In Japan, attention has been focused on the evaluation methods and results of probabilistic risk assessment, and it seems that there have been no nuclear experts who have noticed this basic concept. This is very strange. If only safety is strictly enforced, the actual effect will not increase, and only the technical and economic burden will increase. As a result, the burden on the public increases, and electricity rates become arbitrarily high.

In Japan, because the electricity rate system determined by the total cost method has been in place for a long time, both power companies and the public have avoided talking about generation costs. If the concept of cost-benefit had been considered, the new regulatory standards decided after the TEPCO accident might have been more rational. Japan's response, which has stopped nuclear power plants and is implementing new regulatory standards, is not rational from the perspective of risk-benefit. There may be an excuse that public sentiment was taken into account, but as a result, electricity rates, which are a burden on the public, have increased, and issues of stable supply and energy security have emerged. As mentioned in Chap. 2, the risk associated with evacuation (disaster-related death) has become greater than the risk avoided by evacuation. The question "How safe is safe enough" can technically be answered with the concept of cost-benefit.

Probabilistic risk assessment is a means of analyzing the cost (required for measures) and the effect (reducible risk). For example, by comparing the amount of radiation exposure that can be reduced by the measures for that accident and the cost of the measures, the cost-effectiveness of various measures can be evaluated. In U.S. regulations, a conversion factor is used that equates exposure to 1 person-rem (10 millisieverts) to $5200 (in 2014 dollars). In Japan, probabilistic risk assessment is conducted, but no indicator for converting to cost is used. It is not incorporated into the nuclear regulation rules either. Even if probabilistic risk assessment is conducted in a form similar to the U.S., without a conversion factor, cost-effectiveness analysis is impossible, and it cannot be said that probabilistic risk assessment is being utilized. Continuous safety improvement and regulatory improvement are not just about mimicking the U.S. form, the content is what matters.

Firstly, it is important to do things by oneself. By learning from past domestic and international experiences, such as from reports, and refining them through practical work, and by thinking logically, one might see the path to continuous safety improvement in Japan and dialogue and improvement with regulations. Even if we call for human resource development, it is not good to rely on others. There is a model in the United States, so we should be able to devise based on that.

1.9.9 New Construction of Nuclear Power Plants

In the world, many new nuclear power plants are being built in countries such as China and India. The four nuclear power plants ordered by the United Arab Emirates from South Korea have started operations one after another. In these countries, the manufacturing and construction of nuclear power plants have been ongoing, so their experience has been inherited, and there has been no significant delay in the construction period. A huge amount of money is needed to build a nuclear power plant, and the investment is recovered by operation. Delays in construction increase the interest payments on borrowed money and require financial arrangements for extended borrowing periods, posing a large investment risk and business risk.

In Finland, the United States, and France, the new construction of nuclear power plants has been significantly delayed, and construction costs have skyrocketed. In each country, construction continues, operations have started in the United States and Finland, and operations are scheduled to start in France in mid-2024. The cause of the delay is that new construction was not ongoing, so the experience was not inherited. Specifically, delays in the management of building foundations and civil engineering works, manufacturing management of nuclear reactor equipment such as reactor pressure vessels and digital instrumentation control systems, and delays related to nuclear reactor regulation approvals are the causes. Construction delays have led to a significant increase in construction costs, leading to the withdrawal or reorganization of nuclear reactor manufacturers from the project.

In Japan, it is necessary to learn lessons from the delay in new construction overseas and take measures. In Japan as well, new construction of nuclear power plants has not been carried out for a long period. Nuclear power plants are the largest structures made by mankind. *Only by carrying out new construction can the experience of manufacturing and construction of nuclear power plants be inherited.*

1.9.10 Nuclear Power Technology from the History of Innovation

We often hear the word innovation, but it is good to think about it with reference to the research results of the history of innovation. There is a classic book that analyzes the innovation of industrial products (Utterback 1994). It analyzes product innovation and manufacturing process innovation. Let's consider the innovation of nuclear power technology with reference to this.

Professor Utterback, using the evolution of the typewriter (the act of writing characters with a machine or device) as an example, describes product innovation as follows:

- Innovation occurs when new elements are added to existing technology. For example, the mechanical typewriter was created by combining existing mechanical parts with a keyboard. The electric typewriter was created by combining a motor with a mechanical typewriter. Word processors (personal computers) were created by combining a TV monitor (CRT), a printed circuit board, memory chips, semiconductors, and a keyboard. The arrangement of the keyboard characters remains QWERTY. Dedicated word processors quickly became obsolete and were replaced by personal computers.

- In each phase, various products appeared on the market, and those with a "dominant design" occupied the market. For example, in the case of personal computers, many competitors who once held important positions withdrew with the advent of IBM's personal computer.
- After the "dominant design" is established, the focus shifts to innovation in the manufacturing process.

As an example of innovation in the manufacturing process, the transition in the manufacturing process of incandescent light bulbs and the transition in the manufacturing process of thick plate glass in the material industry from the late 1800s are mentioned. In the manufacture of thick plate glass, the processes of mixing, melting, molding, slow cooling, grinding, and polishing, which were initially performed independently (in batches), were gradually integrated and transformed into a continuous process, and ultimately became a single manufacturing process.

Innovation does not suddenly emerge from nothing, but is born by adding new elements to existing technology, and it is important to note that it is born from ingenuity in a harsh competitive environment surrounding the market.

Considering nuclear power technology based on this, various types of reactors have been developed in the United States and other countries, but light water (natural water) nuclear power plants (light water reactors) have been built the most, and almost all reactors currently under construction are light water reactors. Since they dominated the market, *the light water reactor is the dominant design of nuclear power generation.* The light water reactor uses the coal-fired power generation technology that was developed and used before it. For example, pumps, steam turbines, generators, cooling pipes, and control systems are existing technologies, and these devices are made based on the design and manufacturing experience of coal-fired power generation equipment. To these, the reactor part, which replaces the boiler of coal fired power generation, is added and plays the role of producing steam. The differences between thermal power generation and nuclear power generation were shown in Fig. 1.17. The method of creating steam differs, with thermal power generation using the combustion of oil or coal, while nuclear power generation uses the nuclear fission of uranium. The feedwater pump, turbine (steam turbine), condenser, and generator operate on the same principle in both thermal and nuclear power generation.

1.9.11 Computer-Aided Design, Construction, and Maintenance Technology

When a dominant design is established, innovation occurs in the manufacturing process. This was also the case with nuclear power generation. Japan led the world in this innovation. In the 1970s, when the manufacture and construction of light water reactors began in Japan, the manufacture of equipment, not limited to nuclear equipment, was done by engineers drawing blueprints by hand, which were then used to manufacture in factories. Numerous blueprints were drawn for each piece of equipment, piping, trays for storing cables, and composite blueprints combining these, and were duplicated for use. An example of a blueprint is shown in Fig. 1.25.

In the 1980s, based on these blueprints, scaled-down plastic models were made to check pipe connections and whether there was maintenance space. A photo of the plastic models lined up in a gymnasium is shown in Fig. 1.26. The plastic models occupy most of the gymnasium floor. Nuclear power plants are still the most complex and massive structures created by mankind.

In the late 1980s, with the improvement of computer capabilities and memory capacity, engineers were able to create blueprints directly on computers. There are

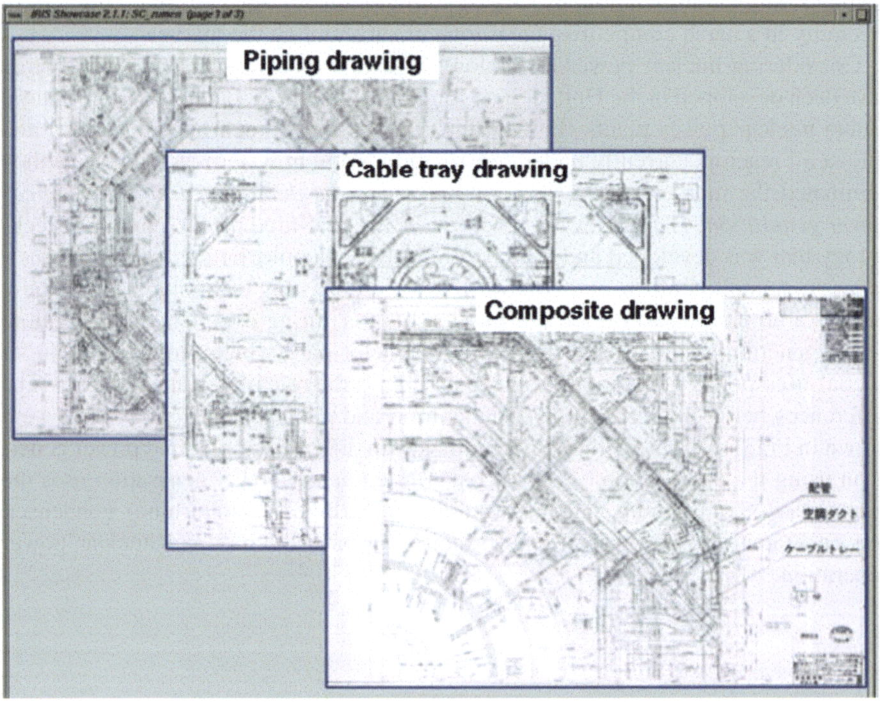

Fig. 1.25 Design by Had drawing (1975). *Source* J. Kawahata: Summer School of Nuclear Power Plants in 2009 at the University of Tokyo

Fig. 1.26 Plastic model of a nuclear power plant (1985). *Source* J. Kawahata: Summer School of Nuclear Power Plants in 2009 at the University of Tokyo

photos of many engineers working with computer terminals at nuclear reactor manufacturers. Figure 1.27 shows the design work using a 3D computer-aided design (CAD) terminal and the model on the computer. At that time in Japan, the development of the improved light water reactor ABWR was being carried out, and the Kashiwazaki Units 6 and 7 of Tokyo Electric Power Company were the first nuclear power plants to be systematically developed and applied using this technology.

Since the blueprints are in the computer, instead of making plastic models to check, you can display equipment, pipes, and buildings in three dimensions on the computer and check whether pipe connections and maintenance space are secured. This kind of work can be done with the help of a computer, not only in the basic design of a nuclear power plant, but also in detailed design, manufacturing, construction, and maintenance work. This is called a computer-aided design and maintenance system.

Furthermore, an innovation called the module method, which involves the manufacturing and construction of nuclear power plant equipment, was developed in Japan. Parts are assembled and pipes are welded at the factory of the reactor manufacturer to create modules for nuclear power plant equipment. Quality control, such as welding, is better managed in a factory than on a construction site. The manufactured modules are transported to the power plant construction site by ship or large vehicles, and the modules are connected and assembled. The foundation of the power plant building is made by combining thick rebar, but the modules are not made at their final location, but in an open space on the power plant site, and are installed in their final position using a large crane. Figure 1.28 shows a module of the steel structure of the body of the boiling water light water reactor containment vessel. Subsequently, the computer database for design and construction is now also used for plant maintenance, as shown in Fig. 1.29.

Fig. 1.27 Design by 3D CAD (after 1990). *Source* J. Kawahata: Summer School of Nuclear Power Plants in 2009 at the University of Tokyo

Fig. 1.28 Modular construction of a nuclear power plant. *Source* TEPCO

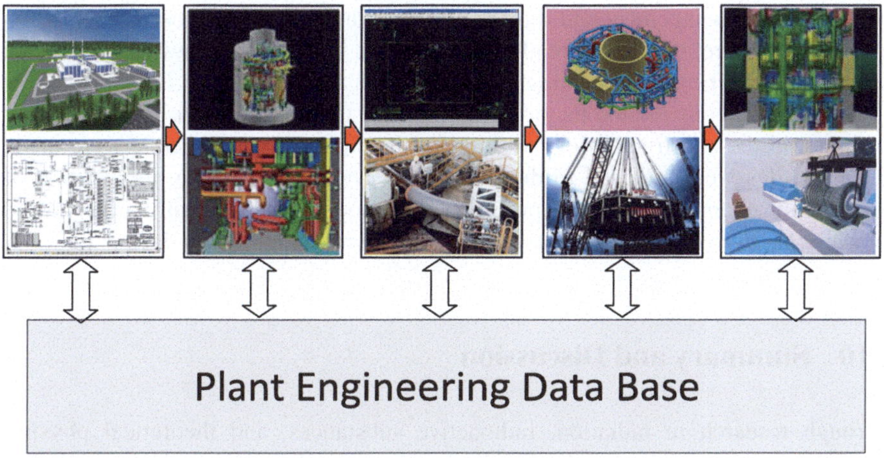

Fig. 1.29 Plant integrated computer-aided engineering system. *Source* J. Kawahata: Summer School of Nuclear Power Plants in 2009 at the University of Tokyo

Another advantage of the module method is that by manufacturing as modules and assembling them at the power plant, the construction period can be shortened. It is also better to manufacture in a factory in terms of quality control during manufacturing, such as welding. Since a power plant does not generate profits until it starts generating electricity, a shorter construction period means less interest payments on loans for construction. The module method has the advantages of quality control and shortening the construction period, and is an innovation in the manufacturing and construction process of nuclear power plants (Kawahata 2011; Oka 2018b). This method has gradually spread to other nuclear power countries, and one country held an exhibition of its own nuclear power technology at the International Atomic Energy Agency in the late 2010s to promote this method. In subsequent construction of nuclear power plants in Japan, in addition to the open top method using large cranes, a parallel construction method was also used, in which civil engineering and mechanical work were carried out simultaneously.

The development of CAD was funded by the US NASA (National Aeronautics and Space Administration) space development budget. Numerical fluid dynamics, which calculates the flow of air around aircraft and rockets using the fluid motion equations (Navier-Stokes equations) with a computer, has evolved, and along with it, technology to graphically display the calculation results and terminals for this purpose have been developed. This terminal is a CAD terminal. Subsequently, this technology has evolved not only for displaying scientific and technological calculation results, but also as animation technology and video technology. A dinosaur movie was made using this technology and was a hit. Computer-aided design, construction, and maintenance technology for nuclear power plants was built using CAD terminals. CAD has features such as easier drafting than handwriting, easy modification of drafting data, and easy management and sharing of data. Currently, CAD is used in all industries. Furthermore, with the spread of 3D printers, it is bringing new developments to various industries in the manufacturing field.

The module construction method is not unique to nuclear power generation. For example, in the construction of buildings in the city, walls of concrete were made and then tiles were attached to the walls. However, after a while, the method of construction changed to carrying pre-made wall modules to the construction site and attaching them to the structure, and this evolution continues to this day.

1.10 Summary and Discussion

Through research in radiation, radioactive substances, and theoretical physics, nuclear science has developed, enabling mankind to utilize nuclear energy and radiation. Many Nobel Prizes have been awarded to researchers in nuclear science. The use of nuclear power is a field that applies nuclear science. When X-rays are irradiated on the human body, a transmission image can be obtained. Roentgen, who discovered X-rays, won the first Nobel Prize in Physics.

Spectroscopy is a method of analyzing the components and characteristics of a substance by irradiating light on a substance and decomposing the transmitted, reflected, or absorbed light by wavelength. The ability and application range of spectroscopy have dramatically improved as the light used has expanded from visible light to radiation. Synchrotron radiation is a high-energy electromagnetic wave that is generated when electrons are bent in a magnetic field. Spectroscopy using synchrotron radiation is widely used from basic research to industrial applications.

Radiation is a general term for high-energy (short wavelength) electromagnetic waves and high-speed material particles. Material particles include various things such as electrons, positrons, protons, and ions, and their interactions with matter are also diverse for each particle. By irradiating not only high-energy electromagnetic waves but also high-speed material particles on substances and measuring their reflection or absorption, it has become possible to know the properties of substances from various perspectives. The target and ability of spectroscopy have dramatically expanded as the light used has expanded from visible light to radiation. *Radiation can be considered as a new light*. In addition to spectroscopy, radiation brings various benefits in many fields of our lives and industries.

Humans have prospered and reigned at the top of the food chain by using fire and tools. The use of fire is said to have begun 100,000 years ago. Firewood was burned for heating and cooking. Later in agriculture, not only the use of fire, but also the energy of livestock was used, and hydropower and wind power were also utilized. In the Industrial Revolution that began in Britain in the eighteenth century, the energy of coal was used in the steam engine invented by Watt, dramatically improving convenience and energy consumption. Nuclear energy has been available since the latter half of the twentieth century. Fossil fuels such as coal, oil, and natural gas, as well as natural energy such as sunlight, wind power, and hydropower, all originate from the energy of the sun, so all energy, not just nuclear power, is energy originating from nuclear reactions. The direct use of energy from nuclear reactions, such as nuclear power generation, has only been around for half a century. *Nuclear energy can be considered a new fire that humanity has acquired.*

Energy is essential for activities in all areas of modern life, not just power generation, but also industry, civil life, and transportation. Currently, the majority (about 80%) is covered by oil, natural gas, and coal, which are called fossil fuels. Its supply disruption affects all socio-economic activities. The issue of energy security in Europe became apparent due to Russia's invasion of Ukraine in 2022. The prices of crude oil, natural gas, and coal have risen sharply, threatening people's lives.

Japan's energy import dependency is 89% (self-sufficiency rate is 11%), which is *extremely high among developed countries.* Most of Japan's primary energy is crude oil, natural gas, and coal, which are called fossil fuels, and almost all of it is imported from overseas. In particular, over 90% of crude oil is imported from the politically unstable Middle East. Before the Great East Japan Earthquake, Japan's primary energy self-sufficiency rate was about 20%, but even in 2024, more than ten years later, many nuclear power plants are still shut down, and the primary energy self-sufficiency rate in 2020 is 11%, which is extremely low among developed countries. France has a high self-sufficiency rate of 55% because it covers most of its electricity with nuclear power generation. Figure 1.18 shows that Japan's nuclear power generation supplied 34% of the power generation in the mid-1990s. In 2020, it was only 4%. Energy is deeply related to international politics, economics, and social activities. Securing energy resources and stable, inexpensive supply, and improving the energy self-sufficiency rate are important issues that will determine Japan's future. In the future, it will be necessary to increase the power generation of renewable energy and expand nuclear power generation.

Japan's electricity rates are high compared to other developed countries. South Korea, like Japan, has scarce domestic energy resources, but its electricity rates are about half of Japan's. Due to the surge in natural gas prices caused by Russia's invasion of Ukraine, Japan's electricity rates were significantly increased in 2022. This not only affects household electricity bills, but also impacts consumer prices and industrial electricity rates, affecting the international competitiveness of export industries, stagnating the Japanese economy, and affecting employment. In Japan, for a long period after the war, power supply was carried out by regional power companies. Attention was paid to stable supply, but because electricity rates were determined by the total cost method, which totals the costs necessary for power generation and transmission, the function to lower electricity rates did not work. The economic rationality that can be realized by supplying cheap power is a competitive principle of global modern society, and neglecting this will mislead Japan's future.

In the West, economic liberalization has progressed since the 1980s, and the power business has also been liberalized. In Europe, power companies expanded their businesses beyond national borders. In the United States, mergers and alliances of power companies and sales of power plants progressed. The power business in the United States is under the jurisdiction of the states, and many states have promoted power liberalization, but due to large-scale power outages, power liberalization in the United States was interrupted.

In the investment environment of economic liberalization, private companies bear the investment risk. Even if the power generation cost is high, if the investment amount is small, the investment risk is also small, so it is preferred in the investment environment of economic liberalization. Gas turbine combined cycle power plants, which use natural gas as fuel, had high fuel costs and high power generation costs, but because the construction investment amount (the cost required for construction) was small, they were preferred in the investment environment of economic liberalization, and many new constructions progressed in the West from the 1990s. The reason why new construction of nuclear power plants in the United States has hardly progressed since the 1980s is not due to safety concerns, but the impact of the investment environment of economic liberalization is large. Even if the power generation cost is high, power is essential for modern life, so it can be sold. Even if it is liberalized, there are restrictions on transmission and distribution, and power cannot be freely purchased from anywhere. As a result, consumers in the United States ended up buying expensive electricity. The electricity rates in the states that have liberalized power in the United States are higher than in the states that have not.

Hydropower plants require a huge investment for new construction, but once the investment is recovered, they can generate power extremely cheaply. For example, hydropower plants built in the United States under the New Deal policy in the 1930s after the Great Depression are still supplying power cheaply. Turbine generators and other aging parts are replaced to continue operation. The same is true for the Aswan High Dam hydropower plant in Egypt. Nuclear power plants also require a huge construction cost, but the proportion of fuel costs in the power generation cost is small, so in the United States, they are generating power cheaply, second only to hydropower. Hydropower plants do not have a lifespan, because things like turbine generators, which change over time, can be replaced. It is the same for nuclear power plants. *There is no lifespan. Nuclear power plants are like hydroelectric power plants.*

In an investment environment of economic liberalization, things that require huge investments, like hydroelectric power plants, nuclear power plants, and transmission lines, *will no longer be built*. The country trying to address this issue in a power liberalization country is the UK. The UK has created a system called the RAB model, which improves the fixed price purchase system to reduce the investment risk of new construction of nuclear power plants, and supports the new construction of nuclear power plants. The purpose of this system is not to support power companies, but to ensure that citizens can obtain cheap and stable electricity in the long term.

The levelized cost of electricity (LCOE), which is used to estimate power generation costs, determines the investment recovery period and calculates the power generation cost. Natural gas power plants, wind power plants, and solar power plants are cheaper to demolish and rebuild than to repair, so when the investment recovery period ends, they are demolished and new ones are built. The LCOE is the power generation cost in the case of demolishing and rebuilding after the investment recovery period has ended. However, hydroelectric power plants and nuclear power plants can operate beyond the investment recovery period. In the United States, many coal-fired power plants also operate beyond the investment recovery period. The power generation cost excluding the construction cost amortization is called the production cost. The production cost of hydroelectric power plants and nuclear power plants in the United States is cheap, as shown in Fig. 1.15. *The LCOE is a value calculated by giving the construction cost amortization period, etc., and is not the actual power generation cost. Nuclear power plants are more advantageous the longer they are used.*

Electricité de France built many nuclear power plants in the 1970s and is carrying out large-scale repairs from 2014 to 2025. Extension of the operating period requires the permission of the nuclear regulatory authority in any country. In Japan, after the TEPCO Fukushima accident, new regulatory standards were established, and the extension of the operating period was limited to once every 20 years, so many relatively small output nuclear power plants were abolished by power companies because they could not expect to recover the cost of construction work to comply with the new regulatory standards within the extended operating period. There is no technical basis for the one-time limit every 20 years. If the operation could have continued, it would have been possible to supply cheap electricity, and the abolition of nuclear power plants is a national loss.

In Japan, nuclear power plants are being shut down to implement regulatory standards that incorporate the lessons learned from the TEPCO Fukushima accident. However, Japan is the only country that is conducting resilience tests to see if they can withstand the lessons learned from the accident while shutting down nuclear power plants. If they had scheduled and implemented the new regulatory standards while operating, there would have been no economic loss to the national economy due to the import of fossil fuels to compensate for the power supply lost by the shutdown of nuclear power plants, and the significant increase in electricity rates in 2022 would have been mitigated.

As mentioned in Chap. 2, the concept of balancing risk and benefit is missing in Japan's safety, regulation, and disaster prevention. This is the underlying issue. According to a report by the United Nations Scientific Committee on the Effects of Atomic Radiation, the public who died from radiation due to nuclear power plant accidents were eight in the Chernobyl accident and zero in the TEPCO Fukushima accident. Nuclear power is a safe source of electricity. Over-worrying about radiation exposure causes more harm, such as damage from rumors, psychological instability, increased disaster-related deaths due to delayed evacuation, and collapse of local communities. A British expert, author of "The Paralysis in Energy Decision Making", says, "*I want nuclear power to become normal*".

The liberalization of the electricity market is to relax regulations on entry into the electricity market and introduce market principles. If the generation of renewable energy increases in a deregulated electricity environment, another problem becomes apparent. If wholesale electricity bidding is being conducted due to electricity deregulation, during favorable weather conditions, renewable energy can supply even at zero cost, making other generation methods such as nuclear power, natural gas-fired power, and coal-fired power disadvantaged in the wholesale electricity market, leading to power plants giving up on continued operation. As a result, there may be problems with stable power supply during bad weather conditions.

In addition to the problem of insufficient power generation facilities (kW) and power generation (kWh), there are issues with maintaining the frequency of alternating current in the problem of stable power supply. The problem of frequency maintenance includes the issue of securing frequency adjustment power and inertial synchronization power. The former is a short-term supply-demand adjustment capability, which is the ability to respond to rapid changes in the power flowing through the power system. If the power generation is insufficient, power supply from other power generation facilities within the transmission network becomes necessary. If a large power generation facility suddenly stops, transmission stops to protect the system, resulting in a major power outage. The latter is a problem of maintaining the frequency of alternating current. Motors in factories rotate in sync with the frequency of alternating current, and if the frequency changes significantly, the rotation speed of the motor changes, affecting the quality of the product. Nuclear power plants, hydroelectric power plants, and thermal power plants generate electricity by rotating turbine generators using the steam they generate. The turbine is a huge weight, and the inertia of rotation (the force to continue rotating) is large. If these power plants are responsible for most of the power supply, the inertial synchronization force is secured by this.

There are other issues to be solved for stable power supply. The suitable location for renewable energy generation is remote from large cities where electricity is consumed (in Japan, Kyushu and Hokkaido), so investment in transmission lines is necessary, but the investment amount is huge, so in a liberalized electricity market, investment in transmission line construction is not made. Electricity is also a public good. It is not possible to leave its supply to the free market, and it is necessary to address both competition activation and public interest issues. The goal is to satisfy both cheap and stable power supply.

Nuclear power plants have value in terms of the environment and employment. Nuclear power plants can generate a large amount of electricity on a small site, have a small impact on the ecosystem, and have high environmental value. Furthermore, nuclear power plants are large-scale businesses, so they have economic value for the local area. Since nuclear power plants are located in rural areas, they contribute to local employment and economy, and play a role in regional revitalization. Not only direct employment at nuclear power plants, but also employment for maintenance and construction work, and indirect economic effects such as accommodation facilities and shops for that purpose. For farmers in the suburbs, since the nuclear power plant was built, there is no need to go out to work in the off-season (winter), and it is possible to have a side job in the power plant related work and agriculture. These things are rarely talked about, but if you compare before and after the location of the nuclear power plant, you should be able to understand.

All nuclear power plants in Japan have been built through local solicitation. While some argue that the "consent" of the local community is necessary for the restart of nuclear power plants, there is no legal basis for this. This does not mean that the understanding of the local community can be disregarded, but in a parliamentary democracy, it is the mechanism of the state to consider and deal with issues, not limited to nuclear power generation.

The nuclear fission reaction was successfully sustained by the genius physicist Enrico Fermi in 1942, but we have not yet succeeded in sustaining a nuclear fusion reaction. Power reactors were developed based on military reactors. Various types of reactors have been developed, but the most widely used in the world is the light water reactor developed by the United States. The reason is that the reactor is compact because water is used for cooling, and the experience of thermal power generation equipment could be utilized. Low-enriched uranium is required for fuel, but it has become available from the global competitive market.

The only countries that have not exported nuclear power plants while using nuclear power on a large scale domestically are Japan and India. India made plutonium in a research reactor imported from Canada and made nuclear weapons, so it was ostracized from the world for a long time. Excluding India, the only country with the ability to manufacture nuclear power plants that has not exported them is Japan. China developed a domestic small light water reactor and exported it to Pakistan in the 1990s immediately after building it domestically. South Korea also exported four Korean-type light water reactors to the United Arab Emirates in the 2010s. As shown in Table 1.2, Western countries started exporting soon after they started domestic use. From this perspective, Japan should have considered exporting in the 1970s. Exporting involves various investment risks. Some countries, like Germany, have struggled with the export of nuclear power plants to Iran and Brazil. However, Japanese companies also export power plant equipment such as power turbines and nuclear reactor pressure vessels. Here, the investment risk is small, but the economic benefit is also small. The nuclear power business in Japan is a unique business environment for reactor manufacturers, where the only customers are electric power companies, unlike other industries' free markets, and the ties are strong. In Japan's nuclear industry, there is a term "Shi-nou-kou-shou". "Shi" refers to government agencies, "Nou" to electric power companies, "Kou" to reactor manufacturers, and "Shou" to trading companies. "Shi-nou-kou-shou" was the social hierarchy in Edo Period between 1603 and 1868, the classes of warriors, farmers, artisans and tradesmen.

In Japan, a large number of nuclear power plants were built across the country in response to the oil shocks of 1973 and 1978. France also built a large number of nuclear power plants in response to the oil shock. Since then, the operation rate and output of the built power plants have increased, and the amount of nuclear power generation in the West has been increasing until the 2000s. In particular, in the United States, the amount of power generation increased and the accident rate decreased due to the continuous safety improvement activities (voluntary safety improvement) of power companies from the 1990s and the improvement of regulations in response to them. The author looks forward to the continuous safety improvement activities of Japanese power companies in the future. *Improving safety and economy can coexist.*

In the history of industrial product innovation, after product innovation is carried out, a dominant design of the product that dominated the market is established, and then innovation in the manufacturing process occurs. In the field of nuclear power generation, light water reactors dominate the market, so light water reactors are the dominant design of power reactors. In the innovation of the manufacturing process of light water reactors, Japan was at the forefront of the world. They developed a computer-aided design construction maintenance system for the design, manufacture, and construction of improved light water reactors. The equipment of nuclear power plants was manufactured as modules in factories, transported to construction sites, connected and assembled. Quality control became easier due to factory production, and the construction period was shortened. The ABWR (Advanced Boiling Water Reactor) of Tokyo Electric Power's Kashiwazaki-Kariwa Nuclear Power Plant was the first to be made by this method, and it was then used for the construction of nuclear power plants in various parts of Japan. This method was later followed by nuclear reactor manufacturers around the world.

Exercise

1. Summarize various nuclear science research and their Nobel Prize winners.
2. Summarize the medical applications of various types of radiation.
3. Summarize the history and development of spectroscopy.
4. Summarize the synchrotron radiation facilities in Japan and their uses (from research to industrial applications).
5. Summarize the history of the steam engine.
6. Summarize the history of electricity generation and use.
7. Summarize the development and history of gas turbine use, including power generation applications.
8. Summarize the history of conflicts over Middle Eastern oil resources from before World War II to the present.
9. Summarize U.S. energy security policies from the 1800s to the present.
10. Summarize the policies surrounding energy resources in the UK since the Industrial Revolution.
11. Summarize the history of Russia's economic situation and energy resource exports since the collapse of the Soviet Union.
12. Discuss the history of Japan's dependence on various energy resources from the post-World War II period to the present, and its energy policies.
13. Summarize the history of electricity deregulation in the United States.
14. Discuss the electricity deregulation in Europe.
15. Summarize the history and current situation of electricity deregulation in Japan.
16. Summarize the situation of energy resources and electricity supply in South Korea. Analyze the reasons for the low cost of electricity in South Korea.
17. Summarize the history of energy resources and supply since the 1970s in China, and the history of electricity supply.
18. Discuss the history and current situation of nuclear power generation in China.
19. Summarize the history of continuous safety improvement and regulatory enhancement in nuclear power generation in the United States.
20. Discuss the power output enhancement and operation rate improvement in nuclear power generation in the United States.

21. Investigate the changes in the capacity factor of nuclear power plants in Japan, and consider why there is a difference with Europe and the United States.
22. Discuss the history of innovation in industrial products, giving examples.
23. Summarize the improvements in the construction methods of nuclear power plants carried out in Japan.
24. Summarize the various construction methods and their changes in building construction.
25. Calculate and compare the area of power plant required per 1kW of power generation equipment for solar power plants, wind power plants, and nuclear power plants.

References

Brook BW et al (2014) Why nuclear energy is sustainable and has to be part of the energy mix. Sustain Mat Tech 1–2:8–16

Council on Foreign Relations (2023) 1850-2022 Oil Dependence and U.S. Foreign Policy, https://www.cfr.org/timeline/oil-dependence-and-us-foreign-policy. Accessed January 31, 2023

EIA (2019) Table 8.4. Average power plant operating expenses for major U.S. investor-owned electric utilities, 2009 through 2019 (Mills per kilowatt hour) Energy Information Administration

Jackson SA (1998) Statement submitted by the United States nuclear regulatory commission to the subcommittee on clean air, wetlands, private property and nuclear safety committee on environment and public works, United States Senate, concerning United States nuclear regulatory commission programs and nuclear safety regulatory issues presented by Shirley Ann Jackson, Chairman, July 30. https://www.epw.senate.gov/105th/nrca-7-3.htm. Accessed June 18, 2024

JAEC (2017) Basic thoughts on the use of nuclear power, reference materials, P 98. Japan Atomic Energy Commission, July 20 (in Japanese). https://www.aec.go.jp/kaigi/teirei/2017/siryo25/1-3_haifu.pdf. Accessed June 18, 2024

JAEC (2019) White paper on nuclear energy for the first year of Reiwa, Chapter 7: Development of the use of radiation and radioisotopes, Japan Atomic Energy Commission, August (in Japanese). https://www.aec.go.jp/kettei/hakusho/2019/pdf/zentai.pdf; https://www.aec.go.jp/kettei/hakusho/2019/pdf/gaiyo_e.pdf (abstract in English). Accessed June 18, 2024

Japan Business Federation (2019) Reconstructing the power system that supports Japan. Japan Business Federation, April 16 (in Japanese) https://www.keidanren.or.jp/policy/2019/031.html. Accessed June 18, 2024

Kawahata J (2011) Advanced 3D-CAD and its application to state-of-the-art construction technologies in ABWR plant projects. In: Oka Y et al (eds) Advances in light water reactor technologies. Springer, Chapter 9

Kemeny JG (1979) Report of the President's commission on the accident at three mile Island, October. http://large.stanford.edu/courses/2012/ph241/tran1/docs/188.pdf. Accessed June 18, 2024

Komiyama R (2022) Basic thoughts on the use of nuclear power, February 15. Atomic Energy Commission (in Japanese). https://www.aec.go.jp/kaigi/teirei/2022/siryo06/1_haifu.pdf. Accessed June 18, 2024

Kondo H (2020) Learning from the initiation of the reactor oversight process (ROP) in the United States, January 14. Atomic Energy Commission Document No. 1 (in Japanese). https://www.aec.go.jp/kaigi/teirei/2020/siryo01/1_haifu.pdf. Accessed June 18, 2024
Kondo H, Yamaguchi A (2018). Learning from the initiation and establishment of the reactor oversight process (ROP) in the United States to successfully reform Japan's inspection system. J Atomic Energy Soc Japan 60(10):17–21 (in Japanese). https://www.jstage.jst.go.jp/article/jaesjb/60/10/60_601/_pdf. Accessed June 18, 2024
Loftness RL (1964) Nuclear power plants. Van Nostrand
Matsuo Y (2015) Overview of the evaluation by the power generation cost verification working group, June 2015. Institute of Energy Economics, Japan (in Japanese). https://eneken.ieej.or.jp/data/6143.pdf. Accessed June 18, 2024
McCormick NJ (1981) Reliability and risk analysis. Academic Press
METI (1993) Annual world economic report 1993, Chapter 2: Conditions for sustainable growth and market economization, Section 1.3: Russia's ongoing economic turmoil. December 10. Economic Planning Agency (in Japanese). https://www5.cao.go.jp/keizai3/sekaikeizaiwp/wp-we93/wp-we93-00303.html. Accessed June 18, 2024
METI (2012) Trade white paper 2012, Chapter 1: Trends in the world economy, Section 1.6: Latin American and Russian economies, 2. Russian Economy, Ministry of Economy, Trade and Industry (in Japanese). https://ndlsearch.ndl.go.jp/books/R100000002-I000000046167-i198007. Accessed June 18, 2024
Nomura K (2015) Materials on the difficulty of energy conservation in the future and the impact of rising electricity prices on industrial structure, p 8. Comprehensive Energy Research Council, Long-term Energy Supply and Demand Outlook Subcommittee, 7th Meeting Materials 11, April 22 (in Japanese) https://www.enecho.meti.go.jp/committee/council/basic_policy_subcommittee/mitoshi/007/pdf/007_18.pdf. Accessed June 18, 2024
NRC (1975) Reactor safety study. Executive summary, WASH-1400 (NUREG 75/014). USNRC, October. https://www.osti.gov/servlets/purl/7134133. Accessed June 18, 2024
NRC (1990) Severe accident risks: an assessment for five U.S. nuclear power plants. NUREG-1150, USNRC, December. https://www.nrc.gov/reading-rm/doc-collections/nuregs/staff/sr1150/index.html. Accessed June 18, 2024
NRC (2002) Perspectives on reactor safety, NUREG-CR-6042, SAND93-0971, Rev.2, Appendix 2.6 risk-informed regulation. USNRC, March. https://www.nrc.gov/reading-rm/doc-collections/nuregs/contract/cr6042/index.html. Accessed June 18, 2024
Oka Y (2018a) Challenges and human resource development for nuclear power utilization in Japan. Special Lecture by the Japan Technical Association, June 15, p112 onwards, Example of Dissent, Cabinet Office Atomic Energy Commission Homepage (in Japanese). http://www.aec.go.jp/jicst/NC/about/kettei/180613.pdf. Accessed June 18, 2024
Oka Y (2018b) Utilization of nuclear energy. IFNEC, Keynote speech, November 15, 2018 Tokyo, Atomic Energy Commission Homepage, Activity Introduction, Lecture Materials, pp21-27. http://www.aec.go.jp/jicst/NC/about/kettei/181115.pdf. Accessed June 18, 2024
Okada S (2017) Past, present, and future of quantum beam science and radiation use, January 31. 4th Atomic Energy Commission Document No. 1 (in Japanese). https://www.aec.go.jp/kaigi/teirei/2017/siryo04/1_haifu.pdf. Accessed June 18, 2024
Pate Z, Ellis J (2011) The evolution of nuclear safety. WNA Personal Perspective. https://www.world-nuclear.org/uploadedFiles/org/Archive/WNA_Personal_Perspectives/evolution_of_nuclear_safety(1).pdf. Accessed June 18, 2024
Utterback JM (1994) Mastering the dynamics of innovation. Harvard Business School Press
Waltar AE (2004) Radiation and modern life, fulfilling Marie Curie's dream. Prometeus Books
WNA (2022) Nuclear power reactors (updated August 2022). World Nuclear Association. https://world-nuclear.org/information-library/nuclear-fuel-cycle/nuclear-power-reactors/nuclear-power-reactors.aspx. Accessed June 18, 2024

Yamasaki H (2022) About the efforts of JANSI as a self-regulatory organization. Japan Nuclear Safety Institute, Agency for Natural Resources and Energy, Comprehensive Energy Research Council, Electricity and Gas Business Subcommittee, Nuclear Subcommittee, Document 5, October 13 (in Japanese) https://www.meti.go.jp/shingikai/enecho/denryoku_gas/genshiryoku/pdf/032_05_00.pdf. Accessed June 18, 2024

Yang RL (2014) EPRI risk & safety R & D & human resource planning, November 14. Voluntary safety improvement WG, Ministry of Economy, Trades & Industry, Japan, November 10. https://www.meti.go.jp/shingikai/enecho/denryoku_gas/genshiryoku/jishuteki_anzensei/pdf/003_02_01.pdf. Accessed June 18, 2024

Chapter 2
Safety and Communication

This chapter analyzes the evacuation and health effects caused by the accident at the Tokyo Electric Power Company's Fukushima Daiichi Nuclear Power Plant and considers the lessons of nuclear disaster prevention measures and crisis management. It introduces the experiences and key points of science communication and stakeholder dialogue in the UK. It discusses risk perception and psychology and considers communication about risk and safety. The safety mainly dealt with in this chapter is the safety at the interface between nuclear power generation and society, that is, the safety related to disaster prevention and emergency response when radioactive materials are released into the environment in the event of a reactor accident. It is not about the safety related to the prevention of accidents or the reduction of radioactive material release that the majority of nuclear safety researchers and engineers are working on.

2.1 Great East Japan Earthquake and Fukushima Daiichi Nuclear Power Plant Accident (TEPCO Accident)

The Great East Japan Earthquake and Tsunami that occurred on March 11, 2011, resulted in the death of approximately 15,000 people. Including those missing and disaster-related deaths, over 20,000 people lost their lives. At that time, 470,000 people evacuated nationwide. Even ten years after the disaster, in 2021, over 40,000 people are still living as evacuees. Among them, the largest number of people who have evacuated outside their prefecture are from Fukushima Prefecture, with about 28,000 people (Reconstruction Agency 2021). The number of evacuees from Miyagi Prefecture, which suffered significant tsunami damage, is just under 4,000, indicating the significant impact of the Fukushima Daiichi Nuclear Power Plant accident (hereafter referred to as the TEPCO accident) on evacuation.

Y. Oka, *Nuclear Power and Society*, Nuclear Science and Technology,
https://doi.org/10.1007/978-981-97-8951-1_2

At the Fukushima Daiichi Nuclear Power Plant, the reactors automatically shut down due to the earthquake, but about an hour later, the tsunami caused units 1, 2, and 3 to lose all power for emergency core cooling, causing the nuclear fuel rods in the reactors to melt. The hydrogen generated by the reaction between the metal of the high-temperature nuclear fuel rods and water caused the buildings of units 1 and 3 to explode. Unit 4 was shut down, but hydrogen from unit 3 flowed into the building of unit 4 through the exhaust duct, causing damage to the building of unit 4 due to the explosion. The unit 2 building did not explode because the explosion at unit 1 opened a panel on the building and released hydrogen into the atmosphere.

The TEPCO accident was the first major nuclear power plant accident to occur in the era of advanced Internet, and the situation of the hydrogen explosion in the building was reported on television, re-emphasizing the importance of crisis management and communication. The staff of the Fukushima Daiichi Power Plant struggled to respond to the accident under the direction of the plant manager, but the response of TEPCO's head office and the government was chaotic.

It is not widely known, but the Tohoku Electric Power Onagawa Power Plant in Miyagi Prefecture, which was built on high ground, is safely shut down. Furthermore, it accepted townspeople who evacuated to the power plant due to the tsunami, providing an emergency evacuation site. The Japan Atomic Power Company's Unit 2 in Tokai Village (mura) had installed a flood protection wall in the area necessary for cooling the emergency diesel generator, based on the tsunami evaluation in Ibaraki Prefecture, ensuring emergency power and safely shutting down.

Regarding the TEPCO accident, accident investigation committees have been held by the government, the Diet, and the private sector, and reports have been issued. The Atomic Energy Society of Japan and the International Atomic Energy Agency have also created and published reports. The chairman of the Diet Accident Investigation Committee, Kurokawa, pointed out that they were "captivate regulation", that is, the nuclear power operators had captivated the regulators (Kurokawa 2016). After the accident, an independent Nuclear Regulation Authority (NRA) and Nuclear Regulatory Agency (NRA) were established. Nuclear Regulatory Agency is secretariat of Nuclear Regulation Authority, but it is a big agency. NRA formulated new regulatory standards and is conducting reviews. Many nuclear power plants have remained shut down since the TEPCO accident because they cannot operate unless they comply with and pass the new regulatory standards. Many power companies have chosen to decommission small nuclear power plants because the cost of construction to comply with the new regulatory standards cannot be recovered over a 20-year operating extension period.

In Western countries, when a major accident occurs, stress tests (resilience tests) are conducted to urgently review and implement measures, without immediately shutting down the power plant. The lessons from the TEPCO accident are also being reviewed and addressed through stress tests. Even in the Chernobyl accident, the nuclear power plant adjacent to the accident reactor continued to operate after the accident. In other words, Chernobyl-type nuclear power plants in the former Soviet Union are also implementing measures based on the lessons of the Chernobyl accident, without immediately shutting down at the time of the accident.

In Japan, even if nuclear power plants were not forced to shut down and stress tests and construction work that met the new regulatory standards were carried out in a planned manner, the risk did not change as a major tsunami has not struck since 2011. It would have also reduced the rise in electricity prices and the release of greenhouse gases.

Before the TEPCO accident, the Atomic Energy Commission and the Nuclear Safety Commission were placed as agencies and councils of the Cabinet Office, and the Agency for Natural Resources and Energy, an external bureau of the Ministry of Economy, Trade and Industry, was in charge of energy policy administration including nuclear power, and its subordinate Nuclear and Industrial Safety Agency was in charge of nuclear safety and industrial safety regulatory work. The Ministry of Education, Culture, Sports, Science and Technology was in charge of research and development of nuclear power.

After the TEPCO accident, the Nuclear Safety Commission was abolished, and the Nuclear and Industrial Safety Agency was abolished after transferring its operations to the Nuclear Regulation Authority. The Nuclear Regulation Authority and the Nuclear Regulatory Agency were created as external bureaus of the Ministry of the Environment, and regulation became independent from promotion. They are in charge of regulations related to nuclear reactors and nuclear fuel facilities, radioactive isotopes, etc., nuclear security measures, safeguards (verification activities carried out to ensure that nuclear materials are not diverted to nuclear weapons, etc.), radiation protection, radiation monitoring, and safety research. In the West, nuclear regulatory agencies are independent of administrative agencies that promote them, and Japan has finally caught up with the West.

The Cabinet Office has established the Nuclear Disaster Prevention Council as a permanent institution, and the Nuclear Disaster Countermeasures Headquarters has been established to promote nuclear emergency measures. The Reconstruction Agency was established as a temporary organization with the aim of reconstruction from the Great East Japan Earthquake, and the Expert Discussion on the Future Image of 12 Cities and Towns in Fukushima and the Fukushima Reconstruction and Regeneration Council from the Nuclear Disaster have been held, and efforts are being made for reconstruction. The Ministry of the Environment has held an expert meeting on the way of health management of residents due to the accident at the Tokyo Electric Power Fukushima Daiichi Nuclear Power Plant and announced an interim summary in December 2014. The Ministry of the Environment has created a unified basic document on the health effects of radiation and is trying to understand the public and victims (Ministry of the Environment 2022). The Ministry of Health, Labour and Welfare has compiled a draft standard for radioactive substances in food at a subcommittee of the Pharmaceutical Affairs and Food Sanitation Council in February 2012.

A cost-benefit analysis of radiation-contaminated food regulations has been reported. It is stated that "compared to the benefit of extending life expectancy by one year obtained from various surveys, 20 million yen, everything except vegetables is inefficient", and a cost-effective standard value has been proposed (Oka 2012). In Fukushima Prefecture, all rice bags were inspected. "3.1 trillion yen was spent to avoid one year of lost life expectancy. This is not a measure to reduce risk, but a measure to prevent anything exceeding the standard from being found later", it is stated. (Oka 2015). There is a word, blood tax. Isn't it necessary to decide on a rational standard value based on the results of cost-benefit analysis in advance?

2.2 Health Effects of the TEPCO Accident and Its Lessons

2.2.1 Health Effects of the TEPCO Accident

The health effects of radiation include deterministic effects (acute effects) where the causal relationship with exposure is clear, and stochastic effects (late effects) such as cancer and leukemia. The United Nations Scientific Committee on the Effects of Atomic Radiation (UNSCEAR) published "Levels and Effects of Radiation Exposure due to the Nuclear Accident after the 2011 Great East Japan Earthquake and Tsunami" in April 2014, evaluating the effects of radiation due to the TEPCO accident. It states the following about the health effects of public exposure to radiation (UNSCEAR 2013).

- In the short term, no deterministic health effects due to radiation exposure after the FDNPS accident were observed among members of the public and none are expected in the long term because the doses were far below the threshold values for such effects.
- The most important and manifest health effects of the nuclear accident in the short term would appear to be on mental and social well-being. This aspect was exacerbated by the enormous impact of the earthquake and tsunami, the uprooting of thousands of people as a consequence of the evacuation and their displacement to unfamiliar surroundings, and the fear and stigma related to radiation exposure.
- Furthermore, the evacuation following the accident caused immediate aggravation of the condition of already vulnerable groups; for example, it was reported that more than 50 hospitalized patients died either during or soon after the evacuation, probably because of hypothermia, dehydration or deterioration of underlying medical problems, and that upwards of 100 elderly people may have died in subsequent months because of a variety of conditions linked to the evacuation.
- Understanding the full health impact of the accident forms an important context for the Committee's commentary on health implications directly related to radiation exposure.

After the Tokyo Electric Power Company's Fukushima Daiichi Nuclear Power Plant accident (hereinafter referred to as the TEPCO accident), a doctor who worked at a hospital in the northern part of Hamadori in Fukushima Prefecture said the following (Ochi 2015a).

- Most residents who heard the order to stay at home left their homes. Disaster-vulnerable people (elderly and sick people who were alone) were left behind.

- Evacuation worsened the health of those who evacuated. Difficulty walking, diabetes, obesity, and osteoporosis increased.
- Evacuation to avoid low-dose exposure increased other health risks. Evacuation is only effective when other risks do not increase.
- The relative risk of cancer due to lack of exercise and obesity is about 1.2 times, equivalent to exposure to 100–200 millisieverts (mSv).
- Talking only about radiation exposure risk causes fear of radiation. Radiation risk is part of the risk of cancer, and the risk of cancer is part of health risks.
- Preventing the mental and social health effects of nuclear accidents should be the goal of countermeasures.

Furthermore, in "Nuclear Accidents from a Health Perspective" (Ochi 2015b), the health effects of evacuation recommendations, evacuation actions, the health effects of evacuation life, health problems of decontamination workers, and the collapse of medical care in the disaster area are discussed.

- As a reality of hospitals near the nuclear power plant, "The biggest factor that decided the evacuation of patients was the lack of food". "Most outsourcing companies issued orders saying 'Employees should not enter within a 50 km radius.' If the staff had not ignored the company's orders and brought supplies, we would have run out of oxygen". "During the autopsies, there were several elderly people who had died of exhaustion (starvation) at home. The disruption of local logistics and infrastructure (including medical) leads to death, but external companies have a duty to protect their employees and themselves (including media and medical). Can you take responsibility for 'dispatching nurses to the vicinity of the nuclear power plant'?"
- Evacuation actions, especially for the elderly, cause significant mental and physical stress. Comparing survival curves before and after the disaster, the mortality rate increases due to evacuation.
- Long-term evacuation life can lead to health damage such as lack of exercise and deterioration of mental state. The discussion is focused on radiation, and many health damages are overlooked. She stated, "We need a disaster prevention, mitigation, and reconstruction plan with health as the goal" (Ochi 2015b).

In 2020, UNSCEAR published a report on the TEPCO Fukushima Daiichi nuclear power plant accident. Compared to the 2013 report, the public's radiation dose assessment is more detailed, but it states that the conclusions of the 2013 report on the health effects on the public are valid (Batandjieva-Metcalf 2020; UNSCEAR 2020/2021). It states the following about the health effects on the public:

- No increase in thyroid cancer due to radiation exposure can be observed.
- An increase in the occurrence of other cancers is also unlikely to be observed.
- An increase in cardiovascular and metabolic (metabolic) diseases was observed in evacuees. These diseases did not increase in those who were not evacuated. The cause is not the effect of radiation exposure, but the changes in lifestyle and psychological stress caused by evacuation.

2.2.2 *Health Effects of Radiation, Low-Dose Exposure*

The health effects of radiation include deterministic effects (acute effects), such as hair loss, cataracts, skin disorders, and central nervous system disorders when the exposure dose is large, and stochastic effects (late effects), such as cancer and leukemia. Deterministic effects are not observed at low doses.

The late effects have been studied by investigating and comparing the cancer data of groups exposed to radiation and those not exposed over many years. This is an epidemiological research method called cohort study, which is based on observation and investigates the relationship between factors and the occurrence of diseases. The data of the exposed people in Hiroshima and Nagasaki not only covers a large number of people, but also covers a wide range of ages and genders, playing a major role as data for estimating late effects. The cancer occurrence in these exposed individuals and in those from non-exposed areas are observed over a long period, and the late effects are estimated from the differences. There are various types of cancer. Models are fitted to the data for estimation. There are cancers that are more likely to occur in certain ethnic groups and those that are not. The International Commission on Radiological Protection (ICRP) uses cancer mortality data from five countries for estimation. When the exposure dose becomes less than 100 millisieverts, the error becomes large and clear conclusions cannot be obtained. However, the dose rate used in the recommendations is less than 100 millisieverts per year. Hiroshima and Nagasaki were exposed in an instant, but in the case of long-term exposure at a low dose rate, there is a repair effect from the exposure. Therefore, it is necessary to consider the dose-dose rate effect. For these reasons, the error and uncertainty of late effects at low doses are two to three times.

Late effects are not visible at doses below about 100 millisieverts (mSv) because the effects of other cancer-causing factors such as smoking, drinking, and lifestyle habits are too large. However, the ICRP has established a way of thinking for radiation protection that there are effects according to the dose even at doses below that (referred to as the linear no-threshold (LNT) model) (Fig. 2.1).

There is no scientific consensus on whether the LNT model is valid. There are opinions that this model creates misunderstandings and has major drawbacks. The OECD/NEA held a workshop titled "Optimization: Rethinking the Art of Reasonable" in 2020 (NEA 2020). This workshop discussed the overall concept of radiation exposure risk, including the LNT model, and is a reference for understanding the issues.

In addition, the ICRP states in Publication 103 that "collective dose (the total of the radiation doses received by each person in a group) is mainly used for comparing radiation protection measures for occupational exposure, and it should be avoided to calculate the number of cancer deaths based on the collective dose obtained by multiplying the dose by the population in the case of extremely low doses". In other words, it is not appropriate to calculate the number of cancer deaths in a group caused by a large number of people receiving very small doses over a long period of time, but we still see this incorrect usage.

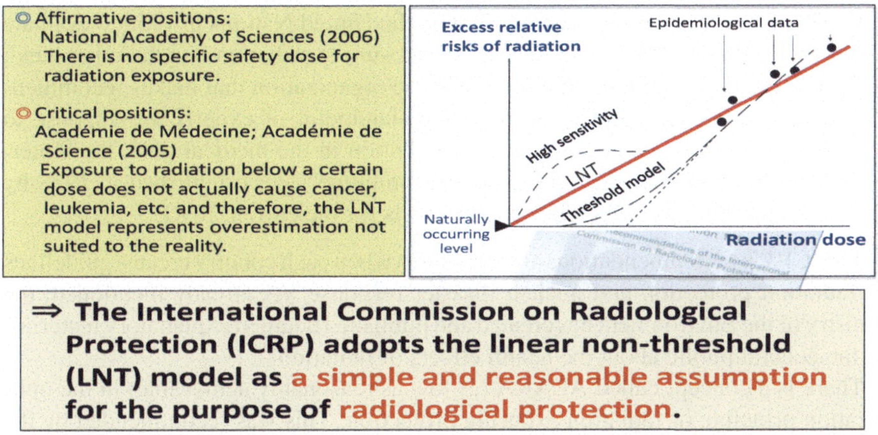

Fig. 2.1 Disputes over the LNT model. *Source* Booklet to Provide Basic Information Regarding Health Effects of Radiation (3rd edition), Chapter 4.1, Ministry of Environment, Government of Japan

At low doses, the LNT model is not a scientific fact. The ALARA concept is a policy for radiation protection.

Nagasaki University Emeritus Professor Shigenobu Nagataki, who has been involved in the work of radiation exposure risk for many years, has stated that it is necessary to distinguish between science (scientific facts) and policy (approach to dealing with the issue) as follows (Nagataki 2011).

- In the case of radiation exposure from atomic bombs, in cases of 100 mSv or more, a proportional relationship is recognized between the exposure dose and the risk of cancer, such as "a 1% increase in cancer risk at 100 mSv, a 2% increase at 200 mSv, and a 5% increase at 500 mSv". However, in cases of less than 100 mSv, no significant correlation (such as a clear survey result of a 0.5% increase at 50 mSv, a 0.1% increase at 10 mSv) can be found. In other words, no evidence of a causal relationship between exposure and cancer can be obtained below 100 mSv. This is a scientific fact ≪Science≫. The agency that announces what has gained international consensus as a scientific fact is UNSCEAR (United Nations Scientific Committee on the Effects of Atomic Radiation), so the report will be "Epidemiologically, the effects of radiation below 100 mSv are not recognized".
- Separately, there is a ≪Policy≫ that "the less radiation exposure, the better". The recommendations issued by the ICRP (International Commission on Radiological Protection) can be said to be a typical ≪Policy≫. ≪Science≫ and ≪Policy≫ are not unrelatedly parallel. The basis for deciding ≪Policy≫ is ≪Science≫. In other words, understanding the scientific impact of radiation recognized by UNSCEAR ≪Science≫, the ICRP recommendation assumes as a ≪Policy≫ that there is an effect even below 100 mSv, and extends the graph of the "linear relationship between dose and effect" above 100 mSv to below 100 mSv, and considers the system of radiation protection.

- UNSCEAR is a committee established by the United Nations in 1955 to evaluate and report on the degree and impact of exposure to radiation. On the other hand, the ICRP is a private international academic organization that makes recommendations on radiation protection from the standpoint of experts. In response to growing concerns about the impact of radiation in the medical field, the "International X-ray and Radium Protection Committee" was established in 1928 by experts in radiology, which became the basis for the current ICRP.

The ICRP's recommendations are referred to when each country creates guidelines for radiation protection and nuclear disaster response. As already mentioned, the Ministry of the Environment has created and published detailed explanatory materials for the general public about the health effects of radiation.

There is a concept called ALARA (as low as reasonably achievable) in the optimization principle of radiation exposure protection. This was recommended by the International Commission on Radiological Protection (ICRP) in 1997, and it means to strive to minimize exposure as much as possible while considering social and economic balance. It does not necessarily mean to minimize exposure as much as possible. How to quantitatively consider the R (reasonably) of ALARA is not mentioned. *ALARA is a policy for radiation protection, not a scientific fact.*

Not limited to nuclear power generation, *there are risks associated with the use of technology. When using it, it is necessary to consider the balance between risk and benefit*. This is stated at the beginning of the International Atomic Energy Agency's handbook on nuclear law as the most basic concept regarding the use of technology, not limited to nuclear power use. In the case of nuclear and radiation accidents, the risk associated with evacuation must be smaller than the risk avoided by evacuation. The U.S. protective action guides (PAGs) consider the risk of traffic accidents during evacuation as a risk associated with evacuation, and other risks of evacuation are not quantitatively considered.

Many of the disaster-related deaths in the TEPCO Fukushima accident were elderly. The author noticed during the writing of this book that their cancer death risk is smaller than the average age, and The author thought that the problem of evacuation during a nuclear accident could be improved, so the author analyzed it, quantified the risk (cost) associated with evacuation and the risk (benefit) avoided by evacuation, and compared them. In Japan, disaster-related deaths were institutionalized in the Great Hanshin-Awaji Earthquake, and data has been published, so the risk associated with evacuation could be quantified. It is a mistake and wrong to interpret the R of ALARA (As low as reasonably achievable) as good if risk communication is performed.

2.2.3 Health Effects and Lessons from Evacuation Due to TEPCO Accident

Disaster-Related Deaths Due to Evacuation

The number of evacuees due to the Great East Japan Earthquake, according to the Reconstruction Agency's announcement data, was highest in June 2012, with about 43,000 in Iwate Prefecture, about 123,000 in Miyagi Prefecture, and about 101,000 in Fukushima Prefecture. Nationwide, about 350,000 people evacuated (Reconstruction Agency 2020a). The transition of the number of evacuees in Iwate, Miyagi, and Fukushima prefectures is shown in Fig. 2.2. The number of evacuees in Fukushima Prefecture continued to be high for several years due to the evacuation caused by the TEPCO accident.

The number of deaths (direct deaths) caused by the Great East Japan Earthquake, according to the National Police Agency's announcement materials, including missing persons, is about 5786 in Iwate Prefecture, 10,758 in Miyagi Prefecture, 1810 in Fukushima Prefecture, and about 18,500 nationwide. Fukushima Prefecture did not suffer as much from the earthquake and tsunami as Miyagi and Iwate prefectures, so the number of deaths due to the earthquake and tsunami is small (National Police Agency 2021).

The number of disaster-related deaths (indirect deaths) in the Great East Japan Earthquake is 3767 nationwide, with Fukushima Prefecture having the most at 2313. The next highest is Miyagi Prefecture with 929. The number of disaster-related deaths includes those who died while evacuating from the earthquake and tsunami. The number of deaths in Miyagi Prefecture exceeded that of Fukushima Prefecture within one month of the earthquake but has rapidly decreased since then. In contrast, as shown in Fig. 2.3, the number of disaster-related deaths in Fukushima Prefecture remains high even after the first year. Over the 10 years following the earthquake, from

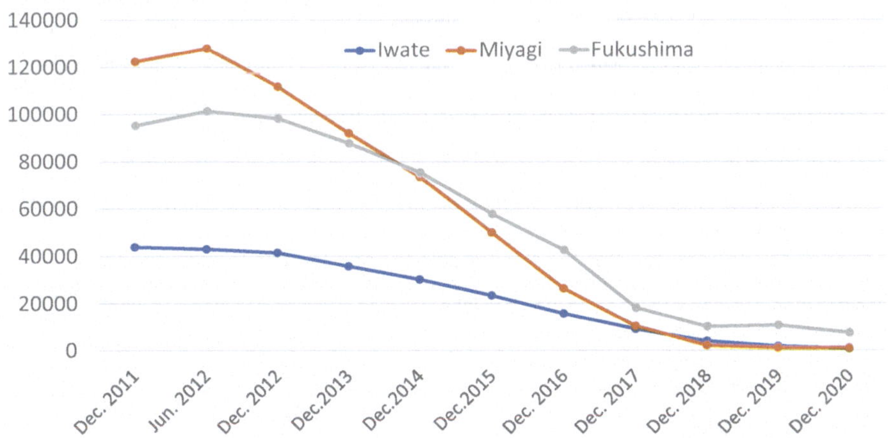

Fig. 2.2 Changes in the number of evacuees due to the Great East Japan Earthquake. *Source* Changes in the number of evacuees, Reconstruction Agency, December 2021 (in Japanese) 20211228_kouhou2.pdf (reconstruction.go.jp)

3 months onwards, the total number of disaster-related deaths in Miyagi Prefecture was 136, compared to 1597 in Fukushima Prefecture. From 1 to 10 years after the accident, the total number was 54 in Miyagi Prefecture compared to 1272 in Fukushima Prefecture. The high number of disaster-related deaths in Fukushima Prefecture is thought to be due to the long-term evacuation caused by the TEPCO accident.

Miyagi and Fukushima Prefectures have similar populations of 2.3 million and 2 million respectively, and the issue of disaster-related deaths due to evacuation from the accident can be analyzed and understood by comparing with Miyagi Prefecture. Approximately 90% of the deceased, as shown in Table 2.1, are over 66 years old (Reconstruction Agency 2020b). In the TEPCO accident, the focus was solely on the risk of radiation exposure, neglecting the occurrence of other health risks, and as a result, it is believed that those who evacuated faced a greater risk.

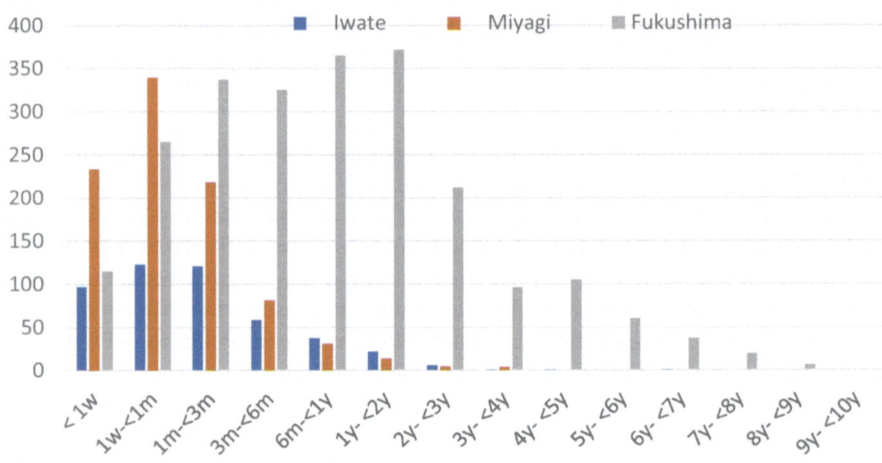

Fig. 2.3 Changes in disaster-related deaths in the Great East Japan Earthquake. *Source* Reconstruction Agency, transition of disaster-related deaths in the Great East Japan Earthquake, December 25, 2020 (in Japanese) 20201225_kanrenshi_teisei.pdf (reconstruction.go.jp)

Table 2.1 Number of disaster-related death by ages of Great East Japan Earthquake and Tsunami (as of September 30, 2020)

	Below 20	21–65	Over 66	Total
Iwate	1	64	404	469
Miyagi	2	118	809	929
Fukushima	3	231	2080	2314
Japan	9	423	3336	3768

Source Number of Disaster related deaths of Great East Japan Earthquake as of September 30, 2020 Reconstruction Agency (in Japanese). https://www.reconstruction.go.jp/topics/main-cat2/sub-cat2-6/20201225_kanrenshi_teisei.pdf (in Japanese)

Disaster-related deaths are defined as "those who died from diseases due to the worsening of injuries caused by the disaster or physical burdens in evacuation life, and were recognized as having died due to the disaster under the law concerning the payment of disaster condolence money", and were first officially recognized at the time of the 1995 Southern Hyogo Prefecture Earthquake (Great Hanshin-Awaji Earthquake). Although the purpose is to provide disaster condolence money, the death is a fact, and the data can be used to analyze the issue of disaster-related deaths. This paper presents the results of an analysis using public data, but tracing back to the cause of each death is a future task for research institutions and universities. However, the overwhelming difference in the number of disaster-related deaths in Fukushima and Miyagi Prefectures after the first year can draw lessons.

While disaster-related deaths are not unique to nuclear accidents, the lesson to be learned is that the number of disaster-related deaths in Fukushima Prefecture exceeded that of Miyagi Prefecture one month after the disaster and accident, and continued for a long period. This will be discussed.

The appropriateness of evacuation can be judged by risk-benefit analysis: the risk avoided by evacuation must not exceed the risk associated with evacuation.

Not limited to nuclear disasters, the risks avoided by emergency measures such as evacuation must not exceed the risks that arise in conjunction with those measures. This is a general principle of emergency measures, and it is also stated in the nuclear disaster response guide of the U.S. Environmental Protection Agency. If it exceeds, the emergency measures are not justified. To make this judgment, it is good to compare the risks to be avoided and the risks that arise in conjunction with them. This method is called risk-benefit analysis or cost-effectiveness analysis. We will conduct a risk-benefit analysis (cost-effectiveness analysis) using data on disaster-related deaths in Fukushima Prefecture.

To state the results first, the impact of prolonged evacuation was calculated using the data released by the Reconstruction Agency, with shortened life expectancy as the risk indicator. The average period until death of disaster-related deaths is 2.2 years, and the average remaining life expectancy of disaster-related deaths if no evacuation had taken place is estimated to be 10.1 years, so the life expectancy shortened by evacuation is 7.9 years per person. The average dose of radiation exposure that could be avoided by evacuation was estimated using the actual measured spatial radiation dose and decay rate announced by the Ministry of Education, Culture, Sports, Science and Technology, and others. When calculating the life expectancy equivalent to the radiation exposure that could be avoided by evacuation, it is 18.5 days per person, and the average life expectancy that could be avoided by the evacuation of 100,000 evacuees is 1.85 million person-days. The average life expectancy loss that occurred with evacuation is 510 million person-days when multiplied by the number of disaster-related deaths in 7.9 years. The average life expectancy loss that occurred with evacuation is 2.8 times greater than the life expectancy extension that was avoided by evacuation. Considering that the lifetime radiation exposure risk of the elderly is one-fifth of the age average (Kai 1999), it is 14 times greater. Evacuation at the time of the TEPCO accident cannot be justified. The risk conversion factor for this calculation is based on the data of the atomic bomb survivors, which is the case of instant exposure. Radiation exposure during evacuation is long-term exposure. The risk factor is twice as small as the instant exposure, because there is damage repair. Taking this into account, the risk of the evacuation increases by 28 times.

In the following, not only the results of the risk-benefit analysis, but also examples of considerations and improvements will be mentioned. Basically, it is a story that administrative organizations related to disaster measures and nuclear disaster prevention are in charge of, and it will be a little specialized, so *the rest of this section is for nuclear and radiation experts and related parties.* The basis of the calculation and other specialized contents were published in an English academic journal (Oka 2022). Since it is open access, anyone can view it. The outline is also explained in Japanese lecture materials (Oka 2023, pp. 42–67). Although this result is calculated based on the government's public data, information such as the age distribution of disaster-related deaths at the start of evacuation is only available to local governments, and there is no information other than for fiscal 2011 to the central government and it is not public. In the future, the author hopes that organizational surveys and research will be conducted, including this data, and the results will be made public.

Disaster-related deaths occur even during evacuations from earthquakes, tsunamis, floods, etc. The recognition of disaster-related deaths and the certification of care levels for the elderly are unique systems in Japan. Research to clarify the relationship between the care level before evacuation and disaster-related deaths should also help reduce health damage associated with evacuation. The United States does not have the concept of disaster-related deaths, and while there are disasters where deaths have been reported due to evacuation, such as hurricanes, it looks that there is no systematically compiled data. There are also things that have not been improved because there is no quantitative data, such as psychological and mental disorders at the time of nuclear accidents. The existence of data is indispensable for the quantitative comparison of risks and benefits associated with evacuation. Research results based on Japanese data on disaster-related deaths should be useful to the world.

Life Shortening Caused by Evacuation

The Reconstruction Agency announces the number of disaster-related deaths caused by the Great East Japan Earthquake every year. There are 12 municipalities in Fukushima Prefecture that received evacuation orders and evacuated. Of these, Hirono Town had its evacuation lifted at the end of September 2011, so the author analyzed the data of the remaining 11 municipalities. The number of disaster-related deaths is shown in Table 2.2 for each municipality, up to the end of September 2020, and for the first year, up to the end of March 2012, and the difference between the two. The total number of disaster-related deaths as of the end of September 2020 is 2,084. The total as of the end of March 2012, about a year after the accident, is 630, and the number of disaster-related deaths that occurred after about a year after the accident is 1,454. The number of disaster-related deaths includes not only those who evacuated due to the nuclear accident but also those who evacuated due to the earthquake and tsunami.

The number of disaster-related deaths in Miyagi Prefecture, as shown in Fig. 2.3, decreased three months after the accident and almost disappeared a year later. The disaster-related deaths in Fukushima Prefecture a year after the accident can be considered to have occurred due to evacuation associated with the accident, not the earthquake and tsunami. However, since Fukushima Prefecture was not as affected by the earthquake and tsunami as Miyagi Prefecture, and the data within one year after the accident includes the number of deaths due to evacuation of nursing homes. The number of deaths after one year shown as a difference in Table 2.2 underestimates the number of disaster-related deaths due to the accident. The number of disaster-related deaths due to accidental evacuation is between 2084 and 1454.

From the data on disaster-related deaths in 11 cities and towns, the average period until death is calculated to be 1.3 years when including data from March 2011 for one year, and 3.1 years when not including it. The average is 2.2 years. 90% of disaster-related deaths are elderly people aged 66 and over. The average remaining life expectancy of disaster-related deaths who did not evacuate is estimated to be 10.1 years from the age distribution of elderly disaster-related deaths at the time of evacuation. Subtracting 2.2 years from this, the remaining life lost due to evacuation is 7.9 years. Using the average of 2084 and 1454 deaths, which is 1769, the reduction in life expectancy is 5.1 million person-days. This is compared with the extension of life expectancy due to radiation exposure avoided by evacuation.

Extension of life expectancy due to radiation exposure avoided by evacuation; prolonged evacuation is not justified.

Table 2.2 Number of disaster-related death of evacuated municipalities

	Until Sep. 30, 2020	Until Mar. 31, 2012	Difference
Tamura	14	1	13
Kawauchi	128	27	101
Naraha	140	33	107
Katsurao	42	9	33
Minami Soma	517	282	235
Iitate	42	35	7
Kawamata	29	0	29
Namie	440	91	349
Tomioka	450	76	374
Okuma	128	38	90
Futaba	154	38	116
Total	2084	651	1454

Source Number of Disaster related deaths of Great East Japan Earthquake as of September 30, 2020 Reconstruction Agency (in Japanese). https://www.reconstruction.go.jp/topics/main-cat2/sub-cat2-6/20201225_kanrenshi_teisei.pdf

This is evaluated from the measurement results of spatial radiation dose after the accident and its decay. Based on the results of the driving survey, the Ministry of Education, Culture, Sports, Science and Technology (MEXT) has published the estimated value of the accumulated radiation dose for one year after the accident (MEXT 2012). The spatial radiation dose decreases over time. Researchers from the Japan Atomic Energy Agency have announced the decay rate based on actual measurement data (Kinase 2015a). The spatial radiation dose decreases faster than the half-life of radioactive materials due to weathering effects. The decay rate is slightly higher in areas where it is difficult to return home than in areas with living restrictions.

Since these are the results of a driving survey, it is considered to be a measurement along roads where there are many residences and can be used to estimate the dose of radiation avoided by evacuation. There are as many as 172 measurement points in total for the 11 cities and towns, so the author decided to use the average value for each city and town. For the 11 cities and towns, the author calculated the average value of the accumulated dose for 10 years until the evacuation was lifted, or for cities and towns where areas where evacuation has not been lifted remain. Table 2.3 shows the results, along with the accumulated dose for the first year.

Table 2.3 Cumulative dose averted by evacuation

	Average (mSv)[a]		Highest (mSv)[b]	
	2011	Total[c]	2011	Total[c]
Tamura	3.3	7.8	7	13.5
Kawauchi	8.65	22.1	10.2	18.7
Naraha	9.1	26.8	14.1	27.7
Katsurao	26.3	102.3	98.2	200
Minami Soma	16.2	63.1	20.3	78.9
Iitate	31	120.4	92.8	360
Kawamata	15.4	50.9	42.5	140
Namie	104.4	355.8	207.4	707
Tomioka	43.4	165.2	110.1	419
Okuma	158.2	538.2	482	1640
Futaba	75.1	254.8	163.7	555
Okuma > 5 km	38.8	132	91.6	312
Futaba > 5 km	73.9	250.7	116.3	394

[a] average of cumulative doses among measured locations in the municipalities,
[b] highest cumulative dose among measured locations in the municipalities
[c] total cumulative dose until lifting evacuation order and in 10 years for Katsurao, Minami-Soma, Iitate, Namie, Tomioka, Okuma, and Futaba where evacuation order is not fully lifted
Attenuation factor for Namie, Okuma, and Futaba is that of difficult to return zone, for other towns is that of residence restricted zone. *Source* https://radioactivity.nsr.go.jp/ja/contents/5000/4766/24/1750_022918.pdf. https://www.pref.fukushima.lg.jp/site/portal/cat01-more.html. https://radioactivity.nsr.go.jp/ja/contents/12000/11995/33/part2.pdf. Fig. 4 (b) and (d). (in Japanese)

From these results, we estimated the number of evacuees in each of the 11 cities and towns by multiplying the number of residents (population) and the proportion of the area evacuated and used this as a weight to calculate the average. The avoided accumulated dose was about 205 millisieverts. The total number of evacuees when adding up the number of evacuees in each city and town is about 100,000. Figure 2.2 shows that the maximum number of evacuees in Fukushima Prefecture as a whole is just over 100,000, and the total number of evacuees in the 11 cities and towns, 100,000, is considered to be a reasonable figure. From this, it is considered that the method of estimating the number of evacuees in each city and town used as a weight in the calculation of the accumulated dose is also reasonable.

The reduction in life expectancy due to exposure to 205 millisieverts (the extension of life expectancy due to avoiding exposure) is, as a coefficient, 0.09 days/mSv (Cologne and Preston 1999, 2000) Using this, it becomes 18.5 days. The life extension avoided by 100,000 people due to evacuation is 1.85 million person-days. This is 2.8 times smaller than the 5.1 million person-days of life-shortening caused by evacuation. Considering that the lifetime radiation risk of the elderly is one-fifth of the average age, it is 14 times smaller. The result is that prolonged evacuation is not justified.

The 5.1 million person-days is a figure calculated based on actual death data, while the 1.85 million person-days is a figure calculated by multiplying the number of evacuees, 100,000, by the life expectancy period converted from the radiation dose, using a conversion factor, based on the average accumulated radiation dose of evacuees over 10 years being 205 millisieverts. It is not good in ICRP103 (2007) to use the number of deaths multiplied by a large population at a low dose as a risk indicator, but it is acceptable to use it for risk-benefit analysis. It should be noted that the 1.85 million person-days is such a figure. The 5.1 million person-days of life shortening calculated from the number of disaster-related deaths is a figure calculated based on actual data.

Furthermore, in the calculation of 1.85 million person-days, the life shortening coefficient of 0.09 days/mSv obtained from the data of atomic bomb victims is used, but this figure is based on the data of atomic bomb victims who were exposed to high dose rates (high doses in a short period). The health effects of long-term exposure at low dose rates are considered to be smaller than those of high dose rate exposure, so considering this, there is a possibility that 1.85 million person-days is an overestimate. Therefore, the risk of evacuation may be even greater than the radiation risk avoided by evacuation (2.8 times or 14 times). The effects of low dose rate exposure are described, for example, in Chap. 3 of the unified document on the health effects of radiation by the Ministry of the Environment. The International Commission on Radiological Protection states that if you are exposed little by little, the impact is half that of being exposed all at once, even if you receive the same dose, so considering this, the risk of evacuation is 5.6 times or 28 times the radiation risk avoided by evacuation.

Based on the result that the life shortening due to evacuation is 2.8 times larger than the life extension avoided by radiation exposure, 14 times larger when considering that the lifetime radiation risk of the elderly is one-fifth of the average age, and 28 times larger when considering the effect of the low dose rate exposure, the following discussion is presented.

First, we discuss inactive lifestyle diseases (disuse syndrome) and point out that it is a mistake to understand that the health effects other than radiation exposure associated with evacuation are only psychological factors as traditionally stated. In the context of radiation protection, the recommended values for radiation protection are given as age averages, and it is pointed out that the fact that the lifetime cancer mortality rate of the elderly is low is not considered. Next, we discuss that other health risks cannot be considered in the framework and scale (dose rate, etc.) of radiation health effects, and present considerations for improving nuclear disaster countermeasures guidelines, etc.

Sedentary Lifestyle Disease (Disuse Syndrome): Health Effects During Evacuation Other than Radiation

If you don't move your body, various functions of the body decline. This is known as sedentary lifestyle disease (also called disuse syndrome), which has been known since the 1980s. Symptoms include decreased blood circulation, obesity, fragility of the musculoskeletal system, depression, and premature aging. It can be experimentally reproduced, and there are treatments through prevention and recovery, which are fortunately said to be inexpensive, safe, simple, and effective (Bortz 1984).

The Ministry of Health, Labour and Welfare has issued a warning about sedentary lifestyle disease during disasters such as earthquakes. If evacuees do not move their bodies for a long time at evacuation centers, their ability to move their bodies decreases (Ministry of Health, Labour and Welfare 2011). During the Great East Japan Earthquake, evacuees from Fukushima Prefecture were forced to evacuate for a long time due to evacuation orders and recommendations from the TEPCO accident. The response taken by the Japanese government followed the ALARA (as low as reasonably achievable) recommendation of the International Commission on Radiological Protection (ICRP), and there is no flaw in the procedure. However, as a result, the evacuation cancellation recommendation value was too low for the elderly, so the evacuation was prolonged, the sedentary lifestyle disease increased, and it is thought that the disaster-related deaths in Fukushima Prefecture became more than in Miyagi and Iwate Prefectures. The warning about sedentary lifestyle disease by the Ministry of Health, Labour and Welfare (Ministry of Health, Labour and Welfare 2011) was issued to the long-term care insurance departments and related organizations of each prefecture and municipality, but ALARA led to an increase in disaster-related deaths in Fukushima Prefecture. From this, we can learn the lesson that it is necessary to clearly state the prevention of disaster-related deaths in the concept of nuclear disaster prevention.

It is a mistake to understand other health risks during nuclear accidents as only psychological aspects.

From the experience of the Chernobyl accident, it seems that nuclear-related organizations and their members, including international organizations, understand that health damage other than radiation during nuclear accidents is mental and psychological, and should be dealt with by risk communication. However, immediately after the Chernobyl accident, there was an increase in artificial abortions in Greece and Italy. The explanation is in the reference material (Ministry of the Environment 2022). It is written in "Chap. 3: Health Effects of Radiation, 3.8 Psychological effects". It is true that the risk of radiation exposure was not properly understood. It is, however, a mistake to think that risk communication can address the problems, even if only from the experience of the Chernobyl accident.

Induced abortions increased in Greece and Italy. The birth rates declined in January 1987, 10 months later the Chernobyl accident. However, in Hungary, where abortion is not allowed unless the fetal radiation dose exceeds 100 millisieverts, no decline of the birth rate was observed. Even if we talk about risk communication, it is not successful because there is a big fuss at the time of the accident. It is learned that preparing standards in advance will be a countermeasure.

Furthermore, the risk communication mentioned here refers to the transmission of short messages explaining risks and safety to the general public, not local stakeholder dialogues or crisis communication during accidents. International organizations have issued many reports on nuclear accidents and emergency responses. For example, the International Commission on Radiological Protection (ICRP) has published Publications 109, 111, 146 in 2009, 2009, and 2020 respectively. The International Atomic Energy Agency (IAEA) has published GS-G-2.1, GCG-2, GSR Part7, GSG-11 in 2007, 2011, 2015, and 2018 respectively as General Safety Guides. The World Health Organization (WHO) also issued a report titled "Health Effects of the Chernobyl Accident and Special Care Plan" in 2006 (WHO 2006).

The WHO report is considered to summarize the health effects of the Chernobyl accident, and on pages 96 to 97 of the report, it recommends "further focusing on mental health maintenance policies and risk communication for the general public". Health effects other than mental, such as sedentary disease, are not mentioned in this report or in the above-mentioned ICRP and IAEA reports. There are no reports on disaster-related deaths in the Chernobyl accident as far as the author has researched.

The Chernobyl accident occurred five years before the collapse of the former Soviet Union, and the country's economic situation was deteriorating. Pensions seem to have been paid, but it was a time of rapid inflation and social change, so those who evacuated may not have been able to sit still in the evacuation centers to survive. Many people seem to have been involved in stabilizing the accident reactor. In the case of the TEPCO accident, an average of 90 million yen per household of four was paid by TEPCO to the residents who lived in the area that became a residence restriction area, for property, inability to work, and mental damage (MEXT 2013). It was not necessarily necessary to work immediately to survive. There was no place to work. The average life expectancy was more than 10 years different between Ukraine at that time and Japan at the time of the TEPCO accident, so the proportion of elderly people was also different. The area affected by radiation was about 6% of the Chernobyl accident in the TEPCO accident, and the population density of the area was also different. For these reasons, disaster-related deaths may not have been manifested in the Chernobyl accident. Disaster-related deaths were officially recognized in Japan from the 1995 Great Hanshin-Awaji Earthquake. This allowed for the compilation of statistical data on disaster-related deaths and made risk-benefit analysis possible. There is no statistical data on disaster-related deaths in the Chernobyl accident. Mental and psychological effects cannot be quantified.

Another difference between the TEPCO Fukushima accident and the Chernobyl accident is that the TEPCO accident was caused by a tsunami, but disaster-related deaths have occurred in Miyagi and Iwate prefectures, which were not affected by the accident, due to evacuation from earthquakes and tsunamis. By comparing disaster-related deaths in Fukushima Prefecture and Miyagi and Iwate Prefectures, we can understand the health effects other than radiation associated with evacuation from nuclear disasters.

The dose rate does not represent the lifetime cancer mortality rate due to radiation exposure.

The International Commission on Radiation Protection (ICRP) makes recommendations on radiation doses for the general public and workers. The Japanese government and other countries use this as a reference to set regulatory values. The dose rate, mSv/y (annual exposure dose), is used for this. Many research papers also use the dose rate for discussion. The dose rate is an annual risk and does not represent the lifetime cancer mortality rate.

Other health effects cannot be considered within the framework of radiation exposure risk.

The author mentioned that the dose rate cannot represent the difference in lifetime exposure risk by age, and that the age average value cannot consider the difference in lifetime cancer mortality rate by age. However, in the first place, within the framework of radiation protection, such as the radiation dose rate, other health risks cannot be considered.

The Japanese nuclear disaster response guidelines need to consider the balance of risks and benefits associated with evacuation.

The U.S. Environmental Protection Agency has created a nuclear protection manual, commonly known as the PAG (protective action guides) manual (EPA 2017).

PAG lists the following three principles on page 3.

(1) Prevention of acute radiation health effects.
(2) Balance protection with other important factors and ensure that actions result in more benefit than harm.
(3) Reduce risk of chronic effects (latent radiation health effects).

The second principle is expanded into the following two principles on page 30.

(2–1) PAGs should not be higher than justified on the basis of optimization of cost and the collective risk of effects on health. That is, any reduction of risk to public health achievable at acceptable cost should be carried out (2–2). Regardless of the above principles, the risk to health from a protective action should not itself exceed the risk to health from the dose that would be avoided.

Based on this principle, PAG is creating guidelines, but only considers traffic accidents during evacuation, which are quantified, as risks associated with evacuation (since psychological damage cannot be quantified). As a result, it is stated that if the benefit of avoiding radiation exposure is greater than the risk of traffic accidents, evacuation can be justified if 5 mSv can be avoided. The disease of inactivity associated with evacuation is not mentioned. Since the data on disaster-related deaths in Fukushima Prefecture is quantified, it should be possible to reflect the results of risk-benefit analysis in the guidelines, taking into account the above principles.

In PAG, risk-benefit analysis is used to determine the dose rate for protective measures, and no analysis of other health risks is performed. In the first place, it should be recognized that other health risks associated with evacuation cannot be considered in the framework of radiation risk. However, if there is data on other health risks, a risk-benefit analysis can be performed.

In Japan, the Basic Disaster Prevention Plan and the Nuclear Disaster Countermeasure Guidelines (abbreviated in English as EPR Guide) have been created after the TEPCO accident. The Basic Disaster Prevention Plan describes the basic plans for various disasters, and Chapter 12 is the Nuclear Disaster Countermeasures Edition. On page 285, in the section on "Consideration for those who need consideration" during wide-area evacuation, it is stated that "Local public bodies should pay sufficient attention to ensure that those who need consideration and temporary residents do not deteriorate their health condition during evacuation, and strive for understanding the health condition at designated evacuation centers, establishing a support system for welfare facility staff, prioritizing entry into temporary housing, and setting up temporary housing for the elderly and disabled. Also, sufficient consideration should be given to providing information for those who need consideration". However, the perspective that the continuation of evacuation worsens the disease of inactivity is missing.

Nuclear Disaster Countermeasure Guidelines (NRA 2022) on page 5 of "Basic Concepts for Radiation Protection Measures", it states, "In accordance with the recommendations of the International Commission on Radiological Protection and the principles of the International Atomic Energy Agency's General Safety Requirements, it is necessary to keep the radiation exposure of residents as low as possible, while also minimizing health impacts not directly caused by radiation exposure". The guidelines do not elaborate further on "minimizing impacts on health and other areas not directly caused by radiation exposure". In the first place, *the idea of reducing radiation exposure as much as possible is contradictory, as it leads to the prolongation of evacuation and increases disaster-related deaths in Fukushima Prefecture.* The guidelines do not discuss the concept of risk-benefit, which is necessary to make a rational judgment on competing risks.

While the U.S. manual requires consideration of health damage separate from radiation exposure due to evacuation, the Japanese guidelines do not specifically require this, so there is no guarantee that the administration will ensure that "the risks associated with protective measures do not exceed avoidable risks". The Japanese nuclear disaster prevention guidelines need to incorporate the concept of balancing the costs of evacuation with the benefits gained from it.

As already mentioned, when comparing the Japanese nuclear disaster guidelines and the U.S. PAG manual, there is a significant difference in the way they are written. The PAG writes about concepts, while the Japanese guidelines write about procedures. Major disasters are infrequent, so it is impossible to predict everything. The PAG's description is more flexible and effective. In the first place, the procedures prepared for the TEPCO accident did not function. For example, the emergency response center was damaged and lost its function, causing confusion. This is one example of a procedure not working well, but shouldn't we fundamentally reconsider the lessons of the TEPCO accident and apply them to the guidelines?

The Shorter the Evacuation Period, the Better

Disaster-related deaths in Miyagi and Iwate Prefectures have been decreasing since one month after the accident, but those in Fukushima Prefecture have continued for a long time. To prevent the onset of inactivity disease in elderly evacuees, the shorter the evacuation period, the better.

The lifting of the evacuation order was delayed in the TEPCO accident. The setting and transition of the evacuation zones after the TEPCO accident are as follows (Fukushima Prefecture 2023).

On the night of the accident day, March 11, 2011, an evacuation order was issued within a 2 km radius from the power plant, which was expanded to 20 km by the next day. On March 15, residents in the area from 20 to 30 km were requested to take shelter indoors. On March 25, voluntary evacuation was called for residents taking shelter indoors. On March 21, the government restricted the distribution of specific foods. On April 24, 2011, the government established alert zones, planned evacuation zones, and emergency evacuation zones, referring to international standard values. Residents of areas included in the planned evacuation zone, such as Iitate Village, were asked to evacuate to other places. Even outside the planned evacuation zone, there were areas (hotspots) where the cumulative dose for one year was expected to exceed 20 mSv, and residents were urged to evacuate (all lifted by December 28, 2014).

On April 1, 2012, some of the alert zones and planned evacuation zones were reviewed according to the situation of the annual cumulative dose and were organized into evacuation order cancellation preparation zones (areas where it is confirmed that the annual dose will be 20 mSv or less), residence restriction zones (areas where the annual cumulative dose exceeds 20 mSv and continued evacuation is required), and difficult-to-return zones (areas where the annual cumulative dose exceeds 50 mSv and is unlikely to fall below 20 mSv even after 5 years). In the evacuation order cancellation preparation zones, temporary return of residents (residence is prohibited) and partial resumption of businesses and farming at hospitals, welfare facilities, stores, etc., became possible. Evacuation orders are being lifted sequentially, but it has taken at least one year from the accident to the start of the lifting. The dose rate to lift the evacuation order is 20 mSv/y.

Page 66 of the U.S. Emergency Response Guide (PAG) states, "The cost of relocation (continuation of evacuation) is constant, but the radioactivity of many radioactive nuclides decays rapidly, so the cost-effectiveness of relocation decreases over time".

Considering the inactive diseases during the evacuation period, the cost of evacuation is not constant, but increases over time.

When it is determined that there is no expansion of the accident, it is better to lift the evacuation order in low-dose areas as soon as possible. In the case of the TEPCO accident, cores were melted down, radioactive substances leaked from the nuclear reactor containment vessel, the reactor buildings were destroyed by hydrogen explosions, and radioactive substances were released into the environment. At that point (around 3:30 pm on March 12 for Unit 1, around 11 am on March 14 for Unit 3, and after 6 am on March 15 for Unit 4, when the hydrogen explosions occurred), in other words, no later than four days after the accident, the massive release of radioactive substances due to the accident had ended. This is the point at which it can be judged that the accident is not expanding (the release of radioactive substances caused by the accident will not increase any further). Confirmation that the melted core could be cooled by external water injection came a little later, but the large-scale release ended at the time of core meltdown. This should be taken as a lesson for disaster management. The environmental radiation dose after the accident is continuously measured at monitoring posts inside and outside the power plants by TEPCO and Fukushima Prefecture, and the above can be understood from the measurement results.

Being conservative about radiation exposure is dangerous in terms of preventing inactive lifestyle diseases.

It needs to be considered in disaster prevention measures. Disaster-related deaths, in terms of the health effects of radiation, are deterministic effects (acute effects), not late-onset (cancerous). The principles of the U.S. PAG prioritize the prevention of acute health effects. The late-onset effects become unclear due to large statistical errors at low doses. The importance of preventing disaster-related deaths is higher than the late-onset effects, especially at low doses (below 100 mSv/y). The principle of radiation protection, ALARA (As low as reasonably achievable), which aims to keep low-dose radiation exposure as low as possible, is a concept (policy) for late-onset effects at low doses. Is it appropriate to apply the concept of ALARA when deciding the evacuation period? It would be better to reconsider with reference to the results of the risk-benefit analysis of the TEPCO accident. Isn't there a need to improve the guidelines for nuclear disaster response? In the TEPCO accident, the government's decision-making tended to be delayed.

The average accumulated dose over the first year for the 11 municipalities that evacuated is summarized in Table 2.3. The average dose for the first year in 9 out of the 11 municipalities is less than 100 mSv. In Sect. 3.7 of Chap. 3 of the Ministry of the Environment's "BOOKLET to Provide Basic Information Regarding Health Effects of Radiation", it is stated that "the risk of cancer incidence is above 100 millisieverts, the risk of cancer death is above 200 millisieverts, and there is a proportional relationship between dose and risk. However, opinions differ among researchers about the relationship at doses lower than 100–200 millisieverts". Considering these points, it might be best to decide in the guidelines for nuclear disaster response that evacuation should be lifted as soon as possible when the radiation dose is less than 100 mSv/y. This would be better, at least for the elderly.

Shelter Indoor is Best for the Elderly in Care Facilities

The issue of the elderly in care facilities is a special case of disaster-related deaths. The reason it's a special case is that while the elderly in care facilities die during or immediately after evacuation, disaster-related deaths in Fukushima Prefecture occur over several years after the start of evacuation. The severely elderly in care facilities and hospitals are the most vulnerable. It is better not to evacuate them. Isolated care facilities and hospitals due to evacuation can request deployment from the Self-Defense Forces with medical teams. Severely elderly people at home can be moved to nearby hospitals after shelter indoors. If evacuation is lifted within a few days of the accident, shelter evacuation is effective not only for these elderly people.

What's important in this case is not the guidelines, but the network of communication with the elderly and their families. The role of local governments is crucial in this. The ones who know each region best are the local governments. They need to maintain a network with the residents of the region during normal times and utilize it during disasters. Top-down emergency measures alone do not work. The importance of the role of local governments is a lesson learned from the 1995 Great Hanshin-Awaji Earthquake.

The issue of what to do with the elderly who need care during a disaster is not unique to nuclear disasters. There are various types of disasters. Earthquakes, typhoons, floods, tsunamis, wildfires, large explosions, etc. There may be disaster-related death measures according to the situation of the disaster, and lessons should have been learned, and it would be good to make use of them and aim for improvement. In the United States, care facilities are required to prepare multiple written evacuation plans in advance for events such as hurricanes. Analysis and evaluation of evacuation plans for care facilities have also been reported (Castle 2008).

In Japan, against the backdrop of an aging population and the progression of nuclear families (families of parents and children), the long-term care insurance system was launched in 2000. This system covers individuals over the age of 40, and through insurance premiums and taxes, provides medical care, long-term care, and support for the elderly, operated by local governments. Care is divided into five levels, and support into two levels, which are certified through an application and review process. Level 4 care is for those who can hardly stand or walk on their own, while level 5 care is for bedridden individuals. Cognitive function is also included in the assessment criteria. Understanding the correlation between the level of care required during evacuation and disaster-related deaths could be useful in analyzing issues related to evacuation and disaster-related deaths. Further discussion on this topic is desired.

Lessons from the Chernobyl Accident for Japan's Nuclear Disaster Prevention Measures

Although not a lesson related to evacuation, an important lesson from the Chernobyl accident for disaster prevention is the need to *limit the intake of food contaminated with radioactive iodine immediately after the accident* (half-life is about 8 days), *especially by young children.* In the Chernobyl accident, the delay in limiting the intake of milk from dairy cows that had eaten contaminated grass led to the accumulation of radioactive iodine in children's thyroid glands, resulting in thyroid cancer. During a nuclear disaster, not only should the distribution and intake of milk and vegetables be restricted, but also the intake of leafy vegetables by self-employed farmers. Radioactive iodine has a relatively short half-life of 8 days, so it is most important to thoroughly disseminate information immediately after core meltdown accidents, which can prevent this.

This important lesson does not seem to be explicitly written in Japan's Nuclear Disaster Prevention Guidelines (as of July 3, 2019 version). Although it does mention avoiding internal exposure through oral intake and taking stable iodine tablets. First and foremost, in the event of an emergency where a core meltdown is likely to occur at a nuclear power plant, it is necessary to restrict the intake of food and beverages, especially by young children, that contain more than the standard amount of radioactive iodine produced by the accident. If people are using stream water as drinking water, its intake should also be restricted.

The guidelines state that iodine intake restrictions should be implemented after radiation measurements have been taken, but when it is anticipated that the reactor core will melt, it is necessary to communicate the intake restrictions to residents within a 30 km radius of the power plant. The wind direction changes, so the direction in which radioactive substances are released changes. Therefore, it is necessary to issue intake restrictions before the release in the circular area around the power plant. In the United States, there are two types of emergency planning zones, and exposure from plumes (masses of smoke) containing radioactive substances released from nuclear power plants is within a 10-mile (approximately 16 km) range of the power plant for emergency planning, but the area to avoid exposure by intake is within a 50-mile radius of the power plant. Taking this into account, the range from the power plant to 80 km becomes the necessary emergency planning area for intake restrictions.

During the TEPCO accident, temporary food regulation values were determined after high concentrations of radioactive substances were detected from plants collected in Fukushima Prefecture on March 15, 2011, and a notification not to provide food exceeding this was made to each prefecture from the Nuclear Safety Commission six days after the accident on March 17 (Government Accident Investigation Report 2011). It seems that it took time to notify because there were no regulation values, but they should have issued a warning immediately, learning from the Chernobyl accident. The administration may not be able to move without regulation values, but rules and procedures alone cannot fulfill the role that the government should play for the people in an emergency. It should inherit lessons and ways of thinking, not procedures.

As already introduced, the Japanese Nuclear Disaster Response Guidelines describe procedures, and the U.S. Nuclear Protective Action Guides (PAG Manual) describes ways of thinking. Since the states are in charge of disaster response in the United States, the federal government's PAG Manual may be showing ways of thinking (guidelines) rather than procedures, but Japan also needs a document that writes ways of thinking. Furthermore, it is problematic that the explanation of the basis for Japan's measures is weak, so to supplement this, it would be good to create training materials for nuclear disaster prevention that summarize the transition of environmental radiation doses, lessons, and ways of thinking based on actual data, understanding the environmental impact at the time of the accident. If you can't imagine the behavior, including various states, the measures won't work. Only Japan has actual experience of a light water reactor core meltdown accident with a large environmental impact. It is the responsibility of the Japanese nuclear industry, which caused the TEPCO accident, to create these training materials based on the available environmental impact data.

After the TEPCO accident, looking at the time change of the spatial dose rate around the power plant (for example, Futaba Town Yamada, 4.1 km west-northwest from the power plant) for one month after the accident, the spatial dose rate after a large release is decaying with a half-life of about 8 days. It is believed that radioactive iodine with a half-life of 8 days, which adhered to the ground, trees, and buildings, was dominant. After the radioactive iodine decayed, it is believed that radioactive cesium with a half-life of about 30 days mainly contributed to the spatial dose rate. Since the energy of the radiation produced by radioactive iodine is low, the exposure dose decreases if you are inside a building.

According to the results of spatial dose rate measurements for several years after the accident, the spatial dose rate decays faster than the physical half-life of iodine and cesium (Kinase 2015b). This is called a weathering effect. It is believed to be due to the effect of soil and leaves contaminated with radioactive iodine and radioactive cesium being washed away and dispersed by wind and rain. Radioactive cesium adheres to particles such as soil. Therefore, if you remove the surface of the soil and bury it underground (decontaminate it), you can reduce the spatial dose rate. The behavior of these radioactive substances also needs to be incorporated into training materials as lessons. Faculty and graduate students of the Faculty of Agriculture, University of Tokyo, conducted research on the situation after the Fukushima accident in a practical manner using their expertise in each agricultural field. The results have been compiled into English documents and have been downloaded many times (Nakanishi and Tanoi 2013, 2016; Nakanishi et al. 2019). Including these, the author would like Japanese nuclear research institutes to create training materials based on practical results.

In a core meltdown accident like the TEPCO accident, not all fission products (radioactive substances) in the fuel are released into the environment. Inert gases and volatile elements are released. Inert gases such as krypton and xenon have no chemical reactivity, so even if they are released, they diffuse and dilute in the atmosphere, and the amount of radiation in the air is temporarily high. Elements that volatilize at high temperatures include iodine and cesium, which are released from the reactor fuel that has melted and become hot, pass through the reactor building, and adhere to surrounding objects. Radioactive iodine with a half-life of 8 days and radioactive cesium with a half-life of about 30 years are the main contributing nuclides to the spatial radiation dose after a core meltdown accident.

The half-life of radioactive cesium, cesium 137, is about 30 years. Because the half-life is long, cesium 137 produced in atmospheric nuclear tests conducted before the 1970s still exists in the global environment. The amount of radioactive substances released into the global environment by atmospheric nuclear tests is 100 to 1000 times more than the amount released by the Chernobyl accident (IAEA 1996). According to the UNSCAER 2013 report, the amount of cesium 137 released into the environment in the TEPCO Fukushima accident is about one-fifth of the Chernobyl accident. After the TEPCO Fukushima accident, cesium 137 in seawater and coasts overseas, such as Indonesia, the Indian Ocean, and the South Pacific countries, it has been reported that cesium has been observed, but this cesium is not derived from the TEPCO Fukushima accident, but from atmospheric nuclear tests. Considering the direction of the currents in the North Pacific, the cesium 137 that is said to have been observed in China and Korea is probably the same. This should be understood by examining observation data from the 1960s. The concentration of cesium 137 in the seawater along the coast of Fukushima Prefecture is decaying much more rapidly than its half-life due to the diffusive action of ocean currents.

The Lifetime Cancer Mortality Rate for the Elderly is One-Fifth of the Age Average

The ICRP's 1991 recommendation report, Publication 60, shows the lifetime cancer mortality rate by age in Figure C-5. The lifetime cancer mortality rate for the elderly decreases with age and becomes zero. Figure C-5 shows the results of two models, but the trend is similar for both models for those over 30 years old. The reason why the lifetime cancer mortality rate decreases with age is probably because the remaining life expectancy of the elderly is short.

There is a Japanese commentary that lists the lifetime cancer mortality rate by age group (Kai 1999). This is shown in Table 2.4. This corresponds to Figure C-5 of ICRP Publication 60. The lifetime cancer mortality rate due to radiation exposure (to be precise, the increased probability of lifetime cancer mortality in the case of exposure compared to the case of no exposure) is 5%/Sv for the age average, while it is 0.9%/Sv for the elderly over 65 years old. In other words, the lifetime cancer mortality rate due to radiation exposure for the elderly is about one-fifth of the age average. The current recommended dose for radiation protection is not given by age.

Table 2.4 Lifetime fatal cancer probability for gender and age at exposure (%/Sv)

Age at exposure	Male	Female	Average
0–19	8.1	12.8	10.4
20–64	3.2	4.2	3.7
65–90	0.8	0.9	0.9
0–90	4.2	5.8	5.0

Source Kai (1999). Available from https://nuce.aesj.or.jp/jnuce/vol6/Jnuce-Vol6-1-p7-12.pdf

The Dose Rate for Lifting Evacuation for the Elderly Should be Five Times Higher

The current dose rate for lifting evacuation is set at 20 mSv/y in accordance with the ICRP's recommendations. This is too low for the elderly. Shouldn't it be five times higher, at 100 mSv/y? The dose rate for lifting evacuation was too low, and the prolonged evacuation was the reason for the high number of disaster-related deaths among the elderly in Fukushima Prefecture. Moreover, even if we calculate the dose rate equivalent to the average life expectancy reduction from the comparison of the risks avoided and incurred by evacuation mentioned above, 100 mSv/y is still too low. This should be examined by experts in public health and public policy.

Examples of Measures Considering the Characteristics of Japanese Agriculture

Aging is progressing in Japan. The national census is conducted every five years, and the results are announced by the country and prefectures. The aging rate in Fukushima Prefecture (the proportion of elderly people aged 66 and over) was 25% in 2010. In Japan, nuclear power plants are located in the regions. The Tokyo Electric Power Company's Fukushima Daiichi Nuclear Power Plant is one of them. The area affected by the radiation from the accident was the Soso region (Soma and Futaba areas), where 85% of the farmers are part-time. 61% of the agricultural workers in this region are elderly people aged 66 and over.

People in the areas with nuclear power plants are engaged in agriculture, as well as working at the power plants, and running canteens for workers and accommodation facilities for long-term residents. Many of the non-elderly (working-age people) are employed and work at local construction companies and power plant maintenance service companies. Many of them are part-time farmers. 90% of agriculture in Fukushima Prefecture is rice farming. Rice farming is busy during the spring planting season and the autumn harvest. Even if they are employed, they may take a vacation during busy periods, or they may do farming work on consecutive holidays. As the term "San-chan agriculture, agriculture by mother, grandfather and grandmother" suggests, the scale of Japanese farms is small. When fathers of working age retire, they devote themselves to farming and jobs in canteens and accommodation facilities. Their parents are also working. In Japan, the age at which pensions start to be paid is 65 (64 in 2011, the year of the accident). Many of the local people who run canteens and long-term accommodation facilities have built apartments and canteens on part of their farmland and are also engaged in agriculture. For these reasons, many of the people aged 65 and over in the areas with nuclear power plants are engaged in farming.

Considering these factors, it would be good to lift the evacuation of local people who voluntarily accept exposure to 100 mSv/y as soon as possible and to enable them to decontaminate their homes and farmland, and to engage in farming and running canteens and accommodation facilities. It would also be possible to have them receive radiation work training. Furthermore, for men and women over the age of 50 who have no chance of having children and who wish to return, it would be good to lift the evacuation and enable them to decontaminate and resume their livelihoods. If we do this, we should be able to prevent the collapse of the local community around the power plant. Even if we do this, we can secure an interim storage site for waste and radioactive waste generated by decontamination.

The present conditions for lifting the evacuation order are.

(1) It is certain that the annual cumulative dose estimated by the spatial dose rate will be 20 millisieverts or less.
(2) Infrastructure essential for daily life such as electricity, gas, water and sewage, major transportation networks, communications, and life-related services such as medical care, nursing care, and postal services are generally restored, and decontamination work focusing on children's living environment is sufficiently progressed.
(3) Sufficient consultation with the prefecture, municipalities, and residents (Nuclear Emergency Response Headquarters 2015).

To prevent disaster-related deaths due to inactive lifestyles, it is better to encourage the participation of residents, rather than leaving all decontamination and infrastructure restoration to the government.

The International Commission on Radiological Protection (ICRP) introduced the "co-expertize" work and "self-protective action" carried out in Fukushima in its 2020 report (Publication 146) as follows (ICRP 2020, p. 122). "Co-expertize work was initiated among different stakeholders such as mayors, residents, health professionals, and scholars, inspired by the ICRP's dialogues and spontaneously, according to the situation in each region. Some were carried out with the help of volunteer experts and remained informal activities, while others became official collaborations with local governments, expert organizations, and universities". Such voluntary activities are not only helpful for residents to understand radiation exposure, but also contribute to the recovery of residents' lives and regional reconstruction through self-help activities. It is considered a good example during the TEPCO accident and a lesson to be passed on. Encouraging voluntary participation of residents in decontamination and infrastructure restoration is a strategy that extends this good example.

Radiological protection in the reconstruction from the TEPCO accident is being promoted with reference to the ICRP's thinking. The land around the TEPCO Fukushima Daiichi Power Station is being acquired by the government as an interim storage site for waste generated by decontamination. There are hardly any cases of disaster-related deaths recognized ten years after the accident. It may not be possible to apply it to the measures against the TEPCO accident, but it is necessary to create a perspective for the elderly, etc., in order to utilize the lessons of the large number of disaster-related deaths caused by the evacuation due to the TEPCO accident for future nuclear emergency measures and nuclear disaster prevention plans.

Many countries in the world, including Japan, China, South Korea, and France, have nuclear power plants located in agricultural areas. Aging is also progressing. These countries should share the lessons of disaster-related deaths at the time of the TEPCO accident.

Japan's Nuclear Disaster Response Guidelines Utilize Indoor Evacuation

The commentary on Japan's nuclear disaster response guidelines describes the calculation of protective measures and radiation doses, lists three lessons from the Fukushima Daiichi nuclear power plant accident, and introduces the response and description of the guidelines (Tanaka 2016).

In the calculation of protective measures and exposure doses, a graph showing the distance from the emission source and the effective dose is presented. Within 5 km from the emission source (PAZ: within the area where preventive protective measures are prepared), the dose reduction effect due to distance is significant, so evacuation is effective as a preventive measure. Beyond 5 km from the emission source, the dose reduction effect due to indoor evacuation can be reliably expected, so indoor evacuation is stated as an effective measure beyond 5 km.

The following are cited as lessons from the Fukushima Daiichi nuclear power plant accident:

(1) Many casualties were caused by unreasonable and unplanned evacuation.
(2) No definitive health effects due to radiation exposure have been recognized.
(3) The environment was contaminated by a large amount of radioactive substances with a long half-life, and evacuation was prolonged.

The lesson from (1) is that many casualties were caused when those requiring care, such as in nursing homes, were forced to travel for long periods due to evacuation. In response to this, the policy is to utilize indoor evacuation and not to evacuate unplanned. The guideline states, "To avoid confusion in evacuation, residents within approximately 5 km (PAZ, area where preventive protective measures are prepared) should prepare and start evacuation from an on-site emergency (a situation where it is expected that the reactor cooling will be lost and the fuel will melt). Priority is given to those requiring care and children".

The response to the lesson from (2) is the evacuation measures from the perspective of reducing radiation exposure (the utility of indoor evacuation), and the guideline states that evacuation before the release of radioactive substances is the principle within the PAZ, while utilizing indoor evacuation as appropriate. Within the UPZ (area where emergency protective measures are prepared, approximately 30 km), indoor evacuation is the principle. Stable iodine tablets are to be used as appropriate.

The lesson from (3) is that "under the new regulatory standards, measures to prevent and mitigate serious accidents are drastically strengthened, and the amount of radioactive substances released into the environment is minimized as much as possible". The evaluation result shows that the amount of cesium 137 released in the maximum accident evaluation of the Ikata nuclear power plant, which has facilities in accordance with the new regulatory standards, is about one-two-thousandth of the TEPCO Fukushima nuclear power plant accident.

In this section of the book, the main issue addressed is the increase in disaster-related deaths due to inactive lifestyle diseases as a result of long-term evacuation. This is a problem corresponding to the above lesson 3. The improvement measures have already been mentioned.

The book did not cover much about evacuation immediately after the accident, so I will express some thoughts on it. Preventive evacuation immediately after an accident in the PAZ (area within 5 km from the accident reactor) (precisely when an emergency situation such as a nuclear power plant losing all power sources occurs) is supposed to be carried out to avoid deterministic radiation effects (exposure of 1 Sv in dose). This is in accordance with the safety guidelines of the International Atomic Energy Agency (IAEA) (JAEA 2014).

The radiation exposure of residents in the TEPCO accident was extremely low (for example, there is a summary on page 5 of the above-mentioned document (Tanaka 2016)). The UNSCEAR also states in its report that it would not be able to recognize not only deterministic effects but also probabilistic effects. Since evacuation was carried out, there is a possibility that the dose was low, so I examined the results of measuring the spatial radiation dose. The measurement results immediately after the accident in Fukushima Prefecture have been reported (Fukushima Prefecture 2011a). Not only the dose measurement values at each point, but also the graphs of their time changes have been published (Fukushima Prefecture 2011b). There are some points where there is no data on time changes due to reasons such as loss of power, but Yamada in Futaba Town, 4.1 km west-northwest from the power plant, has the second highest spatial dose rate next to Kamihatori in Futaba Town, and there is data on time changes. The spatial dose rate is decaying with a half-life of about 8 days. From the data, the dose when staying at that point for one month is estimated to be about 150 mSv, which is lower than the 1 Sv guideline for deterministic effects to be avoided by preventive evacuation. Even if you continue to stay at this point without evacuating, it would have been possible to avoid deterministic effects. In reality, you would not stay outdoors. The dose decreases to about three-quarters in a wooden house and about half in a concrete house due to the radiation shielding effect of the building, so if you had taken shelter indoors at this point, the exposure would have been about 100 mSv. This is not to say that evacuation was unnecessary.

Furthermore, radioactive substances with shorter half lives have greater radioactivity per unit mass, but after a period of 10 times the half-life has passed, the radioactivity decays to about one-thousandth. After that, other radioactive substances with longer half-lives contribute to the spatial dose rate. The main nuclides contributing to the spatial dose rate among the fission products released into the environment in a core meltdown accident are iodine 131 with a half-life of 8 days and cesium-137 with a half-life of about 30 years. The spatial dose rate around the TEPCO Fukushima Daiichi Nuclear Power Plant is decreasing in the long term due to the decay of these nuclides. In reality, the decay rate increases due to the weathering effect. If decontamination is carried out, the spatial dose rate decreases.

The emergency guidelines of international organizations are based on the experience of the Chernobyl accident. The Chernobyl reactor accident was a nuclear runaway accident (where the output of the reactor rapidly increased, the reactor fuel shattered due to instantaneous massive heat generation, reacting with the surrounding cooling water and graphite, and the reactor was destroyed), causing a fire of the graphite of the reactor, and radioactive substances containing core fuel were released into the environment.

The release in the large-scale light water reactor accident, the TEPCO accident, differs in the amount of release. In the TEPCO accident, the nuclear reaction of the reactor stopped, but all power sources lost their function due to the tsunami and the cooling water pump stopped due to a power outage, causing the reactor fuel to lose cooling and melt, and volatile and rare gas radioactive substances such as iodine and cesium contained in it were released. Hydrogen generated by chemical reactions with water explosively reacted with oxygen in the air, and the reactor building was destroyed. Unit 2 did not have a hydrogen explosion because hydrogen escaped into the atmosphere from an opening in the building. It has recently been suggested that not only hydrogen but also combustible gases generated by the high temperature of organic materials such as cable (wire) sheathing contributed to the explosion of the building. Light water reactors are the most commonly used reactors for power generation in the world. For the TEPCO accident, we hope to further advance the analysis and research of environmental data obtained in Japan and contribute to the progress of international knowledge.

Disaster-related deaths in Fukushima Prefecture were high, with 2079 people aged 66 and over, but there were also 231 people aged 21–65. This is more than in Miyagi and Iwate prefectures. 231 is not a small number. There may be lessons to be learned from this as well. The author hopes for a detailed analysis.

There is a need to answer the question, "Is it really good to think conservatively about radiation health risks in emergencies, when there is a conflict with other health damages?" As already mentioned, is it appropriate to apply the concept of ALARA to the continuation of evacuation of evacuated residents during a nuclear disaster involving evacuation? From the lessons of increased disaster-related deaths, was it not rational to apply it? ALARA is a story within the framework of radiation protection. Nuclear disaster measures are considered within the framework of radiation protection both domestically and internationally, but there are fundamental problems with this point as well. With the measure of radiation risk, other health damages cannot be measured. Shouldn't specific guidelines to prevent other health damages be prepared for nuclear disaster measures?

It is Difficult to Respond with Risk Communication in the Event of a Nuclear Accident

The mental and psychological impacts of nuclear accidents and other health damages are thought to be preventable through risk communication to the public, but this is unrealistic. We live in an era where information is conveyed in various ways. It is difficult to prevent other health damages and reputational damage through risk communication when a nuclear accident occurs. This is a lesson from the Chernobyl and TEPCO accidents, and it remains true despite many efforts after the TEPCO accident. For example, there are countries that ban the import of agricultural, forestry, and fishery products from Fukushima Prefecture. What about the release of treated water containing tritium?

Why is it difficult? Because dialogue is essential for risk communication. In his famous paper titled "Perception of Risk" published in Science in 1987, Paul Slovic concluded that "the public's perception of risk is not rational as experts think" and further stated at the end of the paper that "risk communication and risk management efforts are destined to fail unless they are structured as a two-way process" (Slovic 1987). It is impossible to have a dialogue with the entire Japanese public during a nuclear accident and build trust.

Many nuclear experts seem to think that it would be good to educate the public about the safety of nuclear power and radiation by conducting risk communication. This is a pitfall that experts often fall into. The Royal Society, which has been promoting science communication in the UK, expressed concern in its 2006 report that many scientists are participating in communication activities thinking that it is about educating their knowledge, not about having a dialogue (Royal Society 2006).

Risk communication is not education. Nuclear experts may have forgotten how many years it took them to acquire their expertise. In university education, dialogue through question and answer sessions and seminars is essential to acquire specialized knowledge. Without dialogue, understanding does not progress.

In Sect. 2.9, "Risk Perception and Psychology", the author will explain in detail that the purpose of communication is to build trust. It is not easy to build trust by talking about safety and risk related to danger. When discussions about risk and safety are repeated, they become emphasized, resulting in a perception of danger. Negative stories, stories of danger, leave three times stronger impressions. Talking about safety is psychologically the same as talking about danger. In the case of nuclear power, discussions about safety have been repeated. Could this be the reason why nuclear power generation has come to be perceived as extremely dangerous? Nuclear engineers often explain technical safety, and this is frequently communicated to the public through various media. They may think they are talking about safety, but psychologically, it's the same as talking about risk (saying it's dangerous). Not only the safety of nuclear reactors, but what about the geological disposal of high-level radioactive waste? The explanation may be biased towards technical safety.

The technical safety of nuclear reactors and radioactive waste disposal is related to the dissemination of results and acquisition of research funds for nuclear engineers and researchers. With a sense of professional responsibility, there is a consciousness to seek public understanding and financial support, and they may have unconsciously talked about technical safety to the public. This may have led to the public perceiving the safety of nuclear reactors and radioactive waste disposal as extremely dangerous. The parties that the nuclear industry and nuclear-related parties should explain safety to are the Nuclear Regulation Authority and the Nuclear Regulatory Agency. If they want to acquire research funds, it's the ministry in charge of those funds. If they are particularly talking about safety to the public because they want a budget, it's the same as strangling the use of nuclear power themselves.

Communication (Facilitation of Dialogue) Requires Skills, Experience, and Expertise

For example, Steve Robinson, a UK nuclear dialogue expert, states in his materials that "telling the facts is a pitfall" and "encourage joint discovery of facts". The U.S. Regulatory Commission's document titled "Effective Risk Communication" warns that "comparing risks is counterproductive unless there is a certain level of trust with the other party".

The Explanation of Safety and Risk is the Role of the Nuclear Regulatory Agency

In Japan, it seems that many nuclear experts who advocate for risk communication believe it to be sufficient, but they may not be aware of these points. Risk communication is the role of the regulatory side. It is good to create various explanatory materials and commentaries and make them public. After the TEPCO accident, the Ministry of the Environment in Japan has been publishing "BOOKLET to Provide Basic Information Regarding Health Effects of Radiation", Fukushima Prefecture and the government have been disseminating information at "Fukushima Revitalization Information Portal site" in ten languages, and the Agency for Natural Resources and Energy has been creating and publishing "Special Contents-Japan's Energy". The author hopes for the enhancement of various commentaries, reports, training materials, etc. However, it is unreasonable to think that risk communication can somehow prevent other health damages, including mental and psychological aspects, during a nuclear reactor accident.

Even if we talk about safety, we cannot easily alleviate people's anxiety. There may be times when it feels better not to talk about safety unnecessarily.

The international agency's report on the TEPCO accident states that other health effects, such as psychological and mental impacts, should be addressed by strengthening risk communication. The author guesses feel that they are considering the protection from nuclear and radiation accidents only from the perspective of radiation risk prevention. Isn't their perspective too narrow? Isn't it because they are conducting discussions centered on experts in radiation exposure risk and nuclear reactor safety? Along with understanding the past experiences and research results of risk psychology and risk communication, shouldn't we create a report by examining health effects other than radiation exposure risk, centered on public health and social psychology experts?

Risk communication to the entire nation was not successful either at the time of the Chernobyl accident or the TEPCO accident. For example, immediately after the Chernobyl accident, increase in unnecessary abortions were performed in Greece and Italy, and the birth rate decreased in January 1987 (Ministry of the Environment 2022).

In Europe, although the exposure was small, people were mentally and psychologically affected. In the TEPCO accident, not only did many citizens nationwide suffer mental and psychological effects, but disaster-related deaths occurred in Fukushima Prefecture due to the prolongation of evacuation. These are all deterministic effects in terms of the health effects of radiation. The U.S. Nuclear Protective Action Guides (PAG Manual) lists the prevention of deterministic health effects as the first of three principles. The second principle is that the benefits brought about by measures should be greater than the harm. The third principle demands to reduce late effects (cancer).

Trust in the government and nuclear experts was lost during the TEPCO accident. Risk communication does not function when an accident occurs. To prevent other health damages, it is better to decide on a way of thinking in advance. During the Chernobyl accident, in Hungary, abortions were not allowed unless the fetal exposure exceeded 100 millisieverts, so no abortions were performed (Ministry of the Environment 2022). This can be a reference.

It is better not to apply the concept of ALARA to the public in areas that do not exceed 100 mSv/y when a nuclear/radiation accident occurs.

The concept of ALARA is a principle of radiation protection that "radiation exposure should be kept as low as reasonably achievable". By not applying this when an accident occurs to the public in areas with low radiation doses, areas not affected, it may be possible to prevent disaster-related deaths from the TEPCO accident, unnecessary abortions during the Chernobyl accident in Greece and Italy, or effects from psychological and mental stress. These are all deterministic effects.

The R (reasonable) in ALARA has not been considered quantitatively in nuclear disaster prevention and emergency response because there was no quantitative data on evacuation risks. Is it logical to think that R can be addressed by strengthening risk communication? The analysis of disaster-related deaths from the TEPCO Fukushima accident has shown the need to examine the validity of applying R.

ALARA is a concept within the framework of radiation protection. The radiation effects of less than 100 mSv/y are not scientifically settled, and the adoption of ALARA is a policy decision. However, considering other health damages, it is also a policy not to apply ALARA to residents during a major accident. Other health damages are manifesting. Disaster-related deaths due to evacuation in Fukushima resulted in a greater reduction in life expectancy than the radiation exposure avoided by evacuation. For the public who only receive exposure of less than 100 mSv/y during a nuclear/radiation accident, it might be better not to adopt ALARA as a policy. In nuclear reactor accidents, other health damages occur, including mental and psychological aspects. By not applying ALARA, the criteria for lifting evacuation orders will be clearer, and the psychological and mental problems of local residents can be reduced.

The late effects of the Chernobyl accident, such as childhood thyroid cancer, can be prevented by limiting the intake of items (e.g., vegetables, stream water, milk, etc.) that contain or are contaminated with radioactive iodine until it decays, from immediately after the accident to about three months.

Around nuclear power plants, stations that measure the spatial dose rate of radiation are set up to surround the power plants. In the event of an accident, responses can be considered based on these values. Alternatively, it is conceivable to sequentially lift evacuations based on the results of radiation dose distribution measurements by aircraft, drones, and cars. The decision to lift evacuations based on measurement results is also written in the disaster prevention manual.

The conditions for lifting the evacuation during the TEPCO accident include not only the reduction of radiation dose but also the restoration of life infrastructure. To prevent disaster-related deaths due to inactive lifestyle diseases, it is better to enable residents themselves to engage in decontamination and infrastructure restoration, rather than waiting for the government to set the conditions for lifting.

ICRP recommendations are given with a range of 1–20 mSv/y, for example, during the recovery period, so if ALARA is present, attention is focused on the lower value, resulting in a delay in lifting the evacuation. It should be specified as a single value with no range. The idea of avoiding exposure as much as possible is also written in Japan's nuclear disaster prevention guidelines. Compared to the seriousness of other health effects, applying ALARA to the public in areas where the dose is not high at the time of the accident may be more harmful than beneficial.

In the event of a core meltdown accident, evacuation is necessary just like in the case of a major earthquake or tsunami, but in the analysis of the TEPCO accident, the life shortening due to the dose avoided by evacuation is 2.8 times smaller than the life shortening due to disaster-related deaths, and considering that the lifetime exposure risk of the elderly is one-fifth of the age average, it is 14 times smaller. Considering the low-dose rate exposure. It is 28 times smaller. In order to prevent disaster-related deaths due to the prolongation of evacuation, it would be better to review laws and plans related to nuclear disaster prevention. There is no need to treat evacuation during a nuclear disaster as special. It may be possible to learn a lesson that many disaster-related deaths occurred because of the special treatment.

The health impact lessons of the TEPCO Fukushima nuclear accident are shown in Table 2.5.

Table 2.5 Lessons learned from the TEPCO Fukushima nuclear power plant accident (health effects)

Despite no discernible public health effects of radiation exposure, long-term evacuations in Fukushima prefecture resulted in more than 2,000 disaster-related deaths, 90% of whom were elderly and thought to be caused by inactive diseases and mental disorder
There are no radiation protection guidelines that take into account that the lifetime cancer death probability for elderly people is one-fifth of the average age
The radiation exposure risk framework does not take into account other health risks
A cost-benefit analysis shows that the risk of disaster-related death associated with evacuation is 2.8–28 times greater than the risk of radiation exposure avoided by evacuation. Nuclear disaster countermeasure guidelines that take into account the balance between risks and benefits are necessary
To prevent disaster-related deaths, the shorter the evacuation period, the better
In Japan, there are many part-time farmers, and older people are engaged in agricultural work. Enabling them to return home early and engage in decontamination work, agricultural work, etc., will prevent disaster-related deaths and community collapse
These lessons are also important in countries such as China, South Korea, and France, where nuclear power plants are located in agricultural areas and where populations are aging
To prevent other health hazards, ALARA principles should not be applied to members of the public in low-dose areas in the event of a core meltdown

2.3 Crisis Management Lessons from the TEPCO Accident

Training and Workshops

Emergency response requires various training and workshops. Not only comprehensive training with predetermined accident scenarios, but also training without knowing the accident scenarios in advance, training gathering only those responsible for emergency response, training within the company and power plant as well as regular training, learning from past domestic and international lessons is necessary. When the author was managing the research reactor facility at the University of Tokyo, the author conducted training without informing the scenario. The author thinks it was the most effective.

Not only training, but also education is important. It is necessary to have education that allows understanding of the overall picture of severe accidents and their progression, and environmental impacts, not just training on emergency response procedures. Seminar materials on severe accidents in Japan often contain many descriptions of severe accident countermeasure facilities, giving an impression of equipment and technology bias.

There is an opinion that emergency response should be carried out to prevent the worst case, but this is a mistake.

It is correct to focus on resolving the accident while considering the most likely state of the accident. The aforementioned U.S. Environmental Protection Agency's emergency response guidelines do not state that measures should be taken assuming the worst case. Risks and benefits should be considered when taking measures. Radiation exposure is not the only thing to consider in emergency response. If large-scale evacuation is carried out to avoid even a little radiation exposure, disaster-related deaths will increase. The "worst" case can only be considered in terms of a one-dimensional indicator (in the case of the TEPCO accident, the risk of radiation exposure). There are many things to consider at the time of an accident. The indicators are multi-dimensional. Therefore, it is a mistake to respond to emergencies in the worst case.

In the TEPCO accident, there are reports that the case where radioactive materials in the spent fuel in the spent fuel pool of the Fukushima Daiichi Power Station spread is talked about as the worst scenario, and that the evacuation area would extend to Tokyo, but this is a mistake. The heat generation of spent fuel decreases rapidly after the reactor stops, so the heat generation of those left in the pool for a long time is small. It takes about 25 days for the water in the spent fuel pool to evaporate and the spent fuel to be exposed (Gauntt 2012). During this time, it is possible to inject water, restore power, and cool the pool water. In fuel exchange, one-third to one-fourth of the fuel in the reactor is replaced and moved to the spent fuel pool. Therefore, the heat generation of the spent fuel in the pool is smaller than the heat generation of the fuel in the reactor vessel immediately after the reactor stops. As time passes, the heat generation of the spent fuel becomes extremely low, and it can be cooled by natural convection of air. Dry-type (dry) spent fuel storage containers that do not use water have been put into practical use. In the TEPCO accident, a device (called Kirin) that sends liquid concrete from the ground to a high building was used to replenish the water in the spent fuel pool.

The focus on the spent fuel pool was due to confusion of information. The United States believed that the water in the spent fuel pool had disappeared due to damage to the building of Unit 4 from a hydrogen explosion (four days after the accident), and called for a wide-scale evacuation (five days after the accident). In reality, a video taken by a Tokyo Electric Power Company (TEPCO) employee on a helicopter flown by the Self-Defense Forces showed that there was water in the spent fuel pool, and this was confirmed at a TEPCO meeting that night (five days after the accident). This was not communicated domestically (between the TEPCO Fukushima and the TEPCO headquarter in Tokyo and Japanese government) or internationally.

Crisis Management and Emergency Response: The UK was Impressive

The UK was impressive in its crisis management and emergency response (Grimes et al. 2014). The UK had a system in place for the public and media to access scientific and technical information, and a government emergency scientific advisory organization (SAGE, Scientific Advisory Group for Emergencies), drawing on lessons from the government's failed response to bovine spongiform encephalopathy (mad cow disease). During the TEPCO accident, SAGE was activated, collected information, and concluded that radioactive materials would remain in the vicinity of Fukushima, and judged that the impact on British people and embassy staff in Tokyo would be minimal. It was also understood that if evacuation was carried out preventively, there would be other health risks. The Chief Scientific Advisor explained these results in the UK, and a question and answer session was held with other independent scientists, which helped the public understand the government's policy. The Chief Scientific Advisor also directly explained and answered questions to British embassy staff and British citizens in Tokyo. The UK's crisis management capability was impressive. In Japan, a Minister of Foreign Affairs Science and Technology Advisor has since been appointed to the Ministry of Foreign Affairs.

What can be learned from the UK's response is the government's explanation by the Chief Scientific Advisor's single voice, and the system in place to help the public understand it. Furthermore, the assessment of the amount of radioactive material released by the accident was not overestimated, but appropriate. They also kept in mind that there could be other health damages associated with evacuation.

While attention is focused on the role played by the Chief Scientific Advisor, the fact that the assessment of the release amount was not overestimated is not discussed, but it is fundamental and important. Perhaps, the release amount assumed by the UK was one-tenth of that of other Western countries using nuclear powers. To achieve this, in Japan, it is necessary to enhance and systematize the knowledge of core meltdown accidents. It is not enough to refine the calculation of accident probabilities and administrative procedures.

Enhancing Knowledge of Severe Accidents

One of the reasons for the confusion in Japan's response to the TEPCO accident was the lack of sufficient knowledge about the behavior of core meltdown accidents, and the fact that this knowledge was not shared. Since the 1990s, many people involved in nuclear power in Japan have been reluctant and hesitant to talk about, think about, and research core meltdown accidents due to concerns about local and national understanding after nuclear-related troubles and accidents. While it's not that no research was done at all, I believe the TEPCO accident caught us off guard.

The fact that we didn't do what we should have done, using local and national understanding as an excuse, is a big lesson for Japan's use of nuclear power, and the nuclear industry in particular needs to take this to heart.

Furthermore, Japan's unique bureaucratic sectionalism has also affected research on core meltdown accidents. In the 2001 reorganization of government agencies, the regulation of nuclear reactors was transferred from the Science and Technology Agency (at the time) to the Nuclear and Industrial Safety Agency of the Ministry of Economy, Trade and Industry, causing a drastic reduction in the budget for safety research at the Japan Atomic Energy Research Institute (at the time), which was at the center of nuclear safety research. As a result, safety research had to rely on commissioned fees from the Nuclear and Industrial Safety Agency, and the focus of work shifted to areas closer to regulation, such as regulatory rules and probabilistic safety assessment, making it difficult to conduct systematic research on core meltdown accidents. The development of a domestic core meltdown accident calculation code (calculation program) was interrupted, and although some calculation codes have been developed since then, there is still no publicly available severe accident calculation code that graduate students can easily use.

Bureaucratic sectionalism is not unique to this issue or the nuclear field. Every country has sectionalism to clearly define the responsibilities of its administration, but in the West, work is done from the perspective of the people, not for the sake of one's own department. Japan is unique in its sectionalism, including organizations where former bureaucrats are members. The report of the National Diet Accident Investigation Commission states that nuclear power operators have captured regulatory administration, but it is also true that operators and industries cannot oppose the government. This is due to the strong dependence of the Japanese people on the government. Although civil servants are public servants, we hardly hear the term "public servant" these days.

The issues of bureaucratic sectionalism and national dependence are deeply rooted and related to various aspects of Japan's social system, but the author believes they are unavoidable issues for the future of Japan. Companies in Japan that have international competitiveness are not dependent on the country's system and administration. This could be a hint for considering future directions.

South Korea has a strong sense of dependence on the nation among its people, but after the economic crisis of 1997, it carried out administrative reforms and has become a leaner, more competitive society. With a population about half that of Japan, South Korean companies cannot develop without relying on exports. There is a strong overseas orientation, and many citizens are capable of global thinking. This, along with the international expansion of companies, may be the driving force behind the subsequent development of the South Korean economy. Electricity rates in South Korea are also low, with nuclear power contributing to this. The development of South Korea is supported by cheap and stable power supply from nuclear power generation.

After the European Fast Reactor Project ended in the early 1990s, the Karlsruhe Institute in the Federal Republic of Germany conducted systematic research on severe accidents. In France, the Cadarache Institute of the Atomic Energy and Alternative Energies Commission (CEA) is conducting research on core meltdown accidents. In the United States, institutions such as the Sandia National Laboratories have been conducting research on severe accidents and improving calculation codes for many years. The U.S. industry has been conducting research and development from the perspective of accident management (how to resolve accidents), but this alone is insufficient. The research by Sandia National Laboratories and others has produced systematic knowledge of severe accident phenomena, and its results are being utilized in regulation and application. Training materials for regulatory staff are also based on these findings. Although the presentation materials for training are not public, I have seen them and was impressed by how well they were made. The report that served as the basis for the training materials is publicly available (NRC 2002).

After the TEPCO accident, I studied and researched core meltdown accidents, albeit belatedly. This public material explains the progression and phenomena of accidents in relation to the design and structure of the reactor, and was the most helpful. It is written based on knowledge up to the 1990s, but I hope it will be revised to reflect the experience of the TEPCO accident. Severe accidents need to be understood not only as individual phenomena but also in relation to equipment. The release and behavior of radioactive substances into the environment is the most important for ensuring public safety. It is necessary to enhance knowledge with this in mind. It takes time to remove the core meltdown debris (debris) from the accident reactor because the dose is high, but there should be many parts that can be done even at a stage where the inside of the reactor is not fully understood. While attention is focused on the work of removing debris from the accident reactor, the research to elucidate the behavior of the accident and its report is important, and this is the role of the Japan Atomic Energy Agency, a national research institution.

In the future, the areas that should be particularly enhanced in training materials and reports include the environmental impacts such as the diffusion and adhesion of radioactive materials released in accidents, changes in the spatial dose rate over time, radiation exposure, evacuation, disaster-related deaths, psychological and mental impacts, and experiences of decontamination. This is the responsibility of Japan, which caused the TEPCO accident. These points are hardly written in the training materials of the U.S. Nuclear Regulatory Commission mentioned above. The creation of these materials requires the cooperation of relevant ministries and agencies.

While this is a discussion about Japan as a whole and not specific to the nuclear power sector, the direction of Japan's administrative reform should be for each ministry and agency to explain the work of the department in charge on their website, with the signature of the department head, along with the basis of the report for the public. The work of the administration should then be audited by the Board of Audit and the Diet and reflected in the budget and personnel decisions. This is what is being done in Western governments such as the United States and the United Kingdom.

This book is written with reference to these documents. The work of independent administrative organizations such as the Nuclear Regulation Authority can also be checked in this way. There is a system for evaluating the administration and independent administrative corporations in Japan, but it seems to be self-evaluation and evaluation for the sake of evaluation, and I think it is not fulfilling the function it should have like in the West. For example, if the intentions of the evaluated side are reflected in the selection of evaluation items and evaluators, the role of evaluation that should be originally fulfilled cannot be fulfilled. Attention should also be paid to the personnel ties between ministries and independent administrative corporations.

The explanation of the implementation status of policies is not for administrative reform, but the responsibility of the administrative agencies that are working with tax money. Each ministry and agency publish a white paper, but it is an explanation of the progress each year, and it is insufficient for the public to understand the key points of the policy. Furthermore, while a white paper in the West means a document describing a policy, a white paper in Japan is administrative archival and unique to Japan. Japanese white papers are published annually and are not designed to understand the effects of policies over several years. Not limited to white papers, it is not enough to just imitate the form of the West. However, it should be possible to create a policy explanation document using the work of creating a white paper.

Regulatory administration is a procedure, but ensuring safety is not just a procedure. Knowledge and management are important.

Ensuring safety requires a thorough understanding of severe accidents. Speaking of the problem of vertical division of ministries and agencies, both the promotion side and the regulatory side are conducting severe accident research at the Cadarache Research Institute in France. I look forward to the safety research of the Japan Atomic Energy Agency (JAEA). The role of the national atomic energy research and development organization needs to be changed from the development of new types of reactors by the country to the exploration and systematization of knowledge such as severe accidents.

Although not widely known, research on risks due to earthquakes and tsunamis in Japan was conducted at the Japan Atomic Energy Research Institute and then at the Nuclear Safety Research Association (JNES) even before the TEPCO accident. Based on the practical research results of Kyoto University during the Great Hanshin-Awaji Earthquake, a nuclear disaster prevention system against earthquakes and tsunamis was created. In light of the flooding of the seawater pump of an Indian nuclear power plant in the 2004 Sumatra offshore tsunami, the method has been improved. The developed method is reflected in the tsunami standards of the International Atomic Energy Agency. Practical nuclear risk communication is also being conducted. It has been pointed out that in dealing with complex disasters caused by earthquake and tsunami disasters and nuclear disasters, the perspectives of practice and effectiveness, the importance of the role of municipalities managing the basic resident register, etc. are important (Ebisawa 2021). Naturally, the accident scenarios identified by probabilistic risk assessment research coincide well with the TEPCO accident.

After the TEPCO accident, the government as a whole is working and promoting to establish a nuclear disaster prevention system. In peacetime, there is a standing nuclear disaster prevention meeting, and in emergencies, the Nuclear Disaster Response Headquarters is set up in response to the declaration of a nuclear emergency (Disaster Prevention White Paper 2020). Emergency response training is also being conducted.

Nuclear disaster prevention and emergency response are lagging behind in information exchange both domestically and internationally, compared to the safety of the reactor itself and safety standards. There are two reasons for this. One is that most nuclear safety experts are working on accident prevention and reducing the release of radioactive substances into the environment, and there are very few experts on the behavior of radioactive substances released into the environment and emergency response. The other is that the necessary accidents for emergency response and disaster prevention are infrequent in the nuclear field and there is little experience. In the future, wouldn't sharing and exchanging knowledge, experience, and ideas on general disaster prevention and emergency response both domestically and internationally be useful for improvement?

The lessons of crisis management from the TEPCO accident are summarized in Table 2.6.

Table 2.6 Lessons learned from the TEPCO accident (crisis management)

At the time of the accident, the chain of command was confused, and the responses of the Japanese government and TEPCO headquarters caused confusion at the Fukushima Daiichi Nuclear Power Plant site and at home and abroad
The response from the British government and the British embassy was admirable. The unified explanation provided by the chief scientific advisor, based on the UK's experience and knowledge of crisis management, was effective. The estimation of the amount of radioactive material released was also appropriate and not excessive
Japan's nuclear disaster prevention guidelines are top-down, describing procedures, and do not reflect the lessons learned from the confusion in the chain of command. Guidelines and training materials that explain the concept are needed
It is a mistake to conduct crisis management assuming the worst-case scenario. This is because the worst-case scenario can only be considered based on a single indicator: radiation exposure
In crisis management, the role of local governments, which are the point of contact with the field and residents, is important
Regarding the fact that the nuclear regulatory organization was not independent from the promoting organization, see the recommendations of the National Diet Accident Investigation Commission

2.4 TEPCO Accident and Severe Accident Measures

A severe accident is a large-scale accident, like the TEPCO Fukushima nuclear power plant accident, where an abnormal situation beyond the range initially assumed in the reactor design occurs, a large amount of reactor fuel melts, and radioactive substances are released into the environment. It is also called a core meltdown accident.

To ensure the safety of nuclear power plants, it is necessary to shut down the reactor, cool the reactor fuel after shutdown, and confine radioactive substances in the reactor containment vessel (a sealed structure that houses the reactor and other components). In the event of a large earthquake, the reactor is set to automatically shut down based on signals from seismometers, and the reactors did automatically shut down in the TEPCO accident. For cooling after shutdown, power for the pumps that circulate cooling water and the measurement system is required. Electricity from outside the power plant (referred to as external power) was lost at the time of the earthquake due to the collapse of transmission towers, but multiple emergency diesel generators are installed at the power plant. These automatically start when external power is lost, and the emergency cooling pumps are also operated. The heat removed by the pumps was dissipated into the seawater using heat exchangers and pumps.

However, about an hour later, the tsunami arrived, submerging the emergency generators, batteries, distribution boards, and seawater pumps, and all the systems that had been installed lost their safety functions simultaneously. Since both the external power from the transmission lines and the emergency power were lost, it is called a total power loss accident. As a result, it became impossible to cool the reactor core after shutdown, the fuel rods overheated, and water was reduced by the metal-water reaction, generating hydrogen. The fuel melted and fell from the bottom of the reactor vessel to the floor of the containment vessel, the pressure in the containment vessel increased, and gases containing hydrogen and volatile radioactive substances leaked from the packing part where the lid of the containment vessel was bolted and flowed into the reactor building. In the reactor building, oxygen and hydrogen in the air explosively combusted, damaging the buildings of Units 1 and 3. The hydrogen from Unit 3 flowed into the building of Unit 4, which was shut down, through the exhaust system of the reactor building, and the walls of the building were destroyed by a hydrogen explosion. A core meltdown also occurred in Unit 2, but because the panel separating the outside of the reactor building opened due to the shock of the hydrogen explosion in Unit 1, the hydrogen was released into the atmosphere and no hydrogen explosion occurred. Figure 2.4 shows a summary of the progression of the TEPCO accident.

After the TEPCO accident, a mechanism called backfitting was introduced, which applies newly created regulations retrospectively to nuclear power plants that have received permission/approval in the past, while considering large-scale natural disasters and terrorism in safety regulations.

Figure 2.5 shows a comparison of the conventional regulatory standards and the new regulatory standards. Consideration is given not only to earthquakes and tsunamis, but also to natural phenomena such as volcanoes, tornadoes, and forest

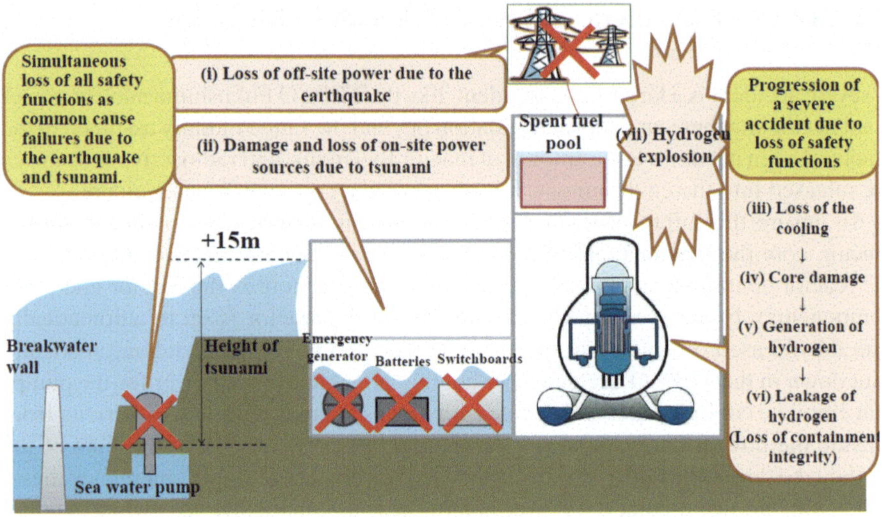

Fig. 2.4 Progression of the TEPCO Fukushima Daiichi Nuclear Power Plant accident. *Source* New regulatory requirements for light-water nuclear power plants—outline—August 2013, Nuclear Regulation Authority, IAEA. https://www.iaea.org/sites/default/files/requirements160913.pdf

fires, severe accident measures to prevent damage to containment vessels, and countermeasures against terrorism such as intentional aircraft collisions. The detailed content will not be explained as it becomes specialized. Explanations can be found on the website of the Nuclear Regulation Authority and other sources.

Since the Chernobyl accident in 1986 facilities and reactor concepts to deal with severe accidents have been considered by nuclear reactor manufacturers worldwide. AREVA of Germany (formerly Siemens KWU) had developed equipment such as a

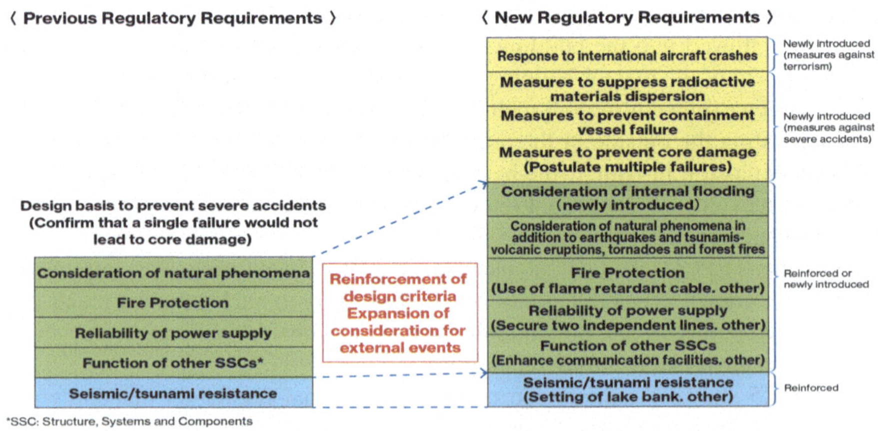

Fig. 2.5 New regulatory requirements of nuclear power plants. *Source* JAERO

hydrogen recombiner that does not require a power source, a device called a filter vent that removes radioactive substances from containment vessels and reduces pressure, a device called a core catcher that spreads and cools the core melt falling from the bottom of the reactor vessel, and devices introduced in Japan as anti-terrorism measures, known as "Specified serious accident response facility" (Oka and Bitterman 2015). Many of the severe accident countermeasures introduced in Japan after the TEPCO accident are either these or were developed with these in mind.

Measures based on the lessons of the TEPCO accident are also being taken at nuclear power plants and nuclear facilities in the United States and Europe. After the 9/11 terrorist attacks in 2001, U.S. nuclear power plants strengthened their counter-terrorism measures, and some of the lessons from the TEPCO accident had already been addressed in the U.S. U.S. nuclear regulation requires a comparison of costs and benefits when considering the introduction of new facilities, so the measures are rational. In Japan, the installation of equipment known as "Specified serious accident response facility" was required as a counter-terrorism measure. This equipment can operate as a backup even in the event of a total power loss accident like the TEPCO accident, so its protection is enhanced, but it is costly.

The concept of cost is missing in Japan's use of nuclear power and its regulation. It is not good because the purpose of power generation, not limited to nuclear power generation, is to supply electricity to the public cheaply, safely, and stably. As mentioned in the section on probabilistic risk assessment in Chap. 1, probabilistic risk assessment is a method for analyzing costs and benefits. By comparing the cost of accident countermeasures and the benefits that can be prevented by them, the cost-effectiveness of the countermeasures can be evaluated. In the United States, a conversion factor is used that equates exposure of 10 person-millisieverts to $5100 (in 2014 dollars). Probabilistic risk assessment is being called for in Japan, but there is no such conversion factor.

Japanese regulations do not have an indicator to consider whether the cost of measures matches the risk (exposure) that can be reduced, so it is not considered whether the facilities established by the new regulatory standards are worth the risk they can prevent.

Not only in nuclear power generation, but technology is used when the benefits outweigh the risks associated with its use. *A cost-benefit indicator should be considered in Japan's nuclear regulation.* This is a future issue. Even if safety measures are provided in large numbers and redundantly, they are not necessarily effective. In any case, it is necessary to consider the goal of nuclear power generation to provide cheap and stable electricity safely to the public. Table 2.7. It demonstrates the lessons of severe accident measures from the Tokyo Electric Power Company's Fukushima nuclear accident.

Table 2.7 Lessons learned from the TEPCO Fukushima nuclear power plant accident (severe accident countermeasures)

Japanese nuclear power officials believed that a core meltdown accident would not occur
The nuclear power industry believed that talking about severe accidents would hinder public and local understanding of nuclear safety, and this awareness was spreading among nuclear power personnel
After the 9/11 terrorist attacks in 2001, the United States strengthened anti-terrorism measures at nuclear power plants and took measures against total power loss, but Japan did not
In Japan, research and improvement of methods for tsunami countermeasures and disaster prevention measures were conducted based on the experiences of the Great Hanshin-Awaji Earthquake in 1995 and the Indian Ocean Tsunami in 2004, but the results were not put to use

2.5 TEPCO Accident and Information Provision to the Public

During the TEPCO accident, the Japanese government was unable to send a unified message to the public. In contrast, the UK was impressive. Japan has traditionally been inadequate in providing and explaining information to the public, including radiation exposure risks. In this state, the TEPCO accident occurred, and as a result of repeated discussions about radiation exposure risks and safety in public, fear of radiation spread among the public.

After the TEPCO accident, various information has been provided to the public. First, a unified basic document on the health effects of radiation (BOOKLET to Provide Basic Information Regarding Health Effects of Radiation), etc., has been created and is published on the Ministry of the Environment's website (Fig. 2.6). Explanations have been created about the situation of the accident, environmental monitoring, radioactive substances in food, efforts towards recovery from the accident, and health management. An English version is also available. A portal site on the health effects of radiation, etc., has also been created. In addition, explanations by a group of experts who have been involved in international activities for many years on the health effects of radiation exposure are disclosed in the section on TEPCO Fukushima Nuclear Power Plant and radiation-related information on the Prime Minister's Office website, for example by Professor Nagataki (Nagataki 2011).

Fukushima Prefecture, together with the Reconstruction Agency, explains radiation and decontamination, evacuation area information, disaster victim support, radioactive substance inspections of water and food, information on TEPCO's Fukushima Daiichi Nuclear Power Plant, etc., along with the status of reconstruction on the "Fukushima Reconstruction Station" reconstruction information portal site. Explanations are also provided in nine languages, including English, Chinese, and Korean (Fukushima Prefecture 2011a, b, 2023).

The Ministry of Economy, Trade and Industry and the Agency for Natural Resources and Energy explain various topics related to energy under the title of "Special Contents" on their website. Among them, explanations are given on topics such as

Vol.1 Basic Knowledge and Health Effects of Radiation
Chapter 1 Basic Knowledge on Radiation
Chapter 2 Radiation Exposure
Chapter 3 Health Effects of Radiation
Chapter 4 Concept of Radiological Protection
Chapter 5 Assessments by International Organizations

Vol.2 Accident at TEPCO's Fukushima Daiichi NPS and Thereafter (Initiatives by Ministries and Agencies)
Chapter 6 Situation concerning the Accident
Chapter 7 Environmental Radiation Monitoring
Chapter 8 Radioactive Materials in Foods
Chapter 9 Efforts toward Recovery from the Accident
Chapter 10 Health Management

Fig. 2.6 BOOKLET to provide basic information regarding health effects of radiation. *Source* Ministry of the Environment, Government of Japan

Fukushima, nuclear power, safety, and disaster prevention (Special Contents 2023). In addition to these, power companies, the Federation of Electric Power Companies, the Nuclear Culture Foundation, and others are posting nuclear-related information.

The nuclear encyclopedia ATOMICA was created with the aim of making a wide range of information related to nuclear power easily accessible to anyone on the Internet and is operated by the Japan Atomic Energy Agency (JAEA). It is designed to be understandable to those who are not specialists in nuclear power (educators, journalists, public servants involved in nuclear administration, university students, etc.).

Furthermore, while not necessarily for the general public, academic journals such as the Journal of the Atomic Energy Society of Japan and the Journal of the Japan Risk Research Society were traditionally only accessible to members of the society. However, they have now become accessible to the general public through the "Science and Technology Information Dissemination and Distribution System" (J-STAGE) operated by the Japan Science and Technology Agency (JST). J-STAGE aims to promote the rapid dissemination and international outreach of scientific and technological information (including humanities and social sciences) from Japan, as well as the promotion of open access. By using it, it is possible to access information based on professional grounds and deepen understanding based on the evidence. It should be noted that *Japan has very few comprehensive reports.* Even if there are, some of them do not have a comprehensive and accurate summary. It is the role of national research and development institutions to create comprehensive reports.

Information on radiation and nuclear power has been enriched, including on the portal sites of the Ministry of the Environment and Fukushima Prefecture. In the future, it would be desirable to move from a state where information on radiation and nuclear power, especially safety and reconstruction-related information, is prominently displayed, to a state where other nuclear-related information and information from other fields can also be found. The UK, which I will introduce next, is in that state. The United States has advanced information provision and disclosure by government agencies, and the public can find various information, including evidence, if they search for it.

The United States and the UK have administrative agencies that explain their policies to the public in a clear and concise manner. Japan has white papers and the like as administrative explanatory documents, but explanations of individual important policies are not made directly. This is not limited to safety, but if we take disaster prevention and emergency response as an example, the U.S. Environmental Protection Agency (EPA) explains on its website what the Protective Action Guide (PAG) manual is and what it contains when it is created. The PAG manual itself and related documents can also be downloaded from the site. This allows the administration to fulfill its responsibility to the public (responsibility for the results). Civil servants are public servants. The author hopes that improvements will be made in Japan as well. This will not progress unless the public monitors the administration, not the public depending on the country. Monitoring is not a simple matter of whether one agrees or disagrees with a policy, or whether it is good or bad. By creating policy explanatory documents, the public can understand how each policy is formulated. The creation of these documents improves the abilities of civil servants. It is a job that should be done with the signature of the section chief. Not only civil servants can improve their abilities by making efforts in their work. This monitoring is the job of the public and the Diet.

The issue of releasing treated water containing tritium from the Tokyo Electric Power Company's Fukushima Daiichi Nuclear Power Plant has attracted public attention and the decision has been prolonged. This is unique to Japan. Not limited to tritium, there are standards set by the Nuclear Regulation Authority for release, and if it falls below that, it can be released. The amount of radioactive substances released from nuclear power plants is disclosed in an agreement with the local government. This is not limited to the Tokyo Electric Power Company's Fukushima Daiichi Nuclear Power Plant. In the West, even if it falls below the standard, everyone does not discuss whether or not to release it. The Nuclear Regulation Authority is a group of experts and is responsible for the regulatory values. It is strange that the situation in Japan has continued for a long time as if there is no responsible party for the regulatory values.

By the way, tritium is universally present in nature, produced when neutrons, which are generated by the reaction of cosmic rays (protons, etc.) with the Earth's atmosphere, react with deuterium contained in atmospheric water vapor. The amount present is far greater from this source. In addition, neutrons have greater penetrating power than charged particles such as protons, and it is known that neutrons from cosmic rays cause malfunctions (soft errors) in semiconductors used in artificial satellites and the like. Table 2.8 summarizes the lessons learned from TEPCO's accident in terms of communication.

Table 2.8 Lessons learned from the TEPCO Fukushima nuclear power plant accident (communication)

At the time of the accident, the Japanese government was unable to send a one-voice message to the people
There was no system in place to provide the public with information on radiation exposure risks (this has now been improved)
Fear of radiation spread among the population due to repeated talk of the risks of radiation exposure
The creation and provision of policy information and scientific information in Japanese remain inadequate
In the United States and the United Kingdom, administrative agencies provide thorough explanations of policies to the public. The basic attitude is that of a public servant
As seen in the issue of the release of treated water containing tritium, even though there are standards set by the Nuclear Regulation Authority, in Japan the person responsible for making decisions is unclear, leading to confusion and prolonging the problem

2.6 UK Science Communication

In the UK, the Royal Society stated in its 1985 report titled "Public Understanding of Science" (Royal Society 1985) as follows:

- Individual decisions about the safety and health of the technologies we use are influenced by an understanding of the science behind them. A statistical understanding related to types of risks and uncertainties is essential.
- Science education in schools is necessary for this.
- In Parliament, the scientific aspects of issues should be discussed more, easy-to-understand reports should be created, and made available to the public.
- The media, especially daily newspapers, should cover more science stories. There should be more contact between journalists and scientists.
- If understanding of what science and technology can achieve is further advanced, the competitiveness of UK industry should improve.
- Scientists must learn to engage with the public. Not only should they do so voluntarily, but it is their duty to do so. The scientific community needs to conduct communication training, improve understanding of the media, give lectures and exhibitions for non-experts, hold science competitions for young people, and brief journalists.
- The Royal Society will promote the improvement of public understanding of science as its main activity.

This report is not only the origin of subsequent activities in the UK, but also a universal point about science communication. In 2006, the Royal Society published a report analyzing factors affecting science communication, emphasizing the need to clarify the definition of "public engagement". There is concern that scientists participate thinking it is necessary to "educate" with the knowledge they have, rather than listening to or dialoguing with the public. The biggest correlation in the motivation of scientists participating in communication is with the potential for budget acquisition of the organization to which they belong (Royal Society 2006).

In the UK, institutions and organizations related to enhancing communication with the public include:

(1) Government and science advisors of each ministry.
(2) Scientific Advisory Group in Emergencies (SAGE).
(3) Science Media Center.
(4) Nuclear Industry Council.
(5) Activities where scholars speak to the media.

Among these, the science advisors and SAGE were explained in the section on the TEPCO accident. The Science Media Center is a private organization aimed at providing journalists with quick, evidence-based, scientifically accurate information on science news of high social interest, and is operated by donations. The themes covered are broad, including diseases, climate change, influenza, genetic technology, drugs, environmental pollution, etc., with nuclear power being a part of it. To ensure neutrality, the proportion of each donation to the total is limited. It started in the UK in 2001 and has spread to Australia, Canada, New Zealand, Germany, etc. In Japan, it was established in 2010 as a follow-up to a public recruitment project, but due to lack of funds, it seems that activities are currently suspended.

The Nuclear Industry Council is a joint forum of the UK government and the nuclear industry, discussing the industrial strategy of nuclear power. In 2014, it issued a report called "In the public eye", stating the importance of public involvement and strategy. Activities where scholars speak to the media have been recommended since the Royal Society's report and are conducted in various fields, not limited to nuclear power. For example, in the field of nuclear power and radiation, a professor of oncology at the Royal College of Medicine speaks to the media. This professor has also come to Fukushima Prefecture after the TEPCO accident. In the UK, a mechanism for public dialogue in policy is an activity called Sciencewise, supported by the government, which has been conducted since 2004. In the UK, it is mandatory to listen to the public's opinion in policy decisions, and this is done through Sciencewise. In the nuclear field, the Department of Energy and Climate Change (DECC) was involved in 2015 and 2016 regarding the geological disposal of high-level radioactive waste.

In the field of nuclear power in Japan, communication is often centered around safety and risk, giving a narrow impression. However, in the UK, nuclear-related topics are treated as part of a broader field of science and technology, and it's worth noting that communication is not solely focused on safety and risk.

2.7 Public Dialogue, Stakeholder Dialogue

In the UK, public dialogue is conducted as a system to hear public opinions in policy-making. In other countries as well, stakeholder dialogues are held in situations such as the siting of radioactive waste disposal sites or nuclear power plants, or the remediation of environmental pollution (Robinson 2018; IAEA 2014). A stakeholder is a party with an interest or stake in a certain situation. There is also a definition of stakeholders as participants who are involved in the project with their own interests. Stakeholders are not predetermined. They need to be identified according to their interests.

The reasons why stakeholders are interested in the project are not only risks but also various. Interest in the scientific and technical aspects is only a small part, and there are interests from social, economic, and political perspectives. For example, employment, business/professional matters, safety/health, security, comfort or specific interests, regulations, media, etc. Therefore, in stakeholder dialogue, it is necessary to clarify issues and problems, consider who is interested in them, ask questions, and match the list of issues with the stakeholders' issues/interests. In the dialogue, it is necessary for each to go beyond their own positions, otherwise, it will turn into a debate. Try to create points of agreement on a common ground. Dialogue helps to discover interests and needs beyond positions. Stakeholders are like bridges. Stakeholders can be used for outreach and feedback.

In stakeholder dialogue, there are common pitfalls (failures) to avoid.

(1) Using a list of stakeholders and limiting stakeholders from the beginning: Stakeholders need to be found by interest.
(2) Showing a pile of documents: Appropriate tools and means need to be prepared for dialogue.
(3) Limiting the meeting place: A place needs to be prepared for the majority.
(4) Sending the wrong staff: Responsible staff need to be sent.
(5) *Telling the facts (this is a pitfall)*: Confirm the facts together. First, agree on who provides the facts and how to find them. The purpose and role of stakeholder dialogue lie beyond the facts.

The purpose of communication and dialogue with local communities regarding nuclear power is to build trust. Building trust is not easy. It's a tough job. Trust is gradually formed, but it can be lost in an instant. Trust is something that arises and develops, and it never reaches a state of completion. Important points for building trust include openness and transparency, professionalism, good listening skills, honesty, and inclusiveness. Respect is an important element in building trust, and it is necessary to approach dialogue participants with a sense of respect.

Facilitation is a technique for smoothly conducting meetings and gatherings. By encouraging participants to speak, organizing the flow of conversation, confirming participants' understanding, expanding discussions, and supporting consensus building and mutual understanding, it promotes the activation, collaboration, and consensus building of organizations and participants. There are organizations in

Japan that aim to popularize facilitation, such as conducting training for improving business skills. The application fields of facilitation include not only normative issues of consensus building, but also utilitarian issues aimed at solving organizational problems such as business improvement and product development, collaboration, and educational learning issues including psychological problems and lifelong education (Hori 2018).

In setting up and operating stakeholder dialogues, facilitators and coordinators who understand facilitation techniques and have practical experience are useful. The public dialogue of Sciencewise in the UK, in the field of nuclear power, was conducted in 2016 on the issue of geological disposal. It involves environmental consultants and experienced dialoguers.

Looking at the history of local dialogues and stakeholder meetings in the UK's nuclear power sector, they were conducted by the then British Nuclear Fuels Limited (BNFL) from 1998 to 2004. Initially dealing with the issue of radioactive liquid waste discharge, they have since accumulated experience in dialogue on more complex themes (McGlynn 2020). Lessons learned include "listening widely to opinions from the community, getting involved as widely and early as possible, maintaining two-way dialogue and transparency. Being ethical and honest towards local organizations". Currently, in the UK, the Nuclear Decommissioning Authority (NDA) operates and manages the country's nuclear-related facilities. Site Stakeholder Groups exist in various places, and as independent forums, they conduct communication between the NDA and the local community. For example, in the West Cumbria region, where the Sellafield nuclear fuel cycle facility and the low-level radioactive waste disposal site near DRIGG are located, the West Cumbria Site Stakeholder Group, which includes local governments, regulators, chambers of commerce, community groups, and the NDA, is active. While the social, cultural, and political environments in the UK and Japan are not the same, there are many commonalities. It is necessary to understand the lessons from countries like the UK and devise communication and stakeholder dialogue that is adapted to the Japanese environment.

2.8 Safety of Nuclear Power Generation

The number of public deaths due to radiation from nuclear power plant accidents is shown in Table 2.9.

In the Chernobyl accident, many children developed thyroid cancer because the restriction on the intake of milk containing radioactive iodine was delayed. Most could be treated by surgery, but eight died. Forty-nine workers who worked on the site to contain the accident died from radiation sickness. They are workers, not the public. The number of public deaths due to radiation from the *TEPCO Fukushima accident is zero*. There were also no deaths among workers due to radiation. The United Nations Scientific Committee on the Effects of Atomic Radiation (UNSCEAR) stated that for the Chernobyl accident, "It would be difficult to observe an increase in cancer or death

Table 2.9 Number of public deaths due to radiation from nuclear power plant accidents

Chernobyl accident: 8 people (pediatric thyroid)[a]
TEPCO Fukushima accident: Zero[b]

[a] In the Chernobyl accident, the majority of pediatric thyroid cancer cases that occurred due to the delay in restricting the intake of radioactive milk were treated with surgery

[a] In the Chernobyl accident, 49 workers died while trying to bring the accident under control. Increases in cancer and death in the public would be difficult to observe in epidemiological studies because the doses are low. *Source* UNSCEAR website

[b] Health damage due to radiation from TEPCO Fukushima accident will not be observed. *Source* UNSCEAR 2013, 2020

in the public due to the low dose in epidemiological studies", and for the TEPCO Fukushima accident, "Health damage due to radiation is unlikely to be observed".

Even if we say that "*nuclear power generation is the safest method of power generation*", many people may not believe it, but that is the case according to statistical data. For example, if you search the internet with words like death, electricity, production, safety, you can find charts and tables comparing the number of deaths per unit of power generated for each method of power generation, including resource mining, transportation, power plant operation, and use. I expect readers to search for themselves, so I will not say more here. For example (Forbes 2012; Brook et al. 2014).

There is a book written by an emeritus professor of nuclear engineering in the United States (a member of the National Academy of Engineering) with the subtitle "The Most Misunderstood Method of Power Generation" (Carbon 1997). Information about the safety of radiation and nuclear power generation and nuclear technology should have been and was supposed to be, on the side of those involved in nuclear power. Yet, why have many people in Japan and around the world come to think of nuclear power generation as the most dangerous method of power generation? Those involved in nuclear power need to think carefully. Could it be due to a lack of understanding of the risk psychology and communication of nuclear engineers, misunderstanding of dialogue activities requiring specialized skills, or the imposition of safety explanations on the public due to collectivism and the dissemination for it, misunderstanding of activities and research results in this field in the West, and a focus on the specific and difficult field of risk communication?

2.9 Risk Perception and Psychology: Communication and Risk Communication

Risk communication was discussed in Sect. 2.2.3, focusing on communication after accidents. In this section, it is discussed in general terms.

With the use of many new technologies in fields such as radiation, nuclear power, chemicals, food, drugs, and household goods, the need to communicate about the risks associated with their use has increased. In the United States, research on risk perception and psychology has been vigorously conducted since the 1970s (Slovic 1987). The National Research Council compiled a report titled "Improving Risk Communication" in 1989 (National Academies 1989). Activities of nuclear industry groups and government agencies have been carried out based on the understanding of these results. In Japan, these have also been introduced, studied, and carried out (Okamoto 1992; Kinoshita 2008).

How are individual attitudes and decision-making about disasters and risks made through what kind of cognitive process? For example, how did those opposed to genetic engineering form their opinions, and why do they think that airplane accidents are a more serious problem than traffic accidents? Fig. 2.7 shows that Slovic evaluated a number of hazards with unknownness and dread as the vertical and horizontal axes, and analyzed how people understand risks. The unknownness factor increases in the case of risks that are unobservable, unknown to familiar people, have delayed effects, are new, and are scientifically unknown. The dread factor increases when the risk is uncontrollable, fatal, globally devastating, unfair, greatly affecting future generations, not easily reduced, increasing, and passive (unavoidable). Risk perception and acceptance are determined by psychological factors. He states that people react to the images they have formed.

As shown in Fig. 2.7, the public understands safety subjectively. We will call this subjective safety. On the other hand, according to reports from international

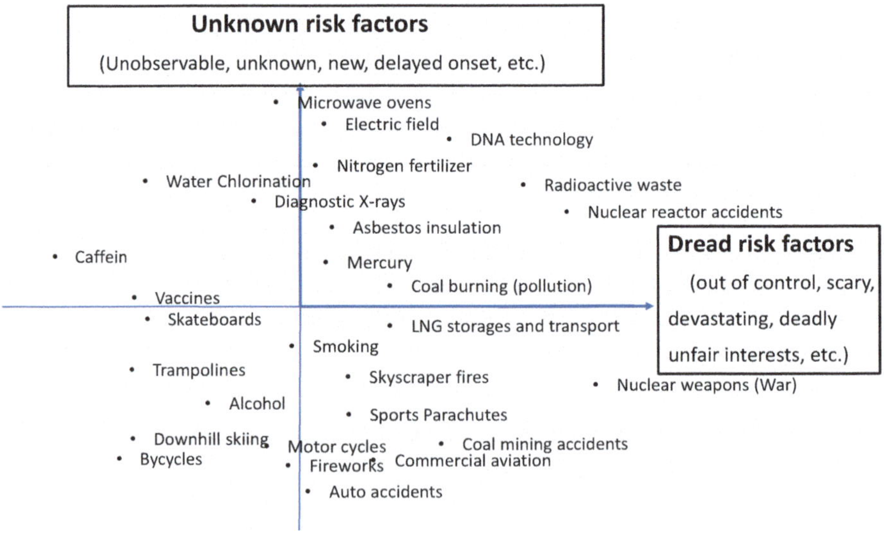

Fig. 2.7 Risk perception factors and awareness distribution of various risks. *Source* Adapted from Fig. 2 of Paul Slovic and Elke U. Weber, "Perception of risk posed by extreme events", Conference of Risk management strategies in an uncertain world, New York, April 12–13, 2002

organizations, eight children died from thyroid cancer caused by radioactive iodine in the Chernobyl accident. In the TEPCO Fukushima accident, there were zero deaths among residents due to radiation. This is the safety of nuclear power generation as understood by experts, which we will call *objective safety. Both subjective safety and objective safety should be considered correct. They are logically different and never intersect.* As will be mentioned later, if we confuse the two and casually talk about safety or safety technology to a large number of the public, it will rather spread the image of danger. The fact that nuclear power stakeholders have been talking about safety to promote nuclear power generation has been spreading the image of danger. Activities to intersect the two will convey the image of danger. Nuclear power stakeholders need to be aware of these points. The nuclear power industry in the West has noticed this and is not particularly explaining safety to the public. For example, Electricité de France (EDF) Energy in UK stopped safety television commercials in 2015. When the author asked a communication expert at the Nuclear Energy Institute in the U.S. What to do if asked about "safety", she said to say "safe". In addition, technology is used based on objective safety, not only for nuclear power generation. This is the responsibility of nuclear engineers.

Slovic stated at the end of his 1987 paper that "*risk communication and risk management will fail if they are not a two-way process*" (Slovic 1987). "Two-way process" is easy to say, but not easy to implement. First, you need skills for dialogue. You need to learn the basics in facilitation training and practically polish it. To implement a two-way process in dialogue, there is a limit to the number of groups. It might be possible to have a dialogue with local stakeholders, but a two-way dialogue with all citizens or prefectural residents is impossible in terms of numbers. It is better to create and publish reports based on evidence for citizens, prefectural residents, and nationals.

In Japan, there is an abbreviation for risk communication called "Riskomi", which is often used. Many nuclear power stakeholders seem to think that risk communication is communication, but the problem of two way dialogue is not only that the number of people who can have a dialogue is limited, but also there are logical breakdowns as mentioned next. In Japan's collectivism, it may be because they don't think for themselves, but only Japan is so biased.

First of all, *the purpose of communication is to build trust.* Here, communication refers to communication about safety issues, environmental issues, and so on. However, *talking about safety is talking about danger* and it is not easy to build trust with the other party by talking about risks. It is necessary to first notice this point.

Another point is that many people involved in nuclear power think that if the public understands the knowledge of nuclear safety and technology they know, their understanding should progress. *Communication and risk communication are not science education. They may have forgotten how many years it took them to acquire knowledge about nuclear power.*

This second point is a misconception that scientists and nuclear experts in the West are prone to. For example, as mentioned earlier, the Royal Society in the UK, which promotes science communication, is concerned in its 2006 report that scientists are participating in dialogue activities thinking that it is necessary to "educate" their knowledge rather than listening to or dialoguing with the public (Royal Society 2006, p. 14). I have also heard a prominent foreigner in the field of nuclear safety say that safety requires "education". It is wrong. The roles and responsibilities in the West are clear. Experts in the technical safety of nuclear power do not necessarily understand communication well. It is a big mistake to think that risk communication is "education" for the public and locals.

"Accepting risk is subjective, so it deviates from risk, and this is called risk perception. Objective risk (science) is closer to safety, and subjective risk corresponds to peace of mind", "Ordinary citizens feel uneasy about both nuclear power plants and genetically modified organisms. Bio experts feel less risk about genetically modified organisms, and nuclear experts feel less risk about nuclear power plants", says a psychologist who has been practically studying risk communication for many years (Kinoshita 2009). After all, the promoters are biased psychologically.

Risk communication is the role of the regulatory side as stated in the documents of the International Atomic Energy Agency (IAEA 2022; Oka 2019). The U.S. Nuclear Regulatory Commission has created and published an excellent manual for the staff titled Effective Risk Communication (NRC 2004). For example, as already mentioned, there are warnings written that it is not good to explain and compare the risks of nuclear power to those who do not have a relationship of trust. Staff of the U.S. Nuclear Regulatory Commission are thought to be conducting risk communication, as they need to explain about risks at public hearings in local areas and at headquarters. For reference materials when considering the practice of risk communication, there are explanations by Kinoshita and Bier (Kinoshita 2016; Bier 1999) and others.

However, in the United States, the *Nuclear Regulatory Commission is not actively conducting risk communication to the public*. This is not written anywhere, and it was an impression I got from observing and hearing about nuclear power in the United States for many years. I confirmed it with a former commissioner. Moreover, the pro-nuclear power generation side, such as the Nuclear Energy Institute, is not conducting risk communication. Media training for power companies and public opinion surveys are conducted by their related organizations. The communication of the Nuclear Energy Institute aimed at "branding" nuclear power. The fact that the NRC is not systematically conducting risk communication to the public can be confirmed by an internet search. Even if you search for NRC, Risk communication, you can only find the literature on Effective Risk Communication. This literature is a manual for NRC staff to conduct risk communication at public hearings and in local areas, and it is not an introduction to activities specifically focused on risk communication to the public. If you search for Public communication, NRC, you will find communication activities such as explanations of the NRC's role to enhance transparency of the NRC.

Even in the United States, there is a need to seek local understanding about safety, so in such cases, communication with the local community and risk communication should be conducted based on the results of past research and practice. This activity is better called stakeholder dialogue. For example, around the year 2000, there was a problem in Hanford, Washington State, where radioactive waste from nuclear development leaked into the ground due to tank corrosion. Communication activities (stakeholder dialogue) were conducted, and now decommissioning measures and waste treatment and disposal are progressing.

In the UK, two public hearings (dialogue activities) on geological disposal have been held. The number of participants is between 30 and 40, which allows for dialogue. The planning of the dialogue activities was carried out by a group of consultants with experience in environmental dialogue activities. In the US and UK, facilitation of dialogue at the local level is being carried out based on practical research results on risk perception and psychology. There are people in Japan who have been carrying out practical activities. However, overall, due to a lack of understanding of risk perception and psychology by the proponents and nuclear power operators, and misunderstandings about risk communication by nuclear engineers, it is rather counterproductive. It is a problem that there are many nuclear experts in Japan who do not notice this point. Japan still has strong lifetime employment, so there are very few experts who can make a living by making dialogue their job. I wonder if the proponents recognize these points as weaknesses.

There is a characteristic of risk perception that anxiety is strengthened by repetition. When we try to discuss individual safety issues with the entire nation, the anxiety of the nation is amplified. It may be only in Japan that the opinion that we have to discuss specific safety issues with the entire nation is widespread. Environmental issues are handled by the Ministry of the Environment, health issues by the Ministry of Health, and nuclear safety by the Nuclear Regulation Authority, and it is their role to create standards and fulfill their responsibilities. If there are opinions, they should be written properly with evidence and published. This is the case in the West.

There is a point that saying things like "We must never repeat such an accident" is unique to nuclear power. In other industries, for example, the oil industry says, "We are making maximum efforts to reduce the impact of the accident in the event of a large-scale crude oil spill from a tanker into the ocean. We can respond better from this experience", "In other industries, we do not say that accidents should never happen again, or that safety is the top priority. Oil spills occur frequently, but they do not lead to strict regulatory requirements like nuclear power generation. By saying this in nuclear power generation, it is spreading the image that it is extremely dangerous", "There have been more than 10 public exposure accidents worldwide that resulted in deaths due to radiation sources, but the symptoms of radiation phobia like nuclear reactor accidents do not last long", as pointed out (Grimston 2014, 2016).

The fact that this kind of language has been repeated in nuclear power generation may be the reason why the actual safety and public perception have greatly differed. Especially in Japan, explanations from nuclear power stakeholders tend to be biased towards risk communication and safety. Many people may not realize the difficulty of risk communication with the public due to their lack of understanding of risk psychology in a collectivist society. There is a strong impression that many nuclear power stakeholders, who are domestically oriented and have a narrow range of experiences, do not realize that domestic information is biased due to their lack of understanding of overseas situations. What I have mentioned here should be understood if one can search in English and think based on evidence. In the internet age, it is easy to access information from the West.

Furthermore, the risk communication I am referring to here is not communication (stakeholder dialogue) with a small number of people who can have a dialogue locally, but rather the dissemination activities by nuclear power promoters and nuclear engineers using verbal and media means, aimed at the entire nation, regarding radiation risk and the safety of nuclear reactors. They should realize that this is inadvertently spreading an image to the public that nuclear power generation and radiation are dangerous. The explanation of radiation risk and the safety of nuclear reactors for the public is the role of the nuclear regulatory agency. A state should be created where interested citizens can find evidence-based information through internet searches and the like, according to their interest.

Moreover, when considering what to do about individual safety issues, I believe that it is best for both nuclear power promoters and the public not to discuss matters below the safety standards. This is the case in the US and UK. There are standards set by the Nuclear Regulatory Commission, an independent group of experts, which also take into account international perspectives. Why should technicians and amateurs, who have no responsibility, need to talk about things below that? This is not done in the West. There is a saying that "Leave it to the experts. Let the baker bake the bread", meaning that everyone is working according to their expertise and responsibility. If there are objections to the established standards, shouldn't the first step be to create and publish a written document stating the reasons? As a parliamentary democracy, discussions and decisions should be made within this framework, which is the rule of Japanese society. It is fundamentally impossible for everyone to agree.

Even below the safety standards, discussing safety cannot alleviate the psychological anxiety of the public, so nuclear engineers should not particularly talk about safety to the public. Information should be disclosed, and written explanations should be created so that the public can find them. It is necessary to think about and discuss risks in stakeholder dialogues with local communities.

Human risk perception is not rational. One becomes aware and concerned about a certain risk once they "notice it". This is a fact, based on the results of risk psychology research. It is necessary to recognize that human risk perception is irrational and think about what should be done. Human risk perception is irrational, but risk management such as regulation and disaster prevention needs to be done rationally. Measures need to consider cost-effectiveness. There are limits to the budget and manpower for measures, and they need to be used effectively and rationally. Considering these, it is better not to talk to the public about risks below the safety standards. This is the case in the West.

If there is a difference of opinion with the Nuclear Regulation Authority, the power company should discuss it with the Regulatory Commission. It would be even better if there is a report on the basis. I heard that in the dialogue between the U.S. Nuclear Regulatory Commission and the industry, there are sometimes heated debates, including meetings to organize the agenda (not open to the public). If you don't exchange opinions to the point of arguing, it's not a real exchange of opinions. I have heard that a private meeting to organize the agenda is held before the public meeting, and there can be heated debates there. It goes without saying that Japanese power companies are required to have the ability to dialogue with the nuclear regulation side and a track record of voluntary safety improvement activities. It is not well known, but in the United States, the safety and operation rate (capacity factor) improvement achievements by the voluntary activities of the nuclear power industry, born from the reflection on the Three Mile Island nuclear accident, preceded the improvement of reactor regulation.

Since the United States does not have lifetime employment, I wonder if safety-related consultants are helping the Nuclear Regulatory Commission and power companies. In Japan, many people conduct nuclear research in narrow specialized fields, but we should be able to nurture a group to do this kind of work. In addition, nuclear disaster prevention and emergency response have a large role for the government and local governments, and it is not left to power companies and reactor manufacturers alone, so the involvement of the country and national research and development institutions is more necessary. Communication with the public in the event of a nuclear emergency is crisis management and emergency response, and as already mentioned, there is a good practice in the UK. Table 2.10 summarizes the communication for nuclear engineers.

In addition, disaster prevention is the responsibility of the country, and I have heard voices saying that power companies have nothing to do with disaster prevention, but this is a mistake. The power company that caused the accident has a responsibility

Table 2.10 Communication (summary for nuclear engineers)

Risk perception and acceptance are determined by psychological factors. The public cannot be called unreasonable
The purpose of communication is to build trust. It is difficult to build psychological trust when talking about risk and safety
Communication is not "education"
Risk communication requires two-way dialogue. This requires facilitation skills
Pay attention to the characteristic of risk perception that anxiety is reinforced through repetition. "I noticed it, so I'm concerned about it."
Nuclear engineers should not talk about safety to the public. Because they can't handle anxiety. It would be a good idea to create and publish various explanations
The phrase "an accident like this must never happen again" is unique to nuclear power, and conveys an image of danger
Risk communication is a regulatory role. However, in the United States, neither the Nuclear Regulatory Commission nor the nuclear industry has discussed risks and safety with the entire nation. Evidence and policy information is created and disclosed
It is best for both nuclear power proponents and the public not to discuss matters that are below safety standards
The Nuclear Regulation Authority is the organization where the nuclear industry talks about safety and risks
British science communication and dialogue on geological disposal can be helpful

to report the situation to the country in real time. Can a power company that doesn't know about disaster prevention convey the necessary information? In the United States, power companies are required to recommend to states and local governments how to take protective measures. The decision on protective measures is made by the state or local government.

2.10 Summary and Consideration

As indicated by the data on the number of public deaths due to radiation exposure from nuclear power plant accidents in the United Nations Scientific Committee on the Effects of Atomic Radiation (UNSCEAR) reports, nuclear power is an extremely safe technology. However, many people around the world believe that nuclear power is extremely dangerous. There are several reasons for this. First, there is a theory that disasters caused by nuclear power are associated with the damage from atomic bombs. However, while the technology of gunpowder is used in wars, it is also used in civil engineering, and civil engineering is not particularly considered dangerous by society. Therefore, the association with atomic bombs and their damage is not the initial cause of society's perception that nuclear power is dangerous.

Nuclear power was not considered dangerous from the beginning of its use. In the 1960s, when the author was a student, expectations were high. The tide turned with the 1974 Three Mile Island nuclear accident in the United States and the international discussions on spent fuel reprocessing and the use of fast reactors and plutonium in relation to nuclear non-proliferation from the late 1970s. Subsequently, the 1986 Chernobyl accident heightened concerns about the safety of nuclear power in Europe, and opposition to nuclear power became apparent. In Japan, since the 1990s, it has become difficult to establish new nuclear power plants, and the fast reactor and nuclear fuel cycle business have attracted public attention. The reason why new nuclear power plants have not been established in the United States since the 1980s is due to the avoidance of construction requiring huge initial investment due to economic liberalization and power liberalization, not due to safety concerns.

There is no technology that has been discussed in society as much as nuclear power in terms of its safety. If you search in English on the Internet, you can find many explanations about the safety of nuclear power. In contrast, if you search for bridges, for example, you will find only a few lines of explanation about their collapse, and you can understand that safety is dealt with in civil engineering.

The scientific knowledge about nuclear power technology and radiation exposure is on the side of nuclear power experts, so it is better to think that the reason why nuclear power has come to be considered extremely dangerous is on the side of nuclear power experts. In this chapter, I introduced the research results on the validity of the application of the ALARA concept based on the linear hypothesis (LNT hypothesis) of radiation exposure and the perception and psychology of risk, and pointed out the lack of understanding of nuclear power experts about the difficulty of risk communication.

The linear no-threshold (LNT) model of radiation exposures, the hypothesis that there is no threshold between radiation exposure dose and the effect, and that a linear relationship is established between the dose and the harm. This has been used to control the risk of radiation exposure. However, the French Academy of Medicine and Academy of Sciences have found successive defensive biological responses specific to low-dose radiation, and in 2005, they announced the view that cancer and leukemia do not actually occur at exposure doses lower than a certain dose, and that the LNT model is overestimated. In contrast, the U.S. National Academy of Sciences announced in 2006 that the LNT model is scientifically valid. The LNT hypothesis has not been scientifically settled.

Based on the LNT hypothesis, even a small amount of radiation is interpreted to have health effects. There are opinions that the LNT hypothesis is the basis for public fear of radiation and nuclear power. There seems to be no other example than the LNT hypothesis that adopts the idea that there is no tolerance for harmful substances, even in small amounts, and that they are harmful. Like the presence or absence of health effects of small amounts of harmful substances, it may be difficult to scientifically determine the validity of the LNT hypothesis. The unknown parts of life phenomena, such as the repair of damage caused by radiation or the elimination of the effects of damage, may be quantified in the future by new scientific means.

The International Commission on Radiological Protection (ICRP) has recommended keeping radiation exposure low since its establishment in 1950. This idea has changed its expression, but in 1977, it became ALARA (As low as reasonably achievable), which is still used today, to reduce radiation exposure as much as reasonably achievable. Following ALARA, reference levels (radiation doses used in emergencies or existing exposure situations) have been used as guidelines for limiting doses. ALARA is not a scientific fact but a policy (way of dealing with it). Unlike the LNT hypothesis, ALARA is a way of dealing with it, so if it is inappropriate, the way of thinking or its interpretation and operation should be able to be revised. Clear evidence has been obtained from the analysis of disaster-related deaths in the TEPCO Fukushima accident, so it should be revised.

Not limited to nuclear power generation, technology is used because it has benefits. The dangers (risks, costs) associated with the use of that technology are evaluated in comparison with the benefits. This is called the cost-benefit or risk-benefit approach. The International Atomic Energy Agency's Handbook of Nuclear Law introduces this approach first. *The way to deal with the risk associated with radiation exposure also needs to be considered in terms of risk-benefit.*

When considering risk-benefit, there are two problems with ALARA. One is that it considers risk within the framework of radiation exposure, without considering risks other than exposure. The other is that it has not quantitatively considered the R (reasonable) of ALARA. Even if risk communication is strengthened, it does not mean that R has been considered.

Risks other than exposure include risks associated with evacuation from accidents, artificial abortion due to excessive concern about radiation exposure, psychological and mental instability, and damage from rumors. These cannot be considered within the framework of radiation exposure. The ICRP is a committee that deals with radiation exposure risks, so these risks cannot be considered within the ICRP framework. The international organization that deals with health risks other than radiation exposure is WHO. A way of thinking that considers both radiation exposure and risks other than radiation exposure is necessary. The public exposure caused by these facilities around nuclear power plants and nuclear facilities in normal times is below the background associated with natural radiation, so the actual problem of public exposure is during a nuclear power plant accident.

To quantitatively consider the R of ALARA, quantified data is necessary. When analyzing the number of disaster-related deaths in municipalities where residents evacuated in the TEPCO Fukushima accident and the radiation dose after the accident, the risk associated with evacuation was about 2.8 times larger than the radiation exposure risk avoided by evacuation. Considering that the lifetime exposure risk of the elderly is one-fifth of the average age, it is 14 times larger. Considering the difference in risk coefficients between instantaneous exposure like in atomic bombings and long-term exposure (about 2 times), it becomes even larger, 28 times.

Since there was data on disaster-related deaths, the risk associated with evacuation could be quantitatively evaluated. In Fukushima Prefecture, the number of disaster-related deaths increased because the lifting of the evacuation was delayed. From the perspective of preventing disaster-related deaths, the shorter the evacuation period, the better. Even in nuclear disasters, like other general disasters, it is better to lift the evacuation at the stage where it is judged that the nuclear accident will not expand. Note that the recognition of disaster-related deaths is a kind of condolence money, so the opinion that the number of disaster-related deaths cannot be used to calculate risk is incorrect. Unlike recognition of occupational accidents due to exposure, death is a fact, and the causal relationship between evacuation due to a nuclear accident is clear from the temporal comparison of the number of disaster-related deaths in Miyagi and Fukushima prefectures. Although a certain percentage of elderly people die in nursing homes, the problem of elderly people with high care needs in nursing homes and the problem of disaster-related deaths at the time of the TEPCO Fukushima accident should not be confused.

After the Chernobyl accident, birth rates declined in Greece and Italy, which is thought to be due to unnecessary abortions performed out of excessive concern about radiation exposure, but this has not been quantitatively analyzed. *The concept of risk-benefit is considered to be higher than ALARA*. Until now, it may not have been possible to examine the validity of ALARA due to the lack of data, but with the data on disaster-related deaths from the TEPCO accident, it has become clear that there is a problem, so it should be corrected.

In the United States, there is a metric for analyzing the cost-effectiveness of safety measures in nuclear power plant safety regulations. That is, there is a conversion factor that equates exposure to 10 people-millisieverts to $5100 (in 2014 dollars). The conversion factor has been revised to keep up with inflation. Probabilistic risk assessment is a method for analyzing costs and benefits and is used for that purpose. Probabilistic risk assessment is often discussed in Japan, but there is no such metric in Japan's reactor regulations. Without a metric to quantify cost-benefit, rational safety regulations cannot be established, and as a result, the burden on the public increases through electricity charges.

In the United States' emergency protection guide (PAG, Protective Action Guides) for nuclear accidents, traffic accidents during evacuation are considered as a risk associated with evacuation. In the United States, evacuations from hurricanes and wildfires occur occasionally, but there seem to be few disasters with data on disaster-related deaths associated with evacuation, and the United States has not quantified evacuation risks other than traffic accidents. Japan's nuclear protection guidelines state, "It is necessary to keep exposure as low as possible while also minimizing health effects not directly caused by exposure". The idea of keeping exposure as low as possible has led to delays in lifting evacuations in Fukushima Prefecture, increasing disaster-related deaths, which is contradictory.

At the time of the Chernobyl accident, children developed thyroid cancer because restrictions on the consumption of milk from dairy cows that had eaten grass contaminated with radioactive iodine were delayed. Most were treated with surgery, but eight died. From this lesson, it is thought that in an emergency situation where a reactor core melts, preventive measures such as restrictions on the consumption of leafy vegetables, drinking water, and milk in areas near the power plant are necessary. However, Japan's nuclear disaster prevention guidelines prioritize thyroid exposure monitoring, and there is a risk that consumption restrictions may be delayed. After the TEPCO accident, safety measures at power plants were strengthened, but relying solely on them contradicts the concept of defense in depth. Japan's disaster prevention guidelines describe procedures, but the U.S. PAG describes concepts. Procedures do not function in an accident if their premises are not met, so the U.S. PAG is superior.

Administrative officers are experts in procedures, but when they work, they need to understand the concepts and insights behind the rules. The creation of materials for this purpose is the role of technical support organizations for regulation (research and development institutions). In the United States, there are excellent training materials on nuclear safety created by researchers at a national laboratory that has been conducting severe accident research for many years. Training materials reflecting the experience of the TEPCO Fukushima accident should be created.

Radiation protection guidelines are given in terms of the annual radiation dose, i.e., the dose rate. The value to determine the need for evacuation is also given in terms of the dose rate. Even if elderly people are exposed to radiation, the proportion of cancer occurring in their remaining life is small. In other words, the cancer risk from radiation exposure in the elderly is smaller than the average age. This is a scientific fact, but the difference in cancer risk by age is not considered in the current radiation protection guidelines.

Most of the disaster-related deaths at the time of the TEPCO Fukushima accident were elderly. Setting the dose for lifting the evacuation at the average age is dangerous from the perspective of preventing disaster-related deaths of the elderly. Nuclear power plants in Japan are located in agricultural areas. The mainstay of agriculture is the elderly. Encouraging the elderly to return early and enabling them to resume farming and engage in decontamination has the effect of preventing disaster-related deaths due to disuse syndrome (inactive lifestyle diseases). This can also prevent the collapse of local communities. Not only in Japan, but there are many countries where nuclear power plants are located in agricultural areas and aging is progressing, for example, South Korea, China, France, etc. It is necessary to make use of the lessons of disaster-related deaths from the TEPCO Fukushima accident worldwide.

In addition, the OECD/NEA held a workshop titled "Optimization: Rethinking the Art of Reasonable" in January 2020. In this workshop, discussions and considerations on radiation exposure risk are being conducted under the title of optimization. Risk-benefit is a method of optimization, and *it is good to quantify the risk associated with evacuation at the time of an accident and quantitatively examine the R of ALARA*. The description in the present book is a proposal to specifically advance the optimization of this workshop. Evacuation during a disaster is often done during hurricanes, tsunamis, and earthquakes, so quantifying that experience should improve the problem of evacuation during a nuclear accident.

The proportion of disaster-related deaths to the total number of deaths from the disaster is 14.3% in the Great Hanshin-Awaji Earthquake, 76.4% in the Niigataken Chuetsu-oki Earthquake, 18.7% in the total of Miyagi, Iwate, and Fukushima prefectures in the Great East Japan Earthquake, and 79.4% in the Kumamoto Earthquake (as of April 2018). These data are in local governments. The Japan Federation of Bar Associations issued an opinion paper in 2018 calling for the "accumulation, analysis, and publication of cases of disaster-related deaths" (Japan Federation of Bar Associations 2018).

It is believed that many disaster-related deaths are among the elderly. Data on the need for care certification for the elderly is also available at local governments. The government should collect data on disaster-related deaths and their need for care from local governments and publish it anonymously. It is necessary to clarify the relationship between the need for care and age at the time of disaster occurrence for each disaster-related death. Analyzing these data can provide lessons on evacuation from various disasters. It is also possible to quantify the R of ALARA. Various natural disasters that require evacuation will continue to occur in Japan. Since consolidated data on disaster-related deaths is only available in Japan, this work is not only important for aging Japan, but also a responsibility of Japan, which caused trouble to the world with the TEPCO Fukushima accident.

In the field of nuclear power, there is a concept of defense in depth for ensuring safety. Up to the fourth layer, it is a concept to prevent accidents and reduce the amount of radioactive material released, and most books on nuclear safety deal with them (Oka and Bittermann 2015). The fifth layer is the reduction of impact by emergency response when radioactive materials are released into the environment during an accident. Disaster prevention drills are conducted in power plants and local governments. This chapter is a description of nuclear safety corresponding to the fifth layer.

Emergency response during disasters is carried out in conjunction with natural disasters. Experience in disaster prevention and emergency response has been accumulated and studied in the field of civil engineering, not nuclear engineering. Researchers and engineers in nuclear safety are working on how to prevent core meltdown accidents, but very few are conducting research on emergency response and disaster prevention, and there are very few actual examples of nuclear accidents where radioactive materials are released into the environment. It is necessary to refer to the experience in the field of civil engineering and apply it to nuclear disaster prevention and emergency response.

Next, the author points out the issues of understanding of nuclear-related personnel regarding risk perception and risk communication. This chapter introduced the results of studies on risk perception and psychology, and public dialogue in the western countries. From the results, it is pointed out that the reason why nuclear power generation has come to be considered extremely dangerous is due to the psychological impact of talking about risks and safety, or in other words, the *lack of awareness that talking about safety is psychologically talking about danger, and the misunderstanding of dialogue activities that require professional skills,* and that nuclear-related personnel have repeatedly communicated explanations of technical safety and radiation risk to the public. It is possible to point out the problem that the focus on specialized explanations about safety and technology, and the concentration of interest in the specific and difficult field of risk communication may be the cause. Japanese nuclear power personnel are highly dependent on groups. The author feels that this tendency is stronger.

Safety technology is not the safety from the perspective of the public. Safety technology refers to the technology used to prevent accidents in nuclear reactors and to reduce the release of radioactive substances into the environment. Safety from the perspective of the public is the safety after radioactive substances have been released into the environment. Most nuclear researchers and technicians are experts in safety technology. They may think that they should explain safety technology to the public, and they may have done so. They may have done so out of a sense of responsibility as experts, but would civil engineering experts verbally communicate the safety of bridges to the public? Nuclear experts may have forgotten how many years it took them to acquire knowledge of safety technology.

In the 1980s, research on risk perception and risk psychology was conducted in the United States, and the results showed that people's perception and acceptance of risk are made by psychological factors, not rationally. There is an introductory book on risk psychology that shows the figure "I'm worried because I noticed it" in the first page. The safety understood by the public is *subjective safety*. The safety of experts is *objective safety*. Both are correct. *Since both are logically different, they never intersect*. Activities to intersect both, not in dialogue, with the public, will ultimately convey the image that nuclear power generation is dangerous.

Communication is the act of conveying feelings, opinions, through dialogue, to the other party. *The purpose of communication is to build trust*. Talking about safety is talking about risk, and it is not easy to build trust with the other party just by talking about safety. This is the first point to notice.

Many nuclear engineers may think that if the public understands the nuclear safety and technology, understanding should progress. Many people may think that they have to do so out of a sense of responsibility as engineers and scientists. However, *communication and risk communication are not education*. They may have forgotten how many years it took them to acquire knowledge of nuclear power in the first place.

This is a misconception that scientists and nuclear experts in the West are prone to. The Royal Society in the UK, which promotes science communication, expressed concern in its 2006 report on science communication that scientists are participating in dialogue activities thinking that they need to "educate" the knowledge they have, rather than listening to or dialoguing with the public's opinions.

A prominent American researcher in risk perception research has said, "*Risk communication and risk management fail if they are not a two-way process*".

"It is a dialogue", they say, referring to the two-way process. Communication, not just risk communication, is an activity carried out through dialogue. It's not easy to execute a bidirectional process; it requires a broad knowledge of the areas of interest to the other party and skills for dialogue. To effectively engage in dialogue, the number of participants in the dialogue is also limited. It is impossible to have face-to-face dialogues with all citizens or all prefectural residents.

Talking about safety is psychologically talking about risk (danger). Negative stories leave an impression three times stronger. Talking about risk is a negative story. There is a characteristic of risk perception that anxiety is reinforced by repetition. Discussing individual safety issues with all citizens amplifies their anxiety, not just in the case of nuclear power. In Japan, there seems to be a common belief that the government has a verbal obligation to explain to its citizens, as the term "accountability" suggests. Not just in the case of nuclear power, if you verbally explain or discuss safety alone, since talking about safety is talking about risk, an image of danger becomes established. Not just in the case of nuclear power, it is difficult to explain and understand the safety of a technology verbally in a short time. It is necessary to read what is written and think for oneself. Building trust is even further ahead. The translation of "accountability" as "responsibility to explain" is a mistranslation; "accountability" refers to the "explanation of policies and responsibility for their results" in administration. Japan has a strong sense of dependence on the country, so many people feel no discomfort with the term "responsibility to explain". U.S. and U.K. administrative agencies are always questioned about their accountability. At the end of Chap. 4, the author explained the accountability of the U.S. Nuclear Regulatory Commission as an example. Without documented information, it is difficult to properly understand the explanation and its basis. In the U.S. and U.K., people do not necessarily trust someone's explanation and do not follow their opinion.

The idea that we need to communicate with all citizens about risk and safety, and that we can communicate, does not exist in the U.S. and U.K. The U.S. nuclear industry does not conduct public relations activities about the safety and risk of nuclear power generation. The party that the nuclear industry should dialogue with about risk is the nuclear regulatory side. The Nuclear Regulatory Commission is the one conducting safety checks on nuclear power plants. The U.S. Nuclear Regulatory Commission explains its activities on its website and public hearings. The U.S. Nuclear Regulatory Commission's dialogue activities at the local level of power plants should not be called risk communication for all citizens, but stakeholder dialogue.

There is a famous high school in Japan that has taken up the safety of nuclear power as a theme for debate. Nuclear power generation is a topic where information about safety (and danger) is extremely prevalent in society. When students bring together their survey results for a debate, many of them end up believing that nuclear power is dangerous. The safety of nuclear power is not an appropriate theme for high school debates. If it were to be used as a practice question, it should be something like "search for the basis of both pros and cons, including English literature, and examine their validity".

In the past, I was asked by a junior high school classmate to give a talk on the safety of nuclear power at a town meeting in metropolitan Tokyo area, along with someone from the opposition. It is impossible to make the general public understand the technical safety of nuclear power in a one-hour talk. I didn't realize it at the time, but talking about safety is talking about danger, so I now think that I may have ended up conveying an image of danger. To begin with, the very act of bringing up the topic of nuclear safety has elements of an anti-nuclear movement through raising public interest.

The phrase "We must never let such an accident happen again" is unique to nuclear power. It is spreading an image of extreme danger. In other industries, even if an accident occurs, they do not use such a phrase. The fact that this kind of language has been repeated in nuclear power generation may be one of the reasons why the actual safety and public perception are greatly different. This was pointed out by a person from the UK, who gave a lecture of the same nature at a convention of the Japanese nuclear industry, but I heard voices from Japanese nuclear industry officials who could not understand the lecture content, saying "He's saying strange things". This is what happens when you understand based on hearsay or written hearsay. In communication, just talking about risks does not deepen understanding about nuclear power. The problem of groupthink and national dependence among Japanese nuclear industry officials is deep-rooted.

One of the reasons for this situation may be that there are no experts or responsible persons who understand the basics of communication and risk psychology in Japanese nuclear industry organizations. The Nuclear Energy Institute (NEI) in the United States has a director in charge of communication and a specialist in communication who supports them. This specialist conducts media response training (press conferences during accidents, etc.) for employees of American power companies and conducts public opinion surveys. They also have contact with the media department of a university in Washington DC for practical research. This person, although a humanities scholar with a doctorate, also understands the scientific aspects of nuclear power. Their ability to understand both the humanities and science aspects of issues is impressive. The Director of Communications at NEI, an important position (director) within NEI, was involved in communication activities such as branding of nuclear power. Both are within NEI, understand NEI's policies and the key points of nuclear policy, and are working together. There are several organizations in Japan's nuclear industry that conduct public relations and publish newspapers of nuclear industry, but they do not have such specialists. As a result, for example, inaccurate and inappropriate understanding of risk communication may have been left unaddressed in the nuclear industry in Japan. The fact that the responsibilities of the person in charge and the department in charge in the United States are clear and that work is being done may be a cause of differences between Japan and the United States.

It is necessary to address both risks and benefits. Not only safety and risks of nuclear power generation, but also its characteristics, benefits, and relationship with security should be explained from a broad perspective and published on the Internet for the public to find. The Ministry of the Environment in Japan has created and published unified explanatory materials on the health effects of radiation exposure. The importance of energy security has become apparent due to Russia's invasion of Ukraine. Not only safety, but also the benefits should be explained and published.

In the event of an accident, it is important to have a single voice for the public. During the TEPCO Fukushima accident, *the UK was outstanding in crisis management and emergency response.* The government's emergency advisory organization was set up, and a unified message was sent to the British people in Japan by the Chief Scientific Advisor. Based on the experience of the UK government's failure to respond to the mad cow disease in the past, preparations were made and functioned.

For those specializing in nuclear engineering, understanding risk psychology and risk communication is not easy. It's not something you can do on the side; you need to dedicate time to it, seek out literature yourself, and study it, otherwise your understanding will not deepen. For reference, I will share my own experience. I have always been interested in the issue of public understanding of nuclear safety and risk communication for many years. In the International Committee of the American Nuclear Society, which I participated in the 1980s, communication was one of the central topics. The Japan Atomic Energy Society also established a Social Environment Committee, and risk communication remains a central topic to this day. Around 2000, I had the opportunity to study in a joint session at the Pacific Basin Nuclear Conference, with the cooperation of the late Professor Yasumasa Tanaka of Gakushuin University, who had long led this field in Japan, and a consultant from the Nuclear Energy Institute (NEI) in the United States. The NEI was conducting media training (training for press conferences) for American power companies, so I misunderstood that communication was simply about conducting media training. Thinking that it would be sufficient to train those who had received the training, I visited the NEI with interested individuals and received media training. At that time, I did not realize that there were no professionals who made communication their full-time job in Japan. I later realized that it was impossible to cultivate experts in this field without full-time job. Furthermore, unless nuclear engineering experts have special talents and experiences, I do not think they can engage in dialogue activities after retiring from their main profession.

When I was a university professor, I was busy with my original job and studied it on the side, but my understanding did not deepen. In 2014, I was appointed as the chairman of the Atomic Energy Commission of the Cabinet Office, and I studied by reading reports summarizing the experiences of stakeholder dialogues on geological disposal in the UK, literature on science communication, and literature on risk communication in the US from the 1980s. When I went to the UK for a Japan-UK dialogue, it was also helpful to talk with experts in communication activities in the UK. The facilitation and presentation materials of the dialogue held at the

British Embassy in Tokyo after the TEPCO accident were also helpful. As for risk psychology, several introductory books introducing American research have been published in Japan in the 1990s. The content written is simple, but it was necessary to understand it within the context of communication as a whole.

Many nuclear engineering experts may feel uncomfortable after reading the description of risk communication in this chapter. The main references used are listed as references for this chapter. I hope they will be helpful for understanding. I recommend searching in English. The lack of logical documentation summarizing the experiences and lessons of actual stakeholder dialogues and communications, to pass on to future generations, also makes understanding this field difficult. In the UK, there are three reports that summarize and evaluate the experience of dialogue activities for geological disposal. The introduction of UK dialogue activities held at the British Embassy is also helpful. To have a dialogue, not only nuclear safety, but also a very wide range of knowledge is required. In the UK, the participation of environmental consulting groups in dialogue activities can be a reference for Japan. If nuclear engineering experts think they can do communication and stakeholder dialogue on the side, isn't that naive?

Dr. A. Weinberg, former director of the Oak Ridge National Laboratory in the United States, is famous for his 1970s remark that "trans-science is needed to solve complex problems brought about by science and technology", but without thinking specifically about the problem of trans-science, there will be no progress. The issue of the health effects of low-dose radiation exposure, which Dr. Weinberg first cites in his paper, is not about scientifically verifying the LNT hypothesis, but about quantifying the R (Reasonably) of ALARA (As Low as Reasonably Achievable) based on cost-benefit thinking, which is a practical way to deal with it.

Social problems need to be thought about inductively, not deductively, based on evidence data and facts. Thinking deductively is a common practice for natural scientists, but there is no "truth of nature" in social problems. For example, even if we think deductively from the proposition that "human life is more important than anything else", we cannot answer the question, "Why do people in Ukraine go to war?" In Japan, there seem to be people in the humanities and social sciences who develop their arguments deductively from propositions. There is an educational scholar who warns that "deductive thinking will destroy Japan", even though it is not about nuclear power. We need to think inductively and logically. Although the author is not familiar with cultural theory, if we consider culture as a matter of the human heart, wouldn't the discussion of nuclear power from the perspective of cultural theory be a matter of a different dimension from the considerations mentioned in this book? Being able to enjoy culture is not unrelated to science and technology. If we consider fashion as a cultural phenomenon, if nuclear power is taken up because it is popular as a topic of conversation in society, wouldn't it result in emphasizing the risk of nuclear power? We live in an era where various social media platforms are available. They are also used as a means to deliver the voices of the people to politics. However, it is impossible to convey various elements related to the point of argument in short messages. Many documents written logically with evidence are necessary for robust policies. Policies are decided under parliamentary democracy.

There are several ways to address social issues and public policy issues, if we refer to the West. Creating and disclosing information, compiling and inheriting/sharing knowledge and experience, improving administrative accountability (responsibility for results), making the audit systems of corporations, corporate groups, administrative agencies, and governments function, for example, mutual checks and balances and management/safety improvements by corporate groups such as external directors, auditors, INPO (Institute of Nuclear Power Operations in the United States), audits and budget amendments on administrative results and plans by the Audit Bureau, Administrative Inspection Bureau, and parliament. In parliamentary democracy, while clarifying the responsibilities and roles of each organization and individual, these should function, but unfortunately, in Japan, the dependence of the people and corporations on the country is large, and audits are not functioning. I feel that the awareness of the cost to the people is also weak in Japan.

The question of how safe is safe enough can be addressed with the concept of risk-benefit in the use of technology. To repeat, in the regulation of nuclear power plants in the United States, indicators for this are determined and used. That is, the relationship between the cost necessary for safety improvement and the amount of radiation exposure that can be reduced by it is defined. Probabilistic risk assessment methods have been developed and used. It is said that if there were anti-terrorism facilities adopted by American nuclear power plants in Japan, the core meltdown of the TEPCO Fukushima accident could have been prevented. The anti-terrorism facilities installed by American nuclear power plants are cost-effective. Here too, the concept of cost-benefit is used. The cost of nuclear power generation in the United States is cheap.

In Japan, I have heard things like "Are you measuring human life with money?" In Japan, many people seem to be insensitive to electricity charges. Power companies, before the liberalization of electricity, were regional monopolies, so they were not in an environment to compete with cheap power supply. They were careful about stable power supply though. From the electric company that manufacture nuclear power plants, I have heard when I was a university professor that the electricity bill for its factories is high, and it is difficult to compete at overseas, but the product, the nuclear power plant has a predetermined power company as a customer, and the nuclear power business is not in a free competition environment in Japan, so the electric companies could not raise their voices. In Japan, isn't it necessary to clarify the responsibilities of organizations and individuals, reduce dependence on the country, and increase cost awareness?

In Japan, neither the nuclear industry nor the nuclear regulatory side has an appropriate concept of safety. There is no concept of cost-effectiveness. Absolute safety can never be achieved, and considering cost-effectiveness, it generates a lot of waste and complicates the equipment, which requires manpower for maintenance and creates unsafe areas. In the past in Japan, risks outside of nuclear technology, such as large tsunamis, were overlooked. The same was true for counter-terrorism measures.

Safety should not be abstracted and considered probabilistically. It is necessary to think concretely from actual data and take measures. Probabilistic risk assessment is a practical technique. It involves evaluating with actual data, considering the reflection of deterministic design, looking at costs, and redoing the tedious and dirty work. If this is not done or cannot be done, and if one is immersed in the atmosphere of risk research, isn't it the same as before the TEPCO accident? This is a lesson in nuclear safety in Japan, and I would like to caution Japanese nuclear safety experts. If the accumulation of objective actual data is insufficient, it is better to downplay the probability of events caused by the plant (internal events). In any case, *the concept of risk benefit and its indicators are necessary for safety and regulation in Japan.*

Exercise

1. Summarize the outline of the stress tests conducted in Europe on operating nuclear power plants after the TEPCO Fukushima accident.
2. Summarize the considerations and measures taken by the U.S. Nuclear Regulatory Commission to learn from the lessons of the TEPCO Fukushima accident.
3. Summarize the health effects of radiation exposure.
4. Summarize the method of evaluating the safety of food. Compare with the method of evaluating the health effects of radiation exposure.
5. Conduct a risk assessment of the radiation-contaminated food from the TEPCO Fukushima accident, and analyze the cost-benefit of regulations and measures.
6. Investigate data on disaster-related deaths associated with evacuation during natural disasters such as hurricanes and wildfires in the United States.
7. Investigate the change in the number of births in Greece and Italy immediately after the Chernobyl accident, and quantitatively compare the risk avoided by abortion during pregnancy.
8. Summarize the changes in evacuation zones in Russia and Ukraine after the Chernobyl accident, and the government's compensation and support for the victims of the accident.
9. Organize and compare the key points of the U.S. Emergency Nuclear Protection Guide (PAG) and Japan's Nuclear Disaster Response Guidelines (including the Basic Disaster Prevention Plan), and discuss.
10. "On the New Basic Concept of Nuclear Disaster Prevention in Our Country—Commentary for Nuclear Disaster Prevention Practitioners—JAEA-Review Compare the "thinking" described in "2013-015" and the U.S. PAG manual, summarize the differences between the two documents, and consider the causes.
11. Summarize the Fukushima Prefecture's requests to the government regarding the decontamination and recovery from the TEPCO Fukushima accident, and the government's subsequent measures.
12. Summarize the findings and lessons from the Great Hanshin-Awaji Earthquake. Consider the man-made and social factors of the earthquake, and examine whether these lessons have been applied to subsequent natural disasters.
13. Read Paul Slovic's "Perception of Risk", Science, Vol. 236, pp. 280–285 (1987) and summarize the key points.

14. Read the U.S. Nuclear Regulatory Commission (NRC)'s "Effective Risk Communication" (NUREG/BR-0308) and summarize the key points.
15. Investigate and summarize the UK government's response to mad cow disease (BSE) since the 1980s and subsequent improvements in crisis management.
16. Investigate and summarize the responses and recommendations of the UK, US, and French governments to their own citizens immediately after the TEPCO Fukushima accident.
17. Read the U.S. NRC's "Perspectives on Reactor Safety", NUREG/CR-6042, Rev.2/SAND 93–0971, March 2002 and summarize the overview.
18. Summarize the history and overview of science communication in the UK.
19. Investigate and summarize the dialogue activities conducted by the UK's BNFL in the early 2000s.
20. Read English literature on stakeholder dialogue and summarize the key points.
21. Select several newspapers (national papers) and compare articles on nuclear power generation around 2010, known as the nuclear renaissance, and after the TEPCO Fukushima accident (for example, editorials). There is no need to discuss differences in opinions between newspapers or the quality of articles.
22. Summarize the key points of the reports from the National Diet Accident Investigation Committee and the Government Accident Investigation Committee on the TEPCO Fukushima accident. Summarize the improvements made by each ministry and power company to apply the lessons of the accident.
23. Criticality accidents occurred frequently in the United States during the early stages of nuclear development. Summarize these incidents. Describe the activities undertaken by U.S. nuclear-related organizations (Department of Energy and National Laboratories, American Nuclear Society) to improve based on these lessons.
24. Summarize the overview and lessons of the JCO criticality accident that occurred in Tokai Village, Ibaraki Prefecture in March 1999. If there have been activities by nuclear-related organizations (including the Japan Atomic Energy Society) to apply these lessons, describe them.
25. Read the report of the U.S. Challenger explosion investigation committee (Rogers Commission), and summarize the key points of the accident and the lessons learned. Also mention Feynman's investigation and opinion. Discuss the improvement activities carried out afterwards to make use of the lessons learned.
26. Investigate and summarize the public deaths due to accidents (including those caused by natural disasters) and the deaths of workers associated with construction and operation, including those associated with fuel mining and transportation, for hydropower, coal-fired power, oil-fired power, natural gas-fired power, nuclear power, solar power, and wind power. The International Commission on Radiological Protection has stated in Report 103 that "it should be avoided to calculate the number of cancer deaths based on the collective dose in the case of extremely low radiation doses", so it is not necessary to include the number of deaths calculated in this way.

27. Summarize the historical background of the public health impact of air pollution caused by exhaust from the use of fossil fuel energy.
28. Compare the change in the capacity factor (operation rate) of Japan's nuclear power plants since the 1990s with other countries that mainly use light water reactors. Consider the reasons why the capacity factor of Japan's nuclear power plants is low.
29. The release of treated water containing tritium from the TEPCO Fukushima Daiichi Nuclear Power Plant into the sea has attracted public attention. Investigate the amount of tritium in the natural world and its origin. Search for and summarize literature on the safety of tritium.
30. Summarize the production amount, half-life, and annual change of the main nuclides of environmental radioactive substances produced in atmospheric nuclear tests. Compare this with the amount and annual change of long-lived fission products released into the environment in the TEPCO Fukushima accident.

References

Batandjieva-Metcalf B (2020) The UNSCEAR 2020 report on Fukushima: implications of information published since the UNSCEAR 2013 report. In: Japan's side event at the 65th IAEA General Conference, September 20. https://www.meti.go.jp/english/earthquake/nuclear/decommissioning/pdf/05_unscear_sideevent.pdf. Accessed June 29, 2024

Bier VM (1999) Summary of the state of the art: risk communication to the public, August. https://www.nrc.gov/docs/ML0116/ML011690134.pdf. Accessed June 29, 2024

Bortz II WM (1984). The disuse syndrome. West J Med 141(5):691–694. https://www.ncbi.nlm.nih.gov/pmc/articles/PMC1011199/pdf/westjmed00183-0115.pdf. Accessed June 29, 2024

Brook BW et al (2014) Why nuclear energy is sustainable and has to be part of the energy mix. Sustain Mat Technol 1–2:8–16. https://doi.org/10.1016/j.susmat.2014.11.001. https://www.sciencedirect.com/science/article/pii/S2214993714000050#aep-article-footnote-id1. Accessed June 29, 2024

Carbon MW (1997). Nuclear power: villain or victim? our most misunderstood source of electricity, Pebble Beach Pub., 1997, nuclear power: is it a perpetrator or a victim? the most misunderstood method of power generation. Energy Information Research Institute, translated by the Energy Information Research Association, 1998

Castle NG (2008) Nursing home evacuation plans. Am J Public Health. 98(7):1235–1240. https://doi.org/10.2105/AJPH.2006.107532. https://www.ncbi.nlm.nih.gov/pmc/articles/PMC2424080/. Accessed June 29, 2024

Cologne JB, Preston DL (1999). Life span of atomic bomb survivors. Radiation Research Report (RR), 8–99 (in Japanese). https://www.rerf.or.jp/library/archives/scientific_pub/rerf_report/rr1999/rr9908/. Accessed June 29, 2024

Cologne JB, Preston DL (2000) Longevity of atomic-bomb survivors. Lancet 356:303–307. https://doi.org/10.1016/S0140-6736(00)02506-X

Disaster Prevention White Paper (2020) 2020 Disaster prevention white paper part 1 chapter 2 status of efforts related to nuclear disasters. Cabinet Office (in Japanese). https://www.bousai.go.jp/kaigirep/hakusho/r02/honbun/index.html. Accessed June 29, 2024

Ebisawa K (2021) Nuclear disaster prevention focused on the risk of natural disasters (2) nuclear disaster prevention system against natural disasters. Japan Atomic Energy Society 2021 Spring Annual Meeting, Abstracts, Risk Division Session 3FPL02 (in Japanese)

EPA (2017) PAG manual: protective action guides and planning guidance for radiological incidents, EPA-400/R-17/001, January. Environmental Protection Agency. https://www.epa.gov/sites/default/files/2017-01/documents/epa_pag_manual_final_revisions_01-11-2017_cover_disclaimer_8.pdf. Accessed June 29, 2024

Forbes (2012) How deadly is your kilowatt? we rank the killer energy sources. https://www.forbes.com/sites/jamesconca/2012/06/10/energys-deathprint-a-price-always-paid/?sh=51d5bf8e709b. Accessed June 29, 2024

Fukushima Prefecture (2011a) Environmental radiation measurement results of nuclear power plants, March 11 to 31 (in Japanese). https://www.pref.fukushima.lg.jp/sec_file/monitoring/etc/post3-20120921.pdf. Accessed June 29, 2024

Fukushima Prefecture (2011b) Reference material: changes in the spatial dose rate at each point, about the measurement results of the spatial dose rate measured at the monitoring post from March 11 to March 31 (after the Great East Japan Earthquake) (in Japanese). https://www.pref.fukushima.lg.jp/sec_file/monitoring/etc/post4-20120921.pdf. Accessed June 29, 2024

Fukushima Prefecture (2023) About the transition of the evacuation zone—commentary—Fukushima reconstruction information portal site, updated February 16 (in Japanese). https://www.pref.fukushima.lg.jp/site/portal/cat01-more.html. Accessed June 29, 2024

Gauntt R et al (2012) Fukushima Daiichi accident study (status as of April 2012). SAND2012-6173, July, Sandia National Laboratories. https://digital.library.unt.edu/ark:/67531/metadc841123/m2/1/high_res_d/1055601.pdf. *Accessed June 29, 2024*

Government Accident Investigation Report (2011) Accident investigation and verification committee at Tokyo electric power company's Fukushima nuclear power plant, interim report, V Mainly dealing with the accident at the Fukushima Daiichi Nuclear Power Plant from outside the power plant, December 26. Cabinet Secretariat, 313 pages (in Japanese). https://www.cas.go.jp/jp/seisaku/icanps/111226Honbun5Shou.pdf. Accessed June 29, 2024

Grimes RW, Chamberlain Y, and Oku A (2014) The UK response to Fukushima and Anglo-Japanese relations, science and diplomacy, June. https://www.sciencediplomacy.org/sites/default/files/the_uk_response_to_fukushima_and_anglo-japanese_relations_science__diplomacy.pdf. Accessed June 29, 2024

Grimston M (2014) Public understanding of nuclear energy: It's not (just) about the science, JAIF Annual Meeting, Tokyo, April 15. https://www.jaif.or.jp/cms_admin/wp-content/uploads/understanding/annual/47th/47-s1_grimston-e.pdf. Accessed June 29, 2024

Grimston M (2016) The paralysis in energy decision making. Whittles Publishing 2016:412–416

Hori K (2018) Introduction to facilitation, 2nd edn. Nikkei Bunko, Nikkei Publishing Inc. (in Japanese)

IAEA (1996) Ten years after Chernobyl: what do we really know? Based on the proceedings of the IAEA/WHO/EC International Conference, Vienna, April, INIS-XA-001, IAEA, p 8. https://inis.iaea.org/collection/NCLCollectionStore/_Public/28/058/28058918.pdf?r=1. Accessed June 29, 2024

IAEA (2014) Communication and stakeholder involvement in environmental remediation projects. IAEA Nuclear energy series No. NW-T-3.5. https://www.iaea.org/publications/10494/communication-and-stakeholder-involvement-in-environmental-remediation-projects. Accessed June 29, 2024

IAEA (2022) Communicating with the public and other interested parties. Nuclear Safety and Security, IAEA. https://www.iaea.org/topics/communicating-with-stakeholders-and-the-public. Accessed June 29, 2024

ICRP (2020) Radiological protection of people and the environment in the event of a large nuclear accident, Update of ICRP publication 109 and 111, ICRP publication 146, Ann. ICRP 49(4):118. https://doi.org/10.1177/ANIB_49_4. Accessed June 29, 2024

JAEA (2014). About the emergency zone (PAZ and UPZ), July 2014, nuclear disaster prevention information. Nuclear Emergency Support and Training Center, Japan Atomic Energy Agency (in Japanese). https://www.jaea.go.jp/04/shien/research/EP017.html. Accessed June 29, 2024

Japan Federation of Bar Associations (2018) Opinion seeking accumulation, analysis, and disclosure of disaster-related death cases, Japan Federation of Bar Associations, August 23 (in Japanese). https://www.nichibenren.or.jp/document/opinion/year/2018/180823_3.html. Accessed June 29, 2024

Kai M (1999) Radiation risk assessment and its uncertainty. Nuclear Backend Res 6(1):Table 1 (in Japanese). https://nuce.aesj.or.jp/jnuce/vol6/Jnuce-Vol6-1-p7-12.pdf. Accessed June 29, 2024

Kinase S (2015a) Development of a predictive model for spatial dose rate distribution, comparison and verification with measured data, Part 2 (in Japanese). https://fukushima.jaea.go.jp/fukushima/try/pdf/pdf08/Part2.pdf. Accessed June 29, 2024

Kinase S (2015b) Development of a predictive model for spatial dose rate distribution part 3 (in Japanese). https://radioactivity.nra.go.jp/cont/ja/docs/reps/rad-dist/aggregate-results-report/2013/part3_sum.pdf. Accessed June 29, 2024

Kinoshita T (2008) Reconsidering risk communication—towards the construction of integrated risk communication. J Japan Risk Res Soc 18(2):3–22, 19(1):3–24 (2009) (in Japanese)

Kinoshita T (2009) Symposium report "risk and risk perception" life and bio plaza 21. Lecture notes by Professor Emeritus Tomio Kinoshita, Kyoto University, June 18 (in Japanese). https://www.life-bio.or.jp/topics/topics378.html. Accessed June 29, 2024

Kinoshita T (2016) The thought and technique of risk communication: techniques of mutual consideration and trust. Nakanishiya Publishing, September (in Japanese)

Kurokawa K (2016) Prisoner of regulation. Groupthink Will Destroy Japan, Kodansha, March (in Japanese)

McGlynn G (2020). An introduction to the BNFL National Stakeholder Dialogue, The Beam nuclear and social research network. The University of Manchester, Dalton Nuclear Institute. https://www.mub.eps.manchester.ac.uk/thebeam/an-introduction-to-the-bnfl-national-stakeholder-dialogue/. Accessed June 29, 2024

MEXT (2012) Announcement of estimated cumulative dose from March 11, 2011 to March 11, 2012 at continuously measured points. Ministry of Education, Culture, Sports, Science and Technology, March 21 (in Japanese). https://radioactivity.nra.go.jp/cont/ja/results/air-dose/estimated-map/1750_032118r.pdf. Accessed June 29, 2024

MEXT (2013) Compensation amount for nuclear damage in model households. Ministry of Education, Culture, Sports, Science and Technology, Nuclear Damage Compensation Dispute Resolution Committee (36th) Distributed Materials Reference 1, October 25 (in Japanese). https://warp.ndl.go.jp/info:ndljp/pid/11373293/www.mext.go.jp/b_menu/shingi/chousa/kaihatu/016/shiryo/__icsFiles/afieldfile/2013/10/25/1340807_6.pdf. Accessed June 29, 2024

Ministry of the Environment (2022) Booklet to provide basic information regarding health effects of radiation, increase in induced abortions in Europe—Chernobyl nuclear accident, 4th edn. Ministry of the Environment. https://www.env.go.jp/en/chemi/rhm/basic-info/1st/03-08-12.html. Accessed June 29, 2024

Ministry of Health, Labour and Welfare (2011) Let's be aware of inactive lifestyle diseases, manual for preventing decline in life functions. Ministry of Health, Labour and Welfare, Bureau of Elderly Health, March 29 (in Japanese). https://www.wam.go.jp/gyoseiShiryou-files/resources/29bd9514-ca47-4656-a3ca-801424184afb/vol.185.pdf. Accessed June 29, 2024

Nagataki S (2011) Distinguishing between science (scientific facts) and policy (approaches to response), September 29. Prime Minister's Office Homepage, TEPCO Fukushima Nuclear Power Plant/Radiation-related Information Nuclear Disaster Expert Group (in Japanese). https://www.kantei.go.jp/saigai/senmonka_g16.html. Accessed June 29, 2024

Nakanishi TM, Tanoi K (eds) (2013) Agricultural implications of the Fukushima nuclear accident. Springer. https://doi.org/10.1007/978-4-431-54328-2. Accessed June 29, 2024

Nakanishi TM, Tanoi K (eds) (2016) Agricultural implications of the Fukushima nuclear accident, The first three years. Springer. https://doi.org/10.1007/978-4-431-55828-6. Accessed June 29, 2024

Nakanishi TM, O'Brien M, Tanoi K (eds) (2019) Agricultural implications of the Fukushima nuclear accident (III) after 7 years. Springer. https://doi.org/10.1007/978-981-13-3218-0. Accessed June 29, 2024

National Academies (1989) Improving risk communication. National Academy Press. https://nap.nationalacademies.org/catalog/1189/improving-risk-communication. Accessed June 29, 2024

National Police Agency (2021) Police activities and damage situation of the Great East Japan Earthquake in 2011, March 10 (in Japanese). https://www.npa.go.jp/news/other/earthquake2011/pdf/higaijokyo.pdf. Accessed June 29, 2024

NEA (2020) Optimisation: rethinking the art of reasonable, workshop summary report, Lisbon, Portugal, 14–15 January. NEA/CRPPH/R(2020)2. https://www.oecd-nea.org/upload/docs/application/pdf/2021-10/nea_crpph_r_2020_2.pdf. Accessed June 29, 2024

NRA (2022) Guidelines for nuclear disaster response. Nuclear Regulation Authority, April 6 (in Japanese). https://www.nra.go.jp/data/000387310.pdf. Accessed June 29, 2024

NRC (2002) Perspectives on reactor safety, NUREG/CR-6042, Rev.2/SAND 93-0971, March. USNRC. https://www.nrc.gov/docs/ML0912/ML091250169.pdf. Accessed June 29, 2024

NRC (2004) Effective risk communication (NUREG/BR-0308), US NRC, January. https://www.nrc.gov/reading-rm/doc-collections/nuregs/brochures/br0308/index.html. Accessed June 29, 2024

Nuclear Emergency Response Headquarters (2015) Requirements for lifting evacuation orders, June 12. Decisions of the Nuclear Disaster Response Headquarters and Cabinet (excerpt) Prime Minister's Office (in Japanese). https://www.kantei.go.jp/saigai/pdf/hinan_youken.pdf. Accessed June 29, 2024

Ochi S (2015a) Potential health and reputation damage caused by radiation education. Energy Rev, 7–10 (in Japanese)

Ochi S (2015b) Nuclear accidents from the perspective of health, materials for the 21st regular meeting of the atomic energy commission, May 19 (in Japanese). https://www.aec.go.jp/kaigi/teirei/2015/siryo21/1_haifu.pdf. Accessed June 29, 2024

Oka T (2012) Risk benefit analysis of radioactive contaminated food regulation. In: Environmental Economic Policy Society 2012 Conference, September (in Japanese). https://www.s.fpu.ac.jp/oka/seeps2012slide.pdf. Accessed June 29, 2024

Oka T (2015) Risk assessment and cost benefit analysis of radioactive contaminated food regulation and measures, structure of rumor damage and regeneration of food and agriculture in the 4th year of nuclear accident Koriyama City Office March 14 (in Japanese). https://www.s.fpu.ac.jp/oka/furesymposium_oka.pdf. Accessed June 29, 2024. Accessed June 29, 2024

Oka Y (2019). Risk communication and public trust in nuclear safety, Atomic Energy Commission Email Magazine, Issue 273, July 26 (in Japanese). https://www.aec.go.jp/mailmagagine/backnumber/2019-0273.html. Accessed June 29, 2024

Oka, Y. (2022). Risks and benefits of evacuation in TEPCO's Fukushima Daiichi nuclear power station accident. Prog Nuclear Energy 148, June. Elsevier. https://doi.org/10.1016/j.pnucene.2022.104222, https://www.sciencedirect.com/science/article/pii/S0149197022001007. Accessed June 29, 2024

Oka Y (2023) Challenges of nuclear power generation in Japan. Lecture materials for the University of Tokyo's Professional Nuclear School, November 28 (in Japanese). http://www.tokai.t.u-tokyo.ac.jp/docs/231128%20%E6%97%A5%E6%9C%AC%E3%81%AE%E5%8E%9F%E5%AD%90%E5%8A%9B%E7%99%BA%E9%9B%BB%E3%81%AE%E8%AA%B2%E9%A1%8C.pdf. Accessed June 29

Oka Y, Bittermann D (2015) Chapter 12, implications and lessons for advanced reactor design and operation, reflections on the Fukushima Daiichi Nuclear Accident. Springer. https://doi.org/10.1007/978-3-319-12090-4_12. Accessed June 29, 2024

Okamoto K (1992). Introduction to risk psychology. Science-Sha (in Japanese)

Reconstruction Agency (2020a) Transition of the number of evacuees (as of December 25, 2020), number of evacuees nationwide—efforts towards reconstruction. Reconstruction Agency (in Japanese). https://www.reconstruction.go.jp/topics/main-cat2/sub-cat2-1/20201225_kouhou2.pdf. Accessed June 29, 2024

Reconstruction Agency (2020b) Number of deaths related to the Great East Japan earthquake (survey results as of September 30), December 25. Reconstruction Agency (in Japanese). https://www.reconstruction.go.jp/topics/main-cat2/sub-cat2-6/20201225_kanrenshi.pdf. Accessed June 29, 2024

Reconstruction Agency (2021) Number of evacuees nationwide. Reconstruction Agency, March 30 (in Japanese). https://www.reconstruction.go.jp/topics/main-cat2/sub-cat2-1/20210330_kouhou1.pdf. Accessed June 29, 2024

Robinson SJ (2018) Public & stakeholder dialog, 9th Regular Meeting of Japan Atomic Energy Commission, Reference Material 1-1-2, March 6. https://www.aec.go.jp/kaigi/teirei/2018/siryo09/1-1-2_sankou.pdf. Accessed June 29, 2024

Royal Society (1985) The public understanding of science. The Royal Society. https://royalsociety.org/-/media/policy/publications/1985/10700.pdf. Accessed June 29, 2024

Royal Society (2006) Science Communication. The Royal Society, UK, p 14. https://royalsociety.org/-/media/policy/publications/2006/1111111395.pdf. Accessed June 29, 2024

Slovic P (1987) Perception of risk. Science 236:280–285. https://www.researchgate.net/publication/271767726_Perception_of_Risk. Accessed June 29, 2024

Special Contents (2023) Here's more about the 6th strategic energy plan nuclear power generation (part 1) enhancement of safety for restart and R&D for advanced reactors. ANRE. https://www.enecho.meti.go.jp/en/category/special/article/detail_176.html. Accessed June 29, 2024

Tanaka S (2016) Nuclear disaster response guidelines and new regulatory standards, December, 9, 13 pp (in Japanese). https://www.nra.go.jp/data/000172848.pdf. Accessed June 29, 2024

UNSCEAR (2013) UNSCEAR 2013 report volume 1, report to the general assembly scientific Annex A: levels and effects of radiation exposure due to the nuclear accident after the 2011 Great East Japan Earthquake and Tsunami, United Nations, p 249, E15, E16. https://www.unscear.org/unscear/uploads/documents/unscear-reports/UNSCEAR_2013_Report_Vol.I.pdf. Accessed June 29, 2024

UNSCEAR (2020/2021) UNSCEAR 2020/2021 report, volume II, scientific annex B: levels and effects of radiation exposure due to the accident at the Fukushima Daiichi Nuclear Power Station: implications of information published after the UNSCEAR 2013 Report, United Nations New York. https://www.unscear.org/unscear/uploads/documents/unscear-reports/UNSCEAR_2020_21_Report_Vol.II-CORR.pdf. Accessed June 29, 2024

WHO (2006) Health effects of the Chernobyl accident and special health care programmes, Report of The UN Chernobyl forum Expert group of Health, Bennet B, Repacholi M, Carr Z (eds). WHO. https://www.crpr-su.se/Chernobyl%20report%20WHO%202006.pdf. Accessed June 29, 2024

Chapter 3
Nuclear Facility Decommissioning and Radioactive Waste

This chapter first discusses the circular (recycling-based) society and waste. Recycling and disposal of waste are methods to realize a circular society. Radioactive waste is a form of waste (industrial waste). In the decommissioning of nuclear facilities, a large amount of industrial waste and a small amount of radioactive waste are generated. Decommissioning and waste disposal should be considered and implemented as a whole. The decommissioning of nuclear facilities and waste disposal in each country are introduced.

3.1 Circular Society and Waste

Before discussing the decommissioning of nuclear facilities and radioactive waste, let's talk about the circular society and waste.

Our lives and social activities utilize resources (materials), producing and consuming products, energy, and food. Waste is also generated. Figure 3.1a, b shows the material flow in Japan for the fiscal year 2000 and 2020, respectively. Resources are not only domestic and foreign, but also include those that are recycled by processing generated waste. The products made are exported, as well as accumulated domestically as durable goods such as civil engineering structures, buildings, and automobiles. Resources are used in power production, transportation, steelmaking, cement manufacturing, etc. These are shown as energy consumption and industrial process emissions. There are also things that are consumed as food and used as fertilizers on farmland. The generated waste is reduced by intermediate processing such as incineration, then finally disposed of in forms such as landfill, naturally returned to farmland as rice straw and livestock manure, or recycled as resources. The amount of recycling has been increasing year by year, and the final disposal amount in fiscal 2017 is less than a quarter of that in fiscal 2000. It should be

Y. Oka, *Nuclear Power and Society*, Nuclear Science and Technology,
https://doi.org/10.1007/978-981-97-8951-1_3

noted that the reduction (weight reduction) by incineration is contributed by carbon-containing waste, and the carbon released into the atmosphere by incineration is not included in this material flow diagram.

Domestic resources are almost 90% non-metallic minerals such as rocks, gravel, and limestone. The reason why the resource input amount in 2017 has decreased compared to 2000 is mainly due to the decrease in this non-metallic mineral amount. The amount of recycling consists of non-metallics such as rubble and mine waste, biomass such as waste paper and manure, metals such as iron scrap and metal scrap, and fossil fuels such as waste plastics. Non-metallics account for about 60% of the total, with rubble accounting for about 25%.

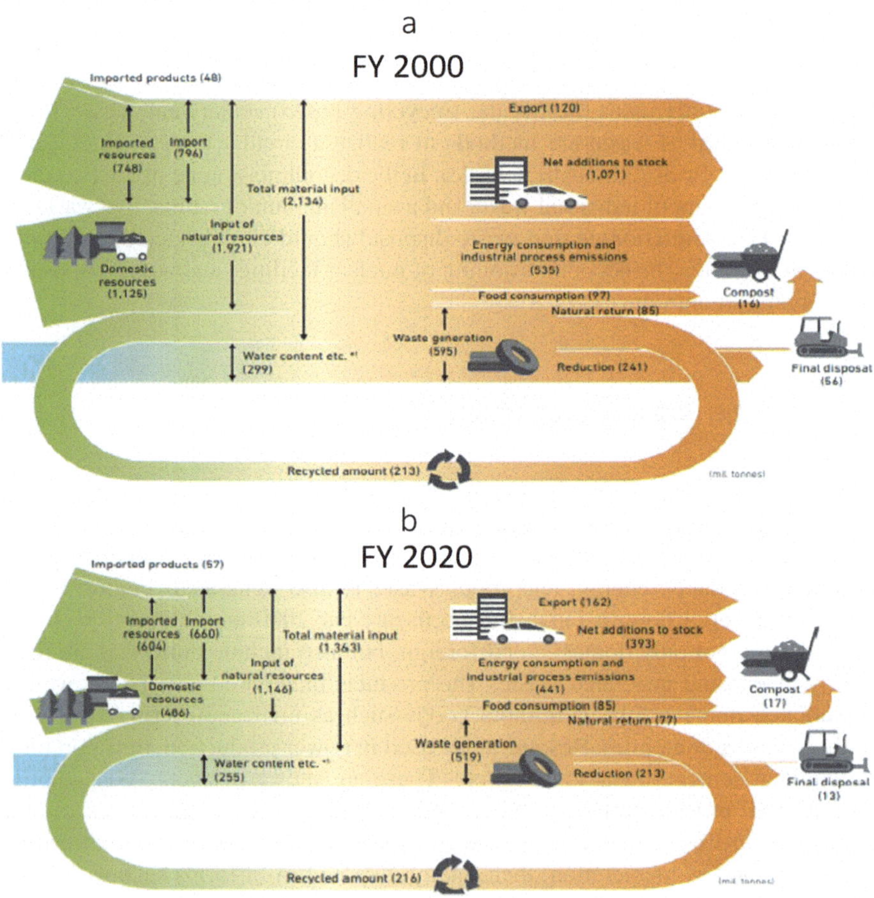

Fig. 3.1 **a** Material flow in Japan, FY 2000 **b** Material flow in Japan, FY 2020. *Source* Annual report on the environment in Japan 2023 Ministry of Environment https://www.env.go.jp/content/000175163.pdf

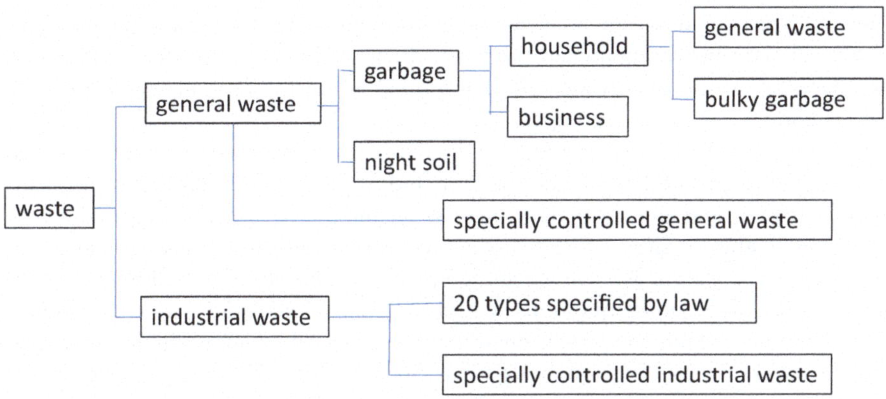

Fig. 3.2 Types of waste in Japan. *Source* MOE, Gov. Japan

The Basic Plan for Establishing a Recycling Society in Japan, established in 2000, sets numerical targets for resource productivity, recycling rate, and final disposal amount as indicators of material flow, with the slogan of Reduce, Reuse, Recycle. Resource productivity is an indicator of how effectively things are used, and is a value obtained by dividing the gross domestic product by the resource input amount. The recycling rate is a value obtained by dividing the recycling amount by the sum of the natural resource input amount and the recycling resource amount. The final disposal amount is the sum of the final disposal amounts of general waste and industrial waste. The recycling rate has numerical targets set at both the input and output sides.

Figure 3.2 shows the classification of waste in Japan. Waste is divided into industrial waste and general waste. General waste includes urine, household waste, and business waste generated from offices and restaurants.

The amount of general waste generated has decreased by more than 20% since around 2000, both in total and per capita. About 70% of general waste is reduced by intermediate processing such as incineration. The amount of resources is about 20%, and about 10% is finally disposed of.

Industrial waste is generated at about 10 times the rate of general waste, and its total volume has hardly changed in recent years. Industrial waste is reduced by about 45% through intermediate processing, about 53% is recycled, and the final disposal amount is about 3% of the total. A large proportion is recycled.

In terms of circular use, not only familiar items such as PET bottles and glass bottles, plastic containers, household appliances like air conditioners and televisions, automobiles and tires, computers, small batteries and mobile phones, but also construction waste and sewage sludge are being recycled. These are large in volume and have a high recycling rate. Sewage sludge is used as cement raw material, construction material, fertilizer, solid fuel, etc.

Both general waste and industrial waste that have the potential to cause harm to human health, such as explosiveness, toxicity, and infectiousness, are designated as specially managed waste. This includes PCB waste, waste mercury, waste asbestos, waste oil, waste acid, waste alkali, waste containing heavy metals and dioxins, and waste that may be contaminated with infectious pathogens. For these, handling, treatment, and disposal methods are determined according to their characteristics, and are handled by authorized operators.

General waste is the responsibility of the municipality to process, while industrial waste is the responsibility of the operator. Industrial waste is defined by government ordinance, and its intermediate processing and final disposal are carried out by authorized operators. General waste is waste other than industrial waste. Construction waste, such as concrete and iron generated from the demolition of buildings over 80 square meters, is legally required to be recycled and is handled by registered demolition contractors. Human waste is general waste, most of which is either discharged into the sewer or treated at human waste treatment facilities. Sludge generated at the end of sewage treatment is processed and recycled as industrial waste.

The remaining years of capacity at the final disposal site are about 21 years for general waste and about 16 years for industrial waste, and the remaining years have been slightly extended over the past 15 years. About 17% of municipalities nationwide do not have a final disposal site and are commissioning landfill to private final disposal sites. The number of illegal dumping cases of industrial waste has significantly decreased over the past 25 years, but it is still occurring, including improper treatment. The cross-border movement of hazardous waste was regulated by domestic laws about 30 years ago based on the Basel Convention. Both exports and imports are carried out for the purpose of metal recovery. Exports include lead scrap (lead batteries), coal ash, tin lead scrap, etc., and imports include scrap of electronic parts and batteries. The counterpart countries are neighboring countries in Asia (Ministry of the Environment 2020).

Japan began to generate a large amount of waste during the period of rapid economic growth from the 1960s. In Tokyo, in the mid-1960s, a problem arose over the construction of a waste disposal plant (incinerator) in Suginami Ward. At that time, the processing capacity could not keep up, and the generated waste was mainly landfilled. However, because the landfill sites were concentrated in Koto Ward, Koto Ward prevented the import of waste from Suginami Ward, and the term "Tokyo Garbage War" was born. A settlement was reached in 1974, a waste disposal plant was built in Suginami Ward, and it has been in operation since 1982. It was rebuilt in 2017 (Suginami Study Club 2014).

It is said that nine waste disposal facilities were planned in the surrounding areas of Tokyo in 1939, before World War II, but now there are waste disposal plants in most of the 23 wards of Tokyo. Some wards, such as Chiyoda Ward in central Tokyo, delegate processing to neighboring wards. Waste collection is the responsibility of the ward, and processing is carried out by the cleaning association composed of waste disposal plants in the 23 wards of Tokyo. Final disposal is the responsibility of Tokyo Metropolitan Government. Landfill disposal sites and bulky waste crushing facilities are located on the central breakwater in Koto Ward, and non-combustible waste crushing facilities are located on the central breakwater and Keihinjima in Ota Ward (Tokyo 23 Wards Cleaning Division Union 2023).

Both general waste and industrial waste were once major social issues, but now they are established in society, and *waste disposal is an important job for realizing a circular society*. It is noteworthy that new industries and employment have been created, mainly in industrial waste disposal. The venues and athletes' village for the Tokyo Olympics used land created by landfill. Even if it is called final disposal, it does not necessarily mean that the land will be unusable forever.

The concept of a circular society, with the slogan of 3R (Reduce, Reuse, Recycle), is expected to become increasingly important. For example, in the field of waste, from the perspective of reducing greenhouse gas emissions, the reuse and recycling of carbon-containing waste should progress. In future, as the decommissioning of nuclear power plants and research and development facilities progresses in Japan, the majority of the waste generated will be equivalent to industrial waste, not radioactive waste. In accordance with the slogan of 3R in a circular society, it is necessary to proceed with its treatment and disposal.

3.2 Decommissioning of Nuclear Facilities and Waste

Factories, buildings, equipment, and pipes that have completed their roles are dismantled, treated as industrial waste, and reused and disposed of. The same applies to nuclear facilities. When their role is completed, they need to stop operation and carry out decommissioning and waste treatment and disposal.

The decommissioning measures for nuclear facilities include the creation of plans such as dismantling, obtaining permits from regulatory authorities, removal and management of spent fuel, decontamination and dismantling of equipment and facilities contaminated with radioactive substances, storage and disposal of generated radioactive waste, disposal of other waste, and preparation for reuse of the facility or the location where the facility was located.

Decommissioning measures for nuclear facilities in Japan include decommissioning measures for nuclear power plants, decommissioning measures for research and development facilities such as research reactors, and decommissioning measures for Tokyo Electric Power Company's Fukushima Daiichi Nuclear Power Plant. In nuclear weapon countries such as the United States, the United Kingdom, and France, in addition to decommissioning measures for nuclear power plants and research and development facilities, there are decommissioning measures for nuclear development facilities used for the production of plutonium for nuclear weapons and decommissioned nuclear submarines.

Decommissioning measures for nuclear facilities are integral to the disposal of generated waste. If the destination of the waste generated by the decommissioning measures is not determined, it is impossible to smoothly proceed with the decommissioning measures. The destination does not necessarily have to be the final disposal site. Here, we will describe the decommissioning measures and the disposal of their waste for each country.

In Western countries that have led in nuclear development and use, decommissioning measures for nuclear facilities and disposal of radioactive waste are progressing. First, we will introduce the United States, the United Kingdom, France, Germany, etc., and then discuss Japan.

3.2.1 Decommissioning Measures and Waste Disposal of Nuclear Facilities in the United States

The decommissioning measures for nuclear facilities in the United States can be broadly divided into decommissioning measures for nuclear power plants, research reactors and nuclear fuel cycle-related facilities, and decommissioning measures for nuclear weapon development-related facilities. The former is the responsibility of the operator to implement, and they apply to the Nuclear Regulatory Commission and are subject to its regulations. The latter is the responsibility of the Department of Energy to implement, and the regulation is also the responsibility of the Department of Energy.

Decommissioning Measures for Nuclear Power Plants

More than 30 private nuclear power plants in the United States, including those with small outputs, have undergone decommissioning measures. There are also power plants that have stopped operation and are planning decommissioning measures. Power operators can choose either immediate dismantling, safe storage, or sealed management. Applications for decommissioning measures are made to the Nuclear Regulatory Commission.

Immediate Dismantling (DECON) is a method of decontaminating or dismantling the relevant parts of the facility, equipment, or structures immediately after the facility is closed to a level below the permissible level of radioactive contamination, and releasing them from nuclear regulation.

Safe Storage (SAFSTOR), also known as delayed dismantling, is a method of managing nuclear facilities until radioactive materials decay, and then dismantling and decontaminating them.

Enclosure management (ENTOMB) is a method of managing and monitoring facilities by covering them with durable materials like concrete until radioactive materials decay to acceptable levels. No nuclear power operators have yet applied this method to the Nuclear Regulatory Commission.

Immediate dismantling and safe storage can also be combined. The presence or absence of a waste disposal site also affects the choice of decommissioning methods. In the United States, it is required to complete decommissioning within 60 years after the power plant stops.

Nuclear power operators in the United States are required to secure the funds necessary for decommissioning and report the situation to the Nuclear Regulatory Commission every two years. It is estimated that decommissioning will require approximately 300–400 M$.

When an application for decommissioning is received from an operator, the Nuclear Regulatory Commission holds a public hearing near the facility scheduled for decommissioning. Public hearings are also held before an application for decommissioning completion is submitted. Public hearings are also held when a correction application for decommissioning is submitted. The Nuclear Regulatory Commission also holds meetings with the applicant operator, and except when confidential information is discussed, the public can participate.

In the United States, the geological disposal site for spent fuel is not proceeding as planned. Therefore, the Nuclear Regulatory Commission allows power operators to sell the land after decommissioning while maintaining the spent fuel storage facility. Independent spent fuel storage facilities are still subject to regulation by the Nuclear Regulatory Commission (NRC 2019).

In the United States, as of July 2019, 30 nuclear power facilities have applied for decommissioning, 3 have completed decommissioning, 7 have completed decommissioning leaving only independent spent fuel storage facilities, 8 are in immediate dismantling, and 12 are in safe storage. In addition, nine nuclear power plants have announced their transition to decommissioning. For example, the Yankee Row power plant, the first commercial nuclear power plant in the United States, started operation in 1960 and, after generating power stably for 32 years until 1992, completed decommissioning in 2007, leaving only an independent spent fuel storage facility, and the land where the power plant was located has been greened.

In the United States, there are private companies that specialize in decommissioning nuclear power plants. For example, the Three Mile Island Unit 2 (TMI-2), which suffered a core meltdown accident in 1979, had 99% of its melted core removed and by 1993 had transitioned to a state of monitored storage with the fuel removed. At the end of 2020, it was sold (ownership transferred for a fee) to a company specializing in decommissioning. The Nuclear Regulatory Commission has approved this company to carry out the decommissioning. The removed melted core has been transported to the Idaho National Laboratory.

Although the United States has not yet been able to operate the Yucca Mountain geological repository, which is a disposal site for spent fuel, it is advancing the decommissioning of nuclear power plants by creating independent spent fuel storage facilities, disposing of industrial waste at disposal sites in the states where the power plants are located, and disposing of low-level radioactive waste at private disposal sites in four locations nationwide. It is necessary to note that the United States has established conditions for decommissioning nuclear power plants as a private business.

The Nuclear Regulatory Commission is also responsible for regulating the decommissioning of facilities dealing with radioactive substances generated by activities other than nuclear weapons development, such as research reactors and uranium refining, enrichment, and processing, and other nuclear fuel cycle-related facilities. As the United States is a federal country, the Nuclear Regulatory Commission sets the rules for decommissioning, but the actual decommissioning is regulated by the states, with which the NRC has agreements. In states without an agreement, the local office of the Nuclear Regulatory Commission is in charge. The NRC directly oversees Indian reservations and military bases (NRC 2023).

Decommissioning of U.S. Government Nuclear Facilities

In the United States, large-scale nuclear weapons development has been carried out since World War II at national laboratories under the Department of Energy (DOE). As a result, millions of gallons of liquid radioactive waste, millions of cubic meters of radioactive waste, thousands of tons of spent fuel and special nuclear materials, and a large amount of contaminated soil and water have been generated. The Department of Energy's Environmental Management Division (DOE-EM) was established in 1989 to advance facility decommissioning, radioactive waste disposal, and environmental restoration.

In 1989, many facilities did not meet safety and environmental conditions. The Environmental Management Division entered into compliance agreements with the regulatory side and each site regarding the scope and schedule of cleanup work. In the first six years, the overall picture of the cleanup and situation understanding was carried out. Identification of the most urgent risks, development of infrastructure, processing, transportation, and disposal processes were also carried out. By the mid-1990s, the contamination situation was understood, the full picture of the cleanup was revealed, and the total cost was estimated to be $200 billion-$350 billion over a 75-year plan. A 10-year plan for each site has been made since 1996, and progress management is being carried out. Since 2000, cleanup has been accelerated by improvement activities (DOE 2014). The budget is about $6 billion annually, and a huge budget has been stably allocated over a long period. Documents setting annual priorities along with a 10-year strategic vision are created and published.

The targets for decommissioning and environmental recovery were initially spread across 31 states and 107 locations. The area was as vast as Rhode Island and Delaware combined, but over the past 30 years, many sites have been completed, and work is being carried out at the remaining 16 sites (DOE 2021, 2023a). Management of completed sites is carried out by the Department of Energy's Legacy Management Division (DOE 2023b).

In the United States, large-scale production and enrichment of plutonium for nuclear weapons were carried out. For example, at Hanford in Washington State, plutonium production for nuclear weapons by nuclear reactors and the extraction of plutonium by dissolving the removed fuel (called reprocessing) were carried out. The waste liquid after extracting plutonium was stored in tanks, but over a long period, the tanks corroded, leaked underground, and there was concern about outflow to the Columbia River. There were nine nuclear reactors used for plutonium production, the last of which was stopped after the end of the Cold War, and decommissioning was necessary. The uranium enrichment plant in Oak Ridge, Tennessee, was the largest building under one roof in the world. Construction and operation of a geological disposal site for transuranic elements such as plutonium were also necessary. The most budget-intensive and time-consuming tasks were the decommissioning of facilities involved in plutonium extraction and the treatment of waste liquid. As already mentioned, the regulation of the Department of Energy's facilities is not the

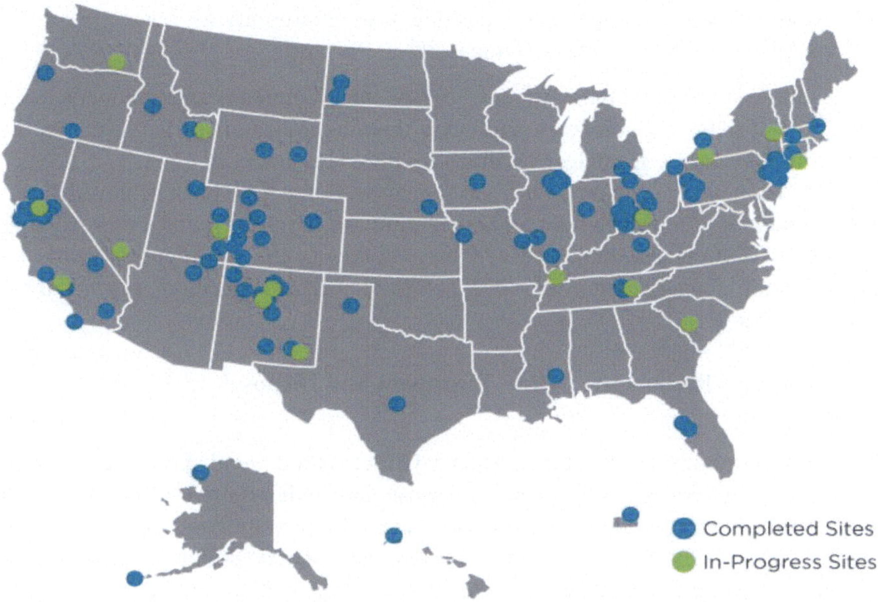

Fig. 3.3 Environmental management Sites of USDOE. *Source* DOE, EM sites map, September 5, 2019

responsibility of the Nuclear Regulatory Commission, but the responsibility of the Department of Energy's Safety Department.

The Department of Energy's decommissioning sites are shown in Fig. 3.3.

The main achievements of the past 30 years are summarized as follows.

- Cleanup of nuclear weapon manufacturing plants such as Rocky Flats in Colorado.
- Operation of WIPP (Waste Isolation Pilot Plant, New Mexico), a geological disposal site for transuranic elements generated by military use.
- Decontamination across 220 square kilometers along the Columbia River in Hanford, Washington.
- Long-term stabilization (safe storage) of six plutonium production reactors in Hanford.
- Dismantling of uranium enrichment plants such as K-25 in Oak Ridge, Tennessee.
- Construction and good operational record of tank waste treatment facilities at the Savannah River Site in South Carolina, and construction and start of operation of salt waste treatment facilities.
- Completion of the mixed waste treatment project at the Idaho National Laboratory and transportation of transuranic waste to WIPP.
- Construction and operation of six depleted uranium hexafluoride conversion plants in Kentucky and Ohio.
- Completion of waste solidification project in New York and dismantling of its facilities.

- Providing over 25,000 acres of land to the local community for reuse.
- Completion of decommissioning measures at 92 sites.

U.S. agencies compile their strategic visions into documents and announce them to the public, defining their work and using them as indicators of achievement. In the decommissioning of nuclear facilities by the Department of Energy, a 10-year strategic vision is created and progress is managed. Some have been achieved ahead of schedule and at a reduced cost. The decommissioning of France's reprocessing plant by ORANO is also managed with an overview of the entire project, with the company's top management involved. If achieved ahead of schedule, it is profitable for the company.

U.S. Waste and Radioactive Waste Treatment and Disposal
General Waste

The U.S. Environmental Protection Agency (EPA) is the federal agency responsible for the treatment and disposal of general waste and industrial waste in the United States. As the U.S. is a federal country, the actual administration is carried out by the states or counties. There are many private companies that deal with waste treatment and disposal. Regulations are carried out in accordance with the standards set by the EPA. The EPA explains waste and its recycling on its website (EPA 2020, 2023a). According to this, about half of municipal solid waste is landfilled, about one third is recycled, and about 12% of the total is incinerated for energy recovery. Although the amount generated has increased by 40% compared to 1990, the amount landfilled has not increased. As landfill sites are limited, efforts are being made to reduce waste and promote the 3Rs (Reduce, Reuse, Recycle) (EPA 2023b). State-by-state waste management reports are also created and can be viewed on the EPA's website (EPA 2023c). Approximately 90% of construction and demolition waste is generated at the time of demolition, and about 10% at the time of construction, with about three-quarters being recycled and the rest being disposed of in landfills (EPA 2023d). The treatment and disposal of non-radioactive waste generated in conjunction with the decommissioning of nuclear power plants may also be handled according to the concept of construction and demolition waste. In the United States, there are private companies that make waste treatment and disposal their business, and handbooks and guidelines on waste treatment and disposal for businesses and states/counties are created and published by the EPA (EPA 2023e). Furthermore, explanations for planning the treatment and disposal of waste generated by natural disasters such as hurricanes are also created and published (EPA 2023f). For hazardous waste, the EPA indicates its identification method and manages the entire process of generation, transportation, recycling, and treatment/disposal (EPA 2023g).

Radioactive Waste in the United States
Radioactive waste in the United States is classified into high-level radioactive waste, transuranic (TRU) waste, and low-level waste. High-level waste is spent fuel from reactors and the waste liquids and glassified bodies generated by its treatment. Both civilian and military generated wastes are intended to be disposed of in geological

repositories. The construction of geological repositories is the responsibility of the federal government, and one has been built at Yucca Mountain in Nevada, but it has not yet become operational.

Transuranic waste is waste containing transuranic elements (elements with atomic numbers larger than uranium), which in the United States was generated in the process of treating solutions containing plutonium for nuclear weapons development. Transuranic waste is disposed of at the WIPP (Waste Isolation Pilot Plant) operated by the Department of Energy (DOE) in New Mexico. WIPP is a disposal site built in an underground salt layer and started operations in 1999.

Low-level radioactive waste generated from civilian uses such as nuclear power generation is classified into four types: Class A, B, C, and greater than Class C, depending on the half-life and concentration of the radioactive nuclides contained. These low-level radioactive wastes are required to be disposed of within the state where they were generated or within the framework of interstate agreements. In Fig. 3.4 the main radioactive waste disposal sites in the United States are shown. The low-level disposal sites in Richland (Washington), Clive (Utah), Texas (Texas), and Barnwell (South Carolina) are all operated by private companies, and the low-level radioactive waste generated by the decommissioning of nuclear power plants in the United States is disposed of at these sites. Although not shown in the figure, the low-level radioactive waste generated at national laboratories and other facilities of the Department of Energy (DOE) is disposed of by the DOE itself at the sites of these national laboratories and other facilities (RWMC 2022).

U.S. Government Accountability Office's Report on Radioactive Waste

The U.S. Government Accountability Office (GAO) has issued reports on the Department of Energy's decommissioning, radioactive waste, and environmental cleanup

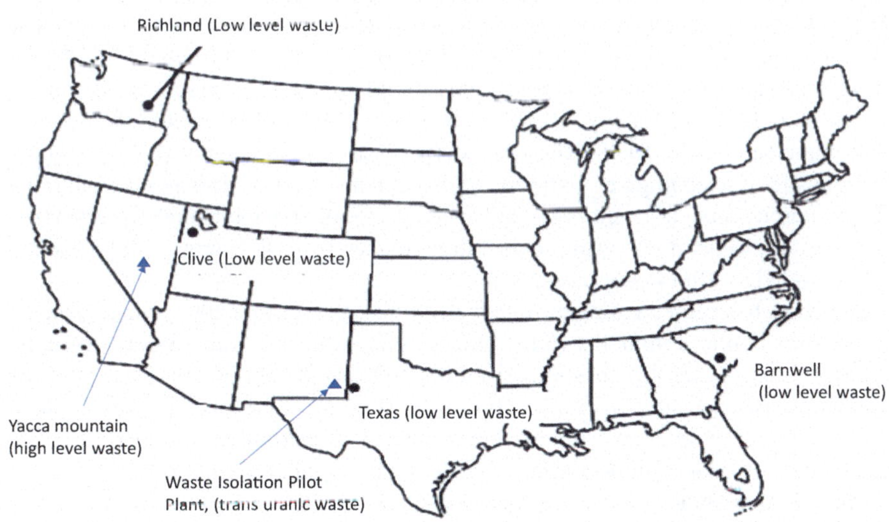

Fig. 3.4 Major radioactive waste disposal sites in the United States. *Source* US DOE

(GAO 2023). For example, GAO has made two recommendations to DOE related to the waste treatment plant under construction at Hanford. Five recommendations have been made for improving the cleanup work at the gas diffusion uranium enrichment plant. In addition to these, recommendations have been made for reducing plutonium and environmental recovery activities. The GAO's reports are referenced in the federal congress and reflected in the budgets and personnel costs of the responsible agencies.

3.2.2 Decommissioning and Waste Treatment and Disposal of Nuclear Facilities in the UK

Decommissioning of Nuclear Facilities

The development of civilian nuclear power in the UK started with the military demand for the production of plutonium for nuclear weapons. The power reactors in the UK were developed based on reactors that produce plutonium. Until around 1990, when economic liberalization took place, electricity in the UK was supplied by the state-owned Central Electricity Generating Board (CEGB). Not only nuclear power generation, but all nuclear facilities in the UK at that time belonged to the public sector, including the CEGB, British Nuclear Fuels Limited (BNFL), the United Kingdom Atomic Energy Authority (UKAEA), and the Ministry of Defence. As a result, the decommissioning of nuclear power plants and nuclear facilities in the UK is a major public debt, characterized by a large responsibility and involvement of the UK government, including its financial aspect.

The decommissioning of nuclear facilities in the UK has been carried out in a unified manner since 2005 by the Nuclear Decommissioning Authority (NDA). The NDA is responsible not only for decommissioning and site environmental recovery, but also for the management of spent fuel, uranium and plutonium and other nuclear materials, and the treatment and disposal of radioactive waste. Although the NDA is not a government agency, it is under the jurisdiction of the Department for Business, Energy & Industrial Strategy (BEIS). It is a public institution outside the government, managed by the BEIS. The Radioactive Waste Management (RWM), which is responsible for geological disposal, is also under the NDA. It is planned that the 15 commercial nuclear reactors (14 advanced gas-cooled reactors and 1 pressurized water reactor, some of which are in operation) owned by EDF Energy will be handed over to the NDA after decommissioning.

The NDA has been creating a strategy every five years since 2005, which is made public after public comments (NDA 2020a, 2021). In the draft report for public comments (NDA 2020a), questions from the NDA to the public are listed after the explanation of each item's strategy. From the strategy, a business plan is created, and semi-annual and annual reports are produced. Reports that succinctly explain the progress of long-term decommissioning are also available (NDA 2019a).

The UK does not yet have a geological disposal site for high-level radioactive waste, but it has infrastructure for general waste, industrial waste, and low-level radioactive waste disposal sites, and decommissioning is progressing at nuclear research and development facilities. Large-scale decommissioning, such as at Sellafield, and decommissioning of UK-type power reactors using graphite, are planned to be carried out over a long period. The cost is about 3 billion pounds per year. Figure 3.5 shows the decommissioning sites in the UK and the proportion of decommissioning costs at each site. Sellafield is the highest. The decommissioning cost of Dounreay at the northern tip of Scotland, where fast reactor development was carried out, is next. The geological disposal site is in the process of site selection, and 8.3% of the cost is allocated. Figure 3.6 shows the estimated amount of decommissioning costs over 100 years. The costs of Sellafield and other decommissioning are shown. The latter's costs are fluctuating because there are stages in the decommissioning of nuclear power plants in the UK (NDA 2019b, 2020b).

The UK's nuclear power plants do not use light water reactors, which are widely used in the world, such as in Japan and the US, but use graphite as a moderator (material that efficiently causes nuclear fission reactions) and pressurized carbon dioxide gas for cooling. In the UK, the operation of the first power plant began in 1956, and 26 MAGNOX-type power plants and 15 advanced gas-cooled reactors (AGR) were built. The UK has developed its own reactor technology using gas cooling and graphite, and it is commendable that it has overcome the challenges associated with graphite and carbon dioxide cooling and used it for long-term power generation.

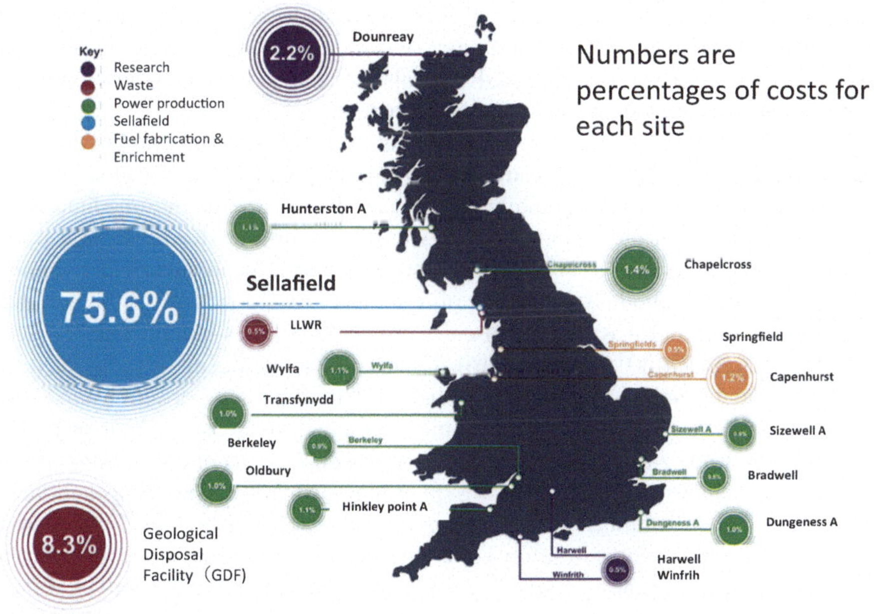

Fig. 3.5 UK nuclear decommissioning sites. *Source* Nuclear provision: the cost of cleaning up Britain's historic nuclear sites, July 4 2019

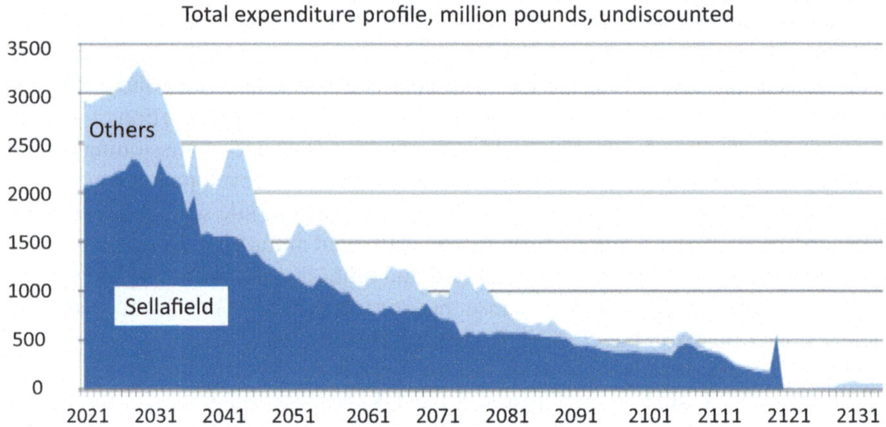

Fig. 3.6 Prospects for decommissioning costs of UK nuclear facilities. *Source* NDA Annual Report and Accounts 2019 to 2020, p. 52, July 2020

Hydropower plants have been used for a long time by replacing generators and water wheels. Light water reactors, including the structures inside the reactor vessel, can be repaired or replaced when they deteriorate, so they have no physical lifespan. However, graphite is a solid and deteriorates over long periods of operation. Since it is installed inside the reactor vessel, it is difficult to replace the graphite. For this reason, all MAGNOX types and three AGRs have already been shut down for economic reasons. The remaining AGRs are also scheduled to cease operation in the 2020s due to graphite degradation and other factors.

In graphite-moderated nuclear power plants, impurities in the graphite become radioactive due to operation, and some of them have long half-lives. Therefore, in their decommissioning measures, graphite needs to be disposed of as intermediate-level radioactive waste, but there is no disposal site for this in the UK yet. Furthermore, because graphite has a large volume, the amount of waste generated by decommissioning measures is about 10 times that of light water reactors, and it is thought that decommissioning and waste disposal costs are about 5 times higher than for light water reactors (IAEA 2010; WNA 2016). The dominant nuclide in the radiation dose of the irradiated graphite moderator is cobalt 60 with a half-life of 5.27 years, so it is thought that the dose of the graphite moderator will decrease by four orders of magnitude over 80 years (Holt 1999). For these reasons, the decommissioning of UK nuclear power plants involves removing spent fuel, dismantling buildings and facilities that can be decommissioned as non-radioactive waste or low-level radioactive waste, and taking measures to safely store the remaining intermediate-level radioactive waste, which is graphite, in the reactor vessels and reactor buildings. After that, safe storage is planned for about 60 years, and the dismantling of the reactor buildings will start about 80 years after permanent shutdown.

The decommissioning of the Magnox power station was carried out using the decommissioning of the Bradwell power station (Bradwell, located a few tens of kilometers northeast of London on the coast) as a precedent. Maintenance work for long-term management and technology development for collecting, adjusting, and boxing intermediate-level waste have been carried out, and the power station is in a maintenance state (NDA 2020c).

The decommissioning of UK nuclear facilities is overseen by the NDA, but the decommissioning of individual sites is undertaken by the PBO (Parent Body Organization), with the SLC (Site License Company) responsible for the work (see reference 4). The PBO is a consortium of private companies, and the SLC is a subcontractor that works under contract with the PBO, and is a local civil engineering and construction company that has been approved by the nuclear regulatory authority. At Sellafield, the PBO method has been used in the past, but due to its large scale and complexity, the NDA currently has the SLC for Sellafield as a subsidiary and is directly responsible (NDA 2023a). The PBO-SLC method is thought to function when there are fewer unknowns related to decommissioning measures, and when the work can be divided and the risks associated with decommissioning and radioactive waste treatment and disposal can be shared by contract.

Decommissioning at Sellafield

The UK built a nuclear facility in Sellafield, in the Cumbria region of northwest England, in the 1940s to produce plutonium and other nuclear weapon materials. Since then, the following facilities have been built at Sellafield.

1. A nuclear reactor called Windscale Piles that produces plutonium for nuclear weapons.
2. Calder Hall nuclear power station, the world's first commercial nuclear power station (started operation in 1956).
3. The prototype reactor of the world's first large-scale Advanced Gas-cooled Reactor (AGR), which started operation in 1963.
4. Storage pools for nuclear reactor fuel and its waste: built in the 1940s and 50s.
5. Nuclear fuel reprocessing plant (a facility that manufactures nuclear fuel for use in reactors).
6. Spent fuel reprocessing plant (a facility that performs chemical processing (reprocessing) to dissolve spent fuel and separate plutonium).
7. Radioactive waste storage facility.

The early facilities at Sellafield, some of which were built during the war, did not consider decommissioning or dismantling, and there are many challenges to overcome for decommissioning, such as the lack of records of what substances were used and the dense concentration of various facilities on a small site (NDA 2019b).

The Windscale Pile No.1 had a fire accident in 1957 and has been placed in a long-term monitoring and maintenance state. An endoscopic survey conducted 50 years later in 2007 provided data on fuel damage and radioactive waste generated by decommissioning. A strategy was decided to remove the fuel with a robotic arm from the top (Ervin 2008). The risk was evaluated to be smaller than initially thought, and the budgetary priority for decommissioning at Sellafield was lowered. The plan is to remove the fuel and put it into a maintenance state by 2030. In 2019, demolition work was carried out on the exhaust tower of Unit 1.

The Calder Hall power station No.2 stopped after 47 years of operation and is in the process of decommissioning. The fuel was removed in 2019.

The AGR prototype reactor No.3 stopped in 1981, and the development of decommissioning technology was carried out.

Concerning items 4, 6, and 7, the stabilization of radioactive waste generated in the process called nuclear fuel reprocessing, which extracts plutonium for military and civilian use, is one of the most difficult and important tasks in the decommissioning measures at Sellafield, and work is continuing.

At Sellafield, nuclear fuel reprocessing plants were built one after another to reprocess fuel taken out from the Windscale Piles, spent fuel from MAGNOX reactors, and even spent fuel from AGR and light water reactors of power companies in Europe and Japan. All of these are now in the process of decommissioning. The waste generated from their operation since the 1950s was thrown into deep pools called ponds at the time, and has accumulated at the bottom. There is also sludge (mud) and water containing radioactive substances. It is necessary to remove and stabilize these, and preparatory work is being carried out. Solid waste that does not need to be stored in water is thrown into concrete storage vaults called silos, six of which are full. Work has begun to remove this with robots and transfer it to containers for future underground disposal (NDA 2020d).

Decommissioning at Dounreay

At Dounreay, at the northern tip of Scotland, development of a fast breeder reactor was carried out. The experimental reactor (DFR) was operated from 1959 to 1977, and the prototype reactor (PFR) from 1976 to 1994. Reprocessing of fast breeder reactor fuel was also carried out. The DFR used a liquid metal alloy of sodium and potassium as a coolant. The contamination by fission products of the reactor coolant (primary coolant) is large, so decontamination work was carried out. The fuel was removed from the reactor, the coolant was treated, and the waste will be treated in future. The removal of fuel from the reactor of the PFR has been completed, and the treatment of the sodium coolant is being carried out (Thompson 2002; Dounreay 2021).

Decommissioning at Harwell

Harwell, located near Oxford in the suburbs of London, is the birthplace of nuclear research in the UK, and had various nuclear research institutions such as five research reactors, plutonium handling facilities, radioactive material laboratories, radioactive waste treatment and storage facilities. Decommissioning of two research reactors has been completed, and decommissioning of the remaining three is underway. Most of the other facilities have been dismantled, and some are in the process of decommissioning. Decommissioning is scheduled to be completed by 2025 (ONR 2023a). Harwell also has a cyclotron accelerator built in 1946. Currently, various research institutions and companies are active as a science park (Harwell 2023).

The decommissioning measures of the above-mentioned site are typical, but on the homepage of the UK Nuclear Regulatory Authority, the names of various sites undergoing decommissioning and environmental recovery are listed. It can be seen that not only the sites of nuclear power plants and research and development institutions, but also commercial sites such as fuel manufacturing, metal recovery, uranium enrichment, and medical isotope production, as well as university research reactors, are involved in decommissioning measures (ONR 2023b). Land environmental management is also explained in conjunction with the Environmental Agency.

Regarding the decommissioning measures and radioactive waste disposal in the UK, the situation at Sellafield, Dungeness Power Station, Harwell, etc., was introduced in the Japan Atomic Energy Commission's email newsletter (Oka 2020a, b).

Monitoring of Decommissioning Measures by the National Audit Office and the UK Parliament

The UK's National Audit Office (NAO) is, together with the UK Parliament, *continuously monitoring the implementation status and budget of decommissioning plans and publishing reports.* For example, it has raised issues with the implementation of the NDA's Magnox reactor decommissioning plan and the losses associated with subsequent changes in contractors, and has called for improvements in the NDA's ability to understand the actual situation of decommissioning (NAO 2020a; UK Parliament 2020). In a report monitoring the progress of Sellafield's decommissioning, it pointed out that there is a large uncertainty in the NDA's future decommissioning cost estimates (NAO 2018). It has also raised concerns about delays and cost increases in plans for nuclear facilities such as the UK Ministry of Defense's nuclear submarines (NAO 2020b). Although not limited to the nuclear field, in the UK and the US, the National Audit Office and Parliament monitor the work of national administrative and public institutions, pointing out whether taxes (including administrative costs) are being used efficiently, and identifying issues in plans, budgets, and operations. This has become a mechanism for improvement. Although this is not well known in Japan, it is a basic principle of democracy, and it would be good if it could be operated in a form closer to that of the West in Japan. The reports of the Western governments' National Audit Offices often provide useful references when considering the issues of nuclear use in Japan.

UK Waste and Radioactive Waste Treatment and Disposal

General Waste

The generation of waste in the UK is 62% from construction, demolition, and excavation. Waste derived from commercial and industrial activities accounts for 18%, household waste for 12%, and other sources for 8%. By material, mineral waste accounts for 37%, soil waste for 26%, and household and equivalent waste for 13%. The recycling rate of household waste is about 45%, and the proportion of urban waste being landfilled has been reduced to one-fifth compared to 1995 (DEFRA 2023).

The UK government, on its homepage's environmental management section, explains to the public about waste handling, disposal, and recycling methods, based on information from various ministries related to a wide range of waste (Gov.UK 2021a). It explains the responsibilities of businesses and the need for registration for waste disposal. It also explains about waste classification, hazardous waste, persistent organic pollutants, reporting of illegal dumping, waste movement, licensing of waste businesses, storage, use, treatment, and disposal of waste, waste from batteries and electrical products, packaging waste, discharge of waste water, and radioactive waste. Links to familiar waste disposal services are also provided. At the end, there is a section on radioactive waste, where the related parts of environmental management are explained.

In the section on radioactive waste, there are explanations on the following eight items:

1. Regulation of radioactive substances in non-nuclear facilities.
2. Radioactive waste from spent fuel and spent radiation sources.
3. Inspections for disposability by RWM (Radioactive Waste Management) and advice on packaging of radioactive waste.
4. Low-level radioactive waste disposal sites.
5. Regulation of radioactive substances in nuclear facilities.
6. Discharge of radioactive waste from nuclear facilities.
7. Examples of incidents.
8. Screening for unsuitability underground.

In 1, radioactive substances are widely used in hospitals, universities for diagnosis and treatment, and manufacturing processes in factories. Therefore, when used in non-nuclear facilities, it is necessary to apply to the Environment Agency for permission to use, and the procedure, including the disposal of radioactive waste, is explained (Gov.UK 2021b).

Within it, there is also an explanation for cases that do not require regulation as radioactive substances. For waste containing naturally derived radioactive substances such as uranium, and waste containing artificially created radioactive substances, based on the annual public exposure dose value (10 microsieverts/y), the standards for exemption are shown in a table for each nuclide for cases that do not fall under radioactive substances, and cases where permission for radioactive substances is exempted. In cases where it does not fall under radioactive substances, the concentration (per gram for solids, per liter for liquids, per cubic meter for gases) is indicated. In cases where permission is exempted, both the total amount and concentration of radioactive substances are required, and the values for each nuclide are shown (Gov.UK 2018).

In item 6, it explains the setting of limit values for the release of radioactive waste from nuclear facilities (Environment Agency 2010). It explains how the legal limit values are determined, citing the thinking and documents of international organizations. Not only the limit values, but also the relationship with the emission values during normal operation that vary, and furthermore, the receipt of reports on the emission results every quarter, including emissions that exceed the normal average value in the short term, which is separate from emissions that exceed the limit value (accidental emissions), are mentioned. In this item, the government explains the concept of limit values and normal emission values on its website, responding to the interest of the public.

In Japan, regulatory values and emission values are published and can be known, but I have never seen such a comprehensive explanation. The UK has been able to explain to the public as an administration related to the release of gases and liquids from nuclear facilities, but Japan is insufficient. Not only for radioactive waste, the UK government has explanations from the perspective of the public on its website, and the basics of a democratic country are excellent.

Treatment and Disposal of Radioactive Waste in the UK

Radioactive waste in the UK is classified as follows (ONR 2015).

1. High-level radioactive waste (HLW): Waste that generates a large amount of heat due to radioactive decay and needs to be considered in storage and disposal; waste liquid containing most of the radioactivity associated with nuclear fuel reprocessing, solidified waste, and other similar radioactive waste.
2. Intermediate-level radioactive waste (ILW): Waste with higher radioactivity than low-level radioactive waste, but heat generation does not need to be considered: for example, fuel cladding generated in the reprocessing of spent fuel, radioactive reactor equipment generated in decommissioning measures, graphite from reactor cores, sludge generated in the disposal of liquid radioactive waste.
3. Low-Level Radioactive Waste (LLW): Waste containing radioactive substances with alpha radioactivity of 4 gigabecquerels or less per ton, and beta-gamma radioactivity of 12 gigabecquerels or less, which can be disposed of in surface facilities. Most low-level radioactive waste is generated from nuclear power plants, reprocessing plants, and decommissioning and cleanup of nuclear facilities. Waste from operating facilities is mainly composed of plastics, paper, and metals, while waste from decommissioning is mainly composed of soil, building materials, and metal equipment.
4. Very Low-Level Radioactive Waste (VLLW): A subcategory of low-level radioactive waste, but as of the 2011 regulation revision, it is now an outdated classification. (It means that VLLW is now subject to different regulations than LLW.)
5. Exempt Waste: Waste whose level of radioactivity is below the set regulations, and is therefore exempt from regulation as radioactive waste. Or waste that is determined as such by environmental regulations. Even if it is exempt from regulation as radioactive waste, it may still be subject to environmental regulations due to other characteristics.
6. High Activity Radioactive Waste (HAW): A broad term that refers to high-level radioactive waste, intermediate-level radioactive waste, and low-level radioactive waste that is currently unsuitable for disposal in low-level radioactive waste disposal sites.

Generated VLLW can be disposed of in permitted landfill sites (not all landfill sites) along with household and industrial waste. The main types are building rubble, soil, and metal objects. These are mainly generated from the dismantling of reactors and nuclear facilities (NDA 2023b).

Very Low-Level Radioactive Waste includes large volume items (bulk disposal): those with a radioactive substance concentration of 4 megabecquerels (MBq) or less per ton, which can be disposed of in designated landfill sites, and small volume items (dustbin disposal): those with 400 kilobecquerels (kBq) or less per 0.1 cubic meter, or a total radioactivity of 40 kilobecquerels or less, which can be disposed of anywhere as urban, commercial, and industrial waste (NDA 2017).

In the UK, it is noteworthy that Very Low-Level Radioactive Waste can be disposed of in permitted disposal sites within waste (non-radioactive waste) landfill sites, and small amounts with low radioactivity can be disposed of in dustbins. As explained in the waste section, the UK has a classification of exempt waste. These are rational and cost-effective in waste regulation administration.

Document from the UK Nuclear Regulatory Authority (ONR 2015) presents the concept of waste management as promoting waste production reduction, reuse, and recycling in order, and disposing of waste that cannot be dealt with in these ways. It states that this approach has been adopted as a policy by the UK government for low-level radioactive waste.

The UK's low-level radioactive waste disposal sites are located near Drigg in Cumbria and Dounreay at the northern tip of Scotland. The Drigg disposal site began operations in 1959 and is operated by LLW Repository Ltd., a Site Licence Company (SLC), under a contract with the NDA. Initially, disposal was carried out by digging trenches in the ground, but since 1988, disposal has been carried out in large storage facilities made of concrete on the bottom and sides. Expansion of the storage facilities requires the submission of an environmental safety assessment document, called an Environmental Safety Case, to the environmental regulatory agency for approval. The bottom of the trench is lined with clay or water-resistant soil called bentonite. The concrete storage facility is made with concrete floors and walls, and iron boxes containing low-level radioactive waste are lined up and stacked. Leachate and rainwater are collected and checked. Once the trench or storage facility is full, it is covered with soil or similar material and monitored and managed for a specified period. The trenches at Drigg that were created in the early stages have already transitioned to this state (LLW Repository LTD 2017a, b, 2018). A document jointly created by regulatory agencies such as the Environment Agency and the Nuclear Regulatory Authority regarding the disposal of solid radioactive waste at existing nuclear sites has been made public (ONR 2021).

The UK has established regulations related to low-level radioactive waste, very low-level radioactive waste disposal sites, and general waste, and has a system in place to process and dispose of these wastes. The government is currently focusing on efforts to locate a geological disposal site for the remaining waste.

The UK's policy is to dispose of waste that cannot be disposed of at existing near-surface disposal sites, including high-level waste, in geological formations. These include intermediate-level radioactive waste, low-level radioactive waste for geological disposal, and used fuel, plutonium, depleted uranium, and recovered uranium generated from reprocessing, which are not currently recognized as waste. If there is no future use for these, they will be disposed of in geological formations, and the UK is making projections of the total amount of waste for geological disposal. The projection of the total amount of waste for disposal is necessary for planning the disposal site. While there is no publicly announced projection of the total amount in Japan, the fact that the UK is making a projection, including those not currently recognized as waste, can serve as a reference.

Geological Disposal of High-Level Radioactive Waste

The site selection process for the UK's geological disposal facilities emphasizes public solicitation, prioritizing active local participation and regional partnerships. The Committee on Radioactive Waste Management (CoRWM), an organization responsible for public debates and policy proposals, was established in 2003. In 2007, the Nuclear Decommissioning Authority (NDA) was designated as the implementing body for geological disposal policies. In the late 2000s, two cities in Cumbria expressed interest, geological screening was conducted by the British Geological Survey, and a report summarizing their opinions, recommendations, and advice to the county and the two cities was compiled. The two cities agreed, but the Cumbria County Council opposed moving to the next stage and withdrew from the site selection process in 2013. Currently, the site selection process is underway again based on the 2014 White Paper (DECC 2014; Yamada 2019).

The UK's geological disposal agency is the Radioactive Waste Management (RWM), a subsidiary of the NDA. RWM has a policy of sequentially promoting awareness, community involvement, and site candidate evaluation. They have created and published general explanatory materials on geological disposal (RWM 2023a) and community involvement guidance documents (RWM 2023b). They have also created materials explaining how to evaluate the site of a radioactive waste disposal facility (Nuclear Waste Services 2020). The general explanatory materials (RWM 2023a) list the benefits that a geological disposal facility can bring to the local community, such as the creation of employment and high-paying skilled workers, investment in public infrastructure, and the provision of support opportunities to the community over the long term until the disposal site is closed. The document discussing site candidate evaluation lists factors to consider in the evaluation, such as safety and security, community desires, environmental protection, technical feasibility, waste transportation, and whether there is value for the cost (Nuclear Waste Services 2020).

The UK's strata are divided into three characteristics: the southern part of England has strata similar to France, the strata of England and Wales, and Scotland has strata close to Sweden and Finland. The British Geological Survey has compiled data on the strata up to 2000 m underground in the UK for geological disposal, dividing England, Wales, and Northern Ireland into 13 regions. This information is intended for strata experts. For example, it shows what era the strata in northern England are, in plan and cross-section (Waters et al. 2018).

RWM provides three types of information on its homepage.

(1) Summaries and videos for non-technical audiences.
(2) Technical information and videos for groups who are not geological experts.
(3) Information for geological experts: Reports from the British Geological Survey.

It is noteworthy that not only information for the general public, but also information for experts in other fields and experts in the relevant field is provided. Some of the reports posted on the website of the Department for Business, Energy & Industrial Strategy (BEIS) have a note under the title stating that they are not intended for experts in related fields (other than nuclear). Specialist reports cannot be written without using specialist terms. By providing information for everyone from the general public to experts, citizens can learn up to the level they want to know. This is not limited to geological disposal. It may also be useful in the nuclear field in Japan.

In 2016, public dialogue activities to raise awareness were conducted in two cities with joint financial support from the then Department of Energy & Climate Change (DECC) and Sciencewise (a UK government program to use public opinion in policy making on science and technology). The dialogue was planned and implemented by a group of environmental consultants, and a report was made and published after the dialogue ended (3KQ 2016). Furthermore, an evaluation of this dialogue activity was conducted by another consulting firm and a report was made (URSUS consulting 2016). In addition, a large research company has produced a report summarizing the lessons and directions based on the investigation of geological disposal in various countries and the past failures of geological disposal in the UK. It is noteworthy that it mentions that past attempts at site selection were biased toward the technical side (Equitis Consulting 2016). Based on these results, in 2018, the Department for Business, Energy & Industrial Strategy (BEIS) created a policy document titled "Implementing Geological Disposal: Working with Communities" (BEIS 2018).

It is noteworthy that this dialogue involves consultants with experience in environmental issue dialogues and experienced individuals who have been involved in nuclear power dialogues for many years (facilitators: dialogue promoters). Experience is necessary for dialogue management and promotion, but even with facilitation training (techniques to smoothly conduct meetings and meetings), it is impossible for many to make a living solely from dialogues related to nuclear power. In the UK, it is believed that dialogues were effectively implemented through the participation of experienced individuals in dialogues outside the nuclear power field, such as environmental issues. In Japan, where lifetime employment is the norm, it is impossible for specific companies or administrations to continue employing these experts, even if dialogue facilitators are trained. The strategy of utilizing experienced consultants and groups in other fields, such as environmental issues, as well as consulting firms and research firms, as in the UK, is considered to be a reference in Japan. Although not a dialogue activity, in the United States, consultants specializing in safety and regulatory issues are active. There are also consulting firms specializing in regulatory compliance, contributing to effective and efficient use of nuclear power. The UK's site selection activity is still in the first stage, but working groups have been formed in two regions in November 2020 and January 2021. The plan is to form community partnerships, conduct local surveys, and further drill holes for investigation. At the local survey stage, an annual budget of 1 million pounds (about 150 million yen), and at the drilling hole stage, an annual budget of 2.5 million pounds (375 million yen) is prepared for the community partnership.

3.2.3 Disposal of Nuclear Facilities and Radioactive Waste in France

In France, the use of nuclear power is characterized by significant national involvement. Nuclear power generation is carried out by EDF (Electricité de France), the nuclear fuel cycle by Orano, and reactor manufacturing by Framatome, all of which are private companies, but the French government owns more than half of the shares. Radioactive waste management is handled by Andra, and research and development by CEA (Commissariat à l'énergie atomique et aux énergies alternatives). Both Andra and CEA are government agencies. From this background, in France, the government is involved in the disposal of nuclear facilities and the treatment and disposal of radioactive waste.

Policy and Overview of Disposal of Nuclear Facilities and Treatment and Disposal of Radioactive Waste

In France, the "Law on Radioactive Waste Management and Research" (commonly known as the Bataille Law, a time-limited law for 15 years) was enacted in 1991, and the Radioactive Waste Management Agency (Andra) was established. In 2006, the "Law on Transparency and Security in the Nuclear Field" (commonly known as TSN Law) was established. It changed the framework of nuclear regulation. Accordingly, efforts are being made for decommissioning measures and the treatment and disposal of radioactive waste. Traditionally, the reactor part was to be dismantled after a safe storage period of about 50 years to wait for the decay of major radioactive nuclides, but after this law, the Nuclear Regulatory Authority (ASN) prefers immediate dismantling in its policy document (ASN 2009). The reasons are.

(1) Staff familiar with the design and operation of the facility will be gone.
(2) The benefits from radioactive decay are minimal.
(3) Delayed dismantling carries the risk of equipment and structures aging.

Even though immediate dismantling is preferred, the period for decommissioning measures is not legally defined.

Following the 2006 law, the Nuclear Regulatory Authority (ASN) regularly (every 3 years) creates and announces the "National Management Plan for Radioactive Substances and Waste (PNGMDR)". This allows for the total amount of radioactive substances and waste generated by decommissioning measures and the like to be understood, and for storage and disposal plans to be coordinated (ASN 2021).

For the disposal of high-level radioactive waste, three types of research and development were conducted for 15 years under the 1991 "Law on Research on Radioactive Waste Management".

(1) Research on group separation of actinides and transmutation from long-lived nuclides to short-lived nuclides.
(2) Testing of recoverability using underground research facilities.
(3) Development of conditioning technology for long-term storage at intermediate storage facilities.

Andra was responsible for research and development on recoverability using underground research facilities. An underground research institute was established in Bure, a comprehensive evaluation was conducted in 2006, and it has been decided to carry out geological disposal considering recoverability for high-level waste and long-lived intermediate-level radioactive waste.

Please note, nuclear facilities in France are classified into two types: Basic Nuclear Installations (BNI) and Installations Classified for Environmental Protection (ICPE). BNI facilities include nuclear reactors, particle accelerator facilities, and nuclear fuel cycle facilities and waste storage, management, and use facilities that contain a greater amount of radioactive substances than a specified value. ICPE facilities are those that possess a level of radioactivity below that of BNI-designated facilities and above a value defined by ministerial order. ICPE facilities are not limited to nuclear facilities, but include all facilities that could potentially pose a risk to the environment and the public. All BNI facilities require approval from the ASN, while ICPE facilities include those that require approval and those that only need to be reported. For example, a very low-level radioactive waste disposal site is an ICPE facility, not a BNI facility. As already mentioned, the UK has a similar approach. Regardless of the amount of radioactive substances handled, it is not rational to make all of them nuclear facilities. What about Japan? *The concept of risk and benefit is necessary for the disposal of radioactive waste and the decommissioning of nuclear facilities in Japan.*

Decommissioning of Nuclear Power Plants

France's electricity is mainly supplied by nuclear power generation, and it is also exported to neighboring countries. Historically, the nuclear fuel cycle business was corporatized by Cogema (the predecessor of Orano) based on the technology developed by CEA, and then merged with the nuclear reactor manufacturing company Framatome and others in 2001 to become Areva. However, due to the delay in the construction of a nuclear power plant in Finland ordered by Areva, which resulted in a loss, the French government stepped in to support it, and in 2018 it was divided into Orano, which is in charge of the nuclear fuel cycle business, and Framatome, which is in charge of the nuclear reactor business. Framatome then became a subsidiary of EDF (Electricite de France).

The French Court of Audit (Cour des comptes) published a report titled "The Cost of the Nuclear Power Sector" in 2012 (Cour des comptes 2012). This report allows us to understand the history and overall picture of civilian nuclear use in France.

Initially, France used a graphite-moderated gas-cooled reactor improved by the Atomic Energy Commission (CEA) for power generation. Six units started operation between 1963 and 1972. In addition, a power reactor using heavy water and a pressurized water light water reactor were each experimentally built. These were shut down between 1973 and 1994, and decommissioning measures were taken (NEA 2018).

Subsequently, Electricite de France built 58 pressurized water light water reactors mainly in the 1970s and 1980s, expanding nuclear power generation. Like Japan, France has scarce fossil fuel resources in its own country and obtains cheap and stable electricity through nuclear power generation. These light water reactors are in operation, except for two units that were shut down in 2020.

Apart from these, the reactors that are being decommissioned include one prototype sodium-cooled fast reactor (Phoenix) and one commercial reactor (Super Phoenix), as well as two graphite-moderated gas-cooled experimental reactors that the French Atomic Energy Commission (CEA) operated at the Marcoule site and stopped in the early 1980s. The graphite used as a material for the reactors is classified as low-level long-lived radioactive waste and is stored until a disposal site is available.

Decommissioning of Nuclear Fuel Cycle Facilities

The nuclear fuel cycle refers to the series of processes that involve mining uranium ore, refining it, converting it to gaseous uranium, enriching the fissile uranium, reconverting it to a solid, processing it into fuel assemblies, using it in a reactor, removing the spent fuel assemblies, shearing them, dissolving the fuel portion into a liquid, separating the plutonium (this part is called reprocessing), mixing the liquid containing radioactive waste with glass to solidify it, and disposing of the radioactive waste. This is also referred to as the nuclear fuel cycle. The process before loading fuel into the reactor is called the front end of the nuclear fuel cycle, and the process after removal from the reactor is called the back end. If reprocessing of spent fuel is not performed, the process excluding that becomes the nuclear fuel cycle.

Plutonium for nuclear weapons is made by reprocessing slightly irradiated metallic natural uranium fuel in a reactor, and it contains a higher proportion of fissile plutonium than plutonium obtained by reprocessing commercial spent fuel. The method of separating plutonium in commercial reprocessing is based on the experience of extracting plutonium for nuclear weapons. The method of making highly enriched uranium for nuclear weapons is the same in principle as making low-enriched uranium for commercial use. For these reasons, from the perspective of nuclear non-proliferation, special consideration for confidentiality (security) is required in commercial nuclear fuel cycle facilities.

Information on nuclear fuel cycle facilities, especially reprocessing facilities, is scarce due to concerns about nuclear proliferation. The French company Orano explains the nuclear fuel cycle with animations on its website, which can be a reference for understanding the concept (Orano 2023a, b).

The decommissioning of reprocessing facilities that extract plutonium from spent fuel, as seen in the UK example, requires the most cost and time in the decommissioning of nuclear facilities. To commercialize the reprocessing technology developed by the Atomic Energy Commission (CEA), Cogema (now Orano) was established in 1976. The produced plutonium is processed into mixed uranium-plutonium oxide fuel (MOX fuel), loaded into 20 units of the 900MWe pressurized water reactors of Electricité de France (EDF) in France, and has been used for power generation.

France, due to its scarcity of fossil fuel resources, has developed and utilized not only nuclear power generation using uranium, but also a nuclear fuel cycle that utilizes plutonium, on a national scale. Power companies in Europe and Japan have commissioned the reprocessing of their spent fuel to nuclear fuel cycle companies in the UK and France, receiving fuel containing plutonium processed into fuel assemblies, which they used in their respective nuclear power plants. Because plutonium is used as fuel in light water reactors, not fast reactors, it is called "pluthermal" in Japanese-English. Currently, both European and Japanese power companies have ceased overseas reprocessing (commissioning reprocessing to companies in France and the UK). The radioactive waste generated by reprocessing is being returned to the country where the company that commissioned the reprocessing is located.

Figure 3.7 shows France's reprocessing facilities and radioactive waste disposal sites. The reprocessing facilities are located in Marcoule in the south and La Hague facing the English Channel in the northwest. In Marcoule, the French Atomic Energy Commission conducted the reprocessing of spent fuel from gas-cooled reactors at a facility called UP1 in 1958, and from fast reactors and light water reactors at a facility called APM from 1974 and 1988. These facilities are currently all being decommissioned. In La Hague, reprocessing of spent fuel from gas-cooled reactors began at a facility called UP2 in 1966, reprocessing of spent fuel from fast reactors began at a facility called AT1 in 1969, and reprocessing of spent fuel from light water reactors began at a facility called UP2-400 in 1976. In 1989, UP3, with a spent fuel processing capacity of 800 tons, is in operation. The construction cost of UP3 was covered by European and Japanese power companies that commissioned reprocessing. Furthermore, UP2-800 has been in operation since 1994. UP2-400 shared some facilities with UP2-800, but it stopped operation in 2004. The decommissioning of AT1's facilities has been completed, and the concrete walls of the decontaminated building remain as a memorial structure. In La Hague, the light water reactor spent fuel reprocessing facilities of UP2-800 and UP3 are in operation. Facilities other than these, such as UP2-400, are being decommissioned. Decommissioning is important for ORANO's management, and the management executives are involved, and progress management is being carried out comprehensively and steadily. There are financial benefits to completing the decommissioning ahead of schedule (Oka 2019a, b).

Information of nuclear fuel cycle from the pro-nuclear side is dominant, but the International Panel on Fissile Materials aims to reduce the amount and production of highly enriched uranium and plutonium. The report of the Institute for Peace and Freedom Movement (IPFM) discusses not only its claims, but also things about the nuclear fuel cycle in France and the UK that are not well known in Japan's nuclear sector (Schneider & Marignac 2008; Forwood 2008). Discussions about reprocessing

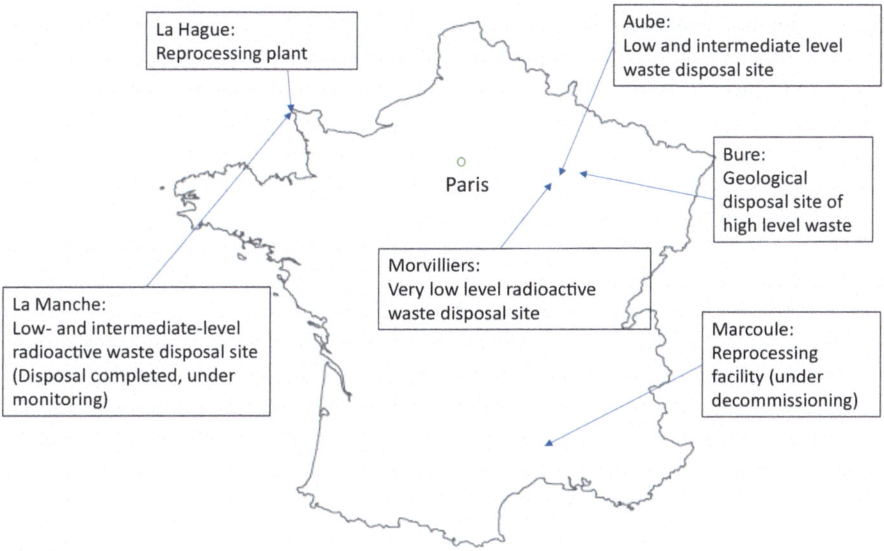

Fig. 3.7 Reprocessing facilities and radioactive waste disposal sites in France

and plutonium tend to be either for or against, but it is necessary to first compile and understand the facts, including history, as a scientific method and for robust (resilient) nuclear use.

The time when foreign power companies planned to reprocess spent fuel was in the 1970s for both the UK and France, but France is ahead in terms of experience in overseas reprocessing and MOX fuel processing. Both countries have the construction costs of plants for overseas reprocessing covered by the foreign power companies that commissioned them. The 2012 French Audit Court (Cour des Comptes) report on the cost of nuclear power introduces the fact that the 1999 French Parliament report on the backend of the fuel cycle (commonly known as the Bataille report) stated that about half of AREVA's investment in uranium enrichment plants and reprocessing plants was covered by foreign customers (power companies), and this is taken into account in cost evaluations.

Nuclear power plants (light water reactors) are built in large numbers, so when they are built, they operate at the designed power output, but this is not necessarily the case for reprocessing plants. For example, the oxide fuel reprocessing plants in both the UK and France have not been operating at their design capacity (for example, 400 tons per year) from the start (Schneider & Marignac 2008). This is also the case for the Tokai Reprocessing Plant of the Japan Nuclear Fuel Services (JNFL, at the time) (with a design capacity of about 200 tons per year).

France built a large-scale uranium enrichment plant from the late 1970s to the early 1980s, which was then COGEMA. This was before the construction of a commercial scale reprocessing plant. A large-scale uranium enrichment plant using the gaseous diffusion method was built in Pierrelatte in southern France. This construction investment was made by a holding company (EURODIF) participated in by power companies from France, Belgium, Italy, and Sweden, and about half of the investment was covered by power companies outside of France. Uranium enrichment by gaseous diffusion consumes a large amount of electricity. Later, the more efficient centrifuge method began to be used for uranium enrichment, and the gaseous diffusion uranium enrichment plant was decommissioned and dismantled.

The dismantling cost is 80% for La Hague (reprocessing plant, etc.), 10% for EURODIF (uranium enrichment plant using gaseous diffusion), and 10% for others (source: French Cour des Comptes report, page 103).

Among other things, there are decommissioning measures for uranium mines. France (now ORANO) conducted uranium exploration domestically and then abroad in Africa and Canada, managed uranium mines, and extracted uranium from ore. According to French regulations, uranium mines are not nuclear facilities but environmental protection facilities. Most of the decommissioning measures for uranium mines in France have been completed.

Decommissioning of Nuclear Research and Development Facilities

Nuclear research and development in France has been conducted by the French Atomic Energy Commission. The name was changed to the French Renewable Energy and Atomic Energy Commission in 2009. Both are abbreviated as CEA. CEA's laboratories are located in Marcoule in the south of France, Cadarache in Provence, Saclay and Fontenay-aux-Roses near Paris, and Grenoble, which has hosted the Winter Olympic Games.

Among these, the Grenoble laboratory was repurposed for research in microelectronics and renewable energy, and decommissioning measures for nuclear-related facilities were taken. The high neutron flux research reactor at the Laue-Langevin Institute in Grenoble is still in operation, and various material property research is being conducted using its neutron beam. The Fontenay-aux-Roses laboratory is the birthplace of nuclear energy research in France, but it is aiming for biotechnology research, and is in the process of being repurposed after taking decommissioning measures for nuclear facilities. Radioactive materials were handled for research such as plutonium chemistry in the early stages of nuclear development, so decommissioning measures are being taken carefully. At the Saclay laboratory, basic nuclear research is being conducted, but the decommissioning of research reactors has been carried out. The Saclay laboratory forms a science city with universities and research institutes in the suburbs of Paris. At the Marcoule laboratory, decommissioning measures have been taken for facilities such as UP1 used for nuclear fuel cycle technology development in the early stages, and for the fast prototype reactor Phoenix. At the Cadarache laboratory, decommissioning measures have been taken for radioactive waste treatment facilities and fast experimental reactors such as Rhapsody (CEA 2016). At the Cadarache laboratory, construction of the nuclear fusion experimental reactor ITER and a new large-scale research reactor (Jules Horowitz reactor) is progressing, and nuclear reactor safety research is being conducted. There are many small facilities where decommissioning has been completed.

Research and development facilities do not handle as much radioactive material as commercial facilities, so generally, the cost of decommissioning is not as high. However, because there are various types of facilities, the experience of decommissioning does not necessarily apply to the next decommissioning as the case of a power reactor. It is necessary to devise and implement appropriate decommissioning measures for each facility. The proportion of the decommissioning costs of each CEA research institute to the total facility dismantling costs is approximately 50% for Marcoule, just over 20% for Cadarache, just over 10% for Saclay, just under 10% for Fontenay-aux-Roses, and less than 2% for Grenoble. The CEA conducted evaluations of the decommissioning of its main facilities in 2001, 2007, and 2010, and the cost estimates have changed (source: 2012 Cour des Comptes report, pages 109 and 110).

The impression is that decommissioning in France is steadily progressing. I have heard from a friend in France that calculation is important in decommissioning. The method of regularly conducting cost evaluations, checking progress and issues, and advancing decommissioning (dismantling) can also be a reference for the decommissioning of research and development facilities in Japan.

The CEA owns shares in Cogema (later AREVA, then Orano) because the nuclear fuel cycle and other technologies it developed were industrialized by Cogema, and it also receives dividend income. The CEA's nuclear research and development costs from 1990 to 2010 averaged about 1 billion euros per year, of which 42% came from the country, 5% from dividends, 13% from other institutions, and 40% from the industrial sector (source: 2012 Cour des Comptes report, page 44). To summarize the decommissioning (dismantling) costs, the costs required for France's decommissioning and spent fuel management, and the management of existing waste, are 78% for EDF (nuclear power related), 13% for AREVA (nuclear fuel cycle related), 9% for CEA (research and development related), and about 0.1% for ANDRA (radioactive waste management related) (source: 2012 Cour des Comptes report, page 164).

This Cour des Comptes report includes joint comments from French ministries, as well as comments from the heads of agencies such as the CEA, AREVA, the French Energy Regulatory Authority, ASN (French Nuclear Regulatory Commission), EDF, and ANDRA (Radioactive Waste Management Agency). France is advancing robust nuclear use by creating and publishing reports at each milestone.

Radioactive Waste Disposal in France

In France, radioactive waste is generated in various fields such as nuclear power generation, research and development in nuclear power, medical biology, nuclear physics, agriculture, military such as nuclear submarines, various industries other than nuclear power generation, and medical fields. The amount of general waste and radioactive waste generated by the decommissioning (dismantling) of nuclear facilities is far greater than that generated during the operation of the facilities.

Radioactive waste in France is classified into very low level, low level, intermediate level, and high-level radioactive waste according to the intensity of the radioactivity of the radioactive substances contained, and into short half-life and long half-life radioactive waste according to the half-life. For example, graphite used in graphite-moderated power reactors is low-level long half-life radioactive waste. There is no distinction by half-life for very low level radioactive waste and high-level radioactive waste (Andra 2023a).

Figure 3.7 shows the radioactive waste disposal sites in France. The La Manche disposal site is a short-lived low and intermediate-level radioactive waste disposal site adjacent to the La Hague reprocessing plant, where radioactive waste from EDF, Orano, and CEA, such as work clothes, filters, concrete, and metals, is disposed of in drums. The waste bodies are placed on a concrete base or in a concrete pit set up on top of it, covered with a waterproof cover, and operations ended in 1994. It is under surveillance for 300 years, and sampling and inspection of drainage, etc. are being conducted. This disposal site is the world's first radioactive waste disposal site to be put under surveillance, and development of means to pass on the records of waste and experience of disposal to the next generation is being carried out.

The Aube disposal site is a short-lived low and intermediate-level radioactive waste disposal site located in central western France, which is in operation and is scheduled to operate for about 60 years. Concrete pits are being made in sequence and disposal is being carried out. After the end of operations, a cover mainly made of clay will be provided, and it will be under surveillance for 300 years.

The Morvilliers disposal site is a very low level radioactive waste disposal site, where very low level radioactive waste generated from the dismantling of nuclear facilities, non-nuclear industries, and radioactive material handling facilities is disposed of in trenches (dugouts) a few meters deep dug into the clay layer. As mentioned above, since it is a very low level radioactive waste disposal site, this disposal site is regulated not as a nuclear facility but as an environmental protection facility. There is a facility to sort and pre-treat waste containing radioactive substances sent from places other than nuclear power plants and large-scale nuclear research and development facilities (Andra 2023b).

Andra, the French radioactive waste management agency, is advancing two projects for the future disposal of long-lived low-level radioactive waste and high-level radioactive waste.

Long-lived low-level radioactive waste is not suitable for disposal at surface disposal sites because of its long half-life, but it is considered that there is no rational reason to dispose of it 500 m underground because it is low level. The main types of long-lived low-level radioactive waste are waste generated in the process of processing ores containing radium to extract rare earth elements used in the radium and electronics industries, and graphite waste generated in the dismantling of graphite-moderated reactors. There are also used radiation sources and products containing radium. Since these wastes were generated in the past, their total amount is not increasing, but they account for 6% of France's radioactive waste.

The disposal method for long-lived low-level radioactive waste has been reviewed every five years in accordance with the Radioactive Waste Management Act of 2006. Based on the results of a preliminary geological survey by Andra and the French Geological Survey, a public call was made in 2008 to 3115 towns and villages believed to have strata suitable for a shallow geological disposal site for this waste. About 40 towns and villages that showed interest were analyzed by Andra, and after advice from the Nuclear Regulatory Authority (ASN) and the National Evaluation Committee (CNE), representatives of each municipality were selected. The government ordered Andra in 2009 to conduct detailed geological surveys in two towns and villages, but both withdrew their applications under pressure from people opposed to the project. However, the government has ordered Andra to continue dialogue with the interested towns and villages. The High Committee for Transparency in Nuclear Safety Information (HTISN) has ordered the continuation of site exploration and recommended the following three principles for site selection in 2011.

(1) Utilize the results of the 2008 application.
(2) Preferentially select regions where nuclear facilities already exist.
(3) Deepen dialogs among a very narrow range of local government stakeholders, with the support of the government and broader local governments.

In accordance with these recommendations, Andra published a report in 2012 and announced the continuation of the search for potential disposal sites.

In 2013, following the declaration by the inter-municipal officials of Soulaines (Aube department) that geological surveys could be conducted, the government ordered Andra to conduct scientific investigations in this region regarding the possibility of a disposal site for very low-level radioactive waste, in addition to a long half-life low-level radioactive waste disposal site. An advisory committee, led by the governor of the Aube department, was established, and geological surveys were conducted in a 50 square kilometer area spanning five municipalities. A report, including previous research results, was published by Andra in 2015. Preliminary design and safety assessments of the disposal site were conducted. At the same time, considerations were made regarding the management of long half-life low-level radioactive waste mixed with reactor graphite and asphalt. The 2015 report identifies suitable areas for the disposal site (Andra 2023c).

The geological disposal of high-level radioactive waste and long half-life medium-level radioactive waste in France is a project called CIGEO (Industrial center for geological disposal), which is being carried out by Andra in Bure in northeastern France. The following introduces the historical background leading up to this.

Andra was established in 1979 and selected four candidate sites to establish an underground research institute for geological disposal. However, as no explanation was given to the residents, strong opposition was encountered. In 1990, the opposition intensified, and the then Prime Minister Rocard ordered Andra to halt its research and instructed a policy reevaluation. Representative Bataille issued a report, and the Radioactive Waste Management Act (Bataille Law) was established in 1991. This law was a time-limited legislation, and research in the following three fields was to be conducted for 15 years.

1. Separation and transmutation (conversion) of long half-life radioactive nuclides: CEA reduces the quantity and toxicity of radioactive waste. First, long half-life ones are separated and converted into short half-life nuclides through nuclear reactions.
2. Long-term storage: CEA stores it on or near the surface for 300 years.

3. Deep geological disposal: Andra conducts research on the identification of geological disposal sites and their safety and feasibility.

The Bataille Law required a report with scientific and technical backing to be submitted to the parliament by 2006.

In accordance with the Bataille Law, an underground research institute was to be established, and Andra proposed three clay layer sites and one granite site as candidate sites for the underground research institute. After consulting with the French government and local governments, considerations were made for the three sites. One of these was concluded to be the most suitable as a disposal site, and the construction of the underground research institute was approved in 1996 and built in Bure in 2000. It reached the target stratum in 2004, and research has been conducted since then.

Andra and CEA each submitted their research results to the French government. As a result, the following conclusions were made.

1. Separation and Transmutation: Separation (partitioning) and transmutation is only possible for a limited number of nuclides, and it cannot eliminate the need for geological disposal. The limited nuclides refer to minor actinides [Americium, Curium, Neptunium]. Furthermore, the process of separation and transmutation produces radioactive waste that requires geological disposal.
2. Long-term Storage: While storage facilities are robust, there is a need to guarantee retrieval after the storage period ends, to recondition the radioactive waste and move it to a new storage facility. This requires monitoring and surveillance during the storage period. This necessitates continuous monitoring by society, imposing a burden on future generations. It is impossible to guarantee this for hundreds of years. For these reasons, the French Nuclear Safety Authority (ASN) has rejected the long-term storage option.
3. Deep Geological Disposal: Based primarily on the results obtained by underground laboratories, Andra concluded that the clay layer in the Bure (Meuse/Haute-Marne) region has suitable characteristics for deep geological disposal. This conclusion and the design and operation of the geological disposal facility were evaluated by the national evaluation committee and the ASN. Andra's work was evaluated by an international group of experts. This confirmed the conclusion reached by Andra.

In 2005, the results of the research conducted under the Bataille Law were presented to a national public hearing at the request of the French Ministry of Environment and Industry. The public hearing was held 13 times with the participation of 3000 people and ended in 2006. The hearing shed light on whether to store or dispose. One was geological disposal considering reversibility. The other was to implement a comparison program between geological disposal and long-term storage and postpone the conclusion until 2020. As a result, the former, geological disposal, was concluded to be better, and in the 2006 law, the French government adopted the option of reversible deep geological disposal for the long-term management of high-level radioactive waste and intermediate-level long-lived radioactive waste. Andra was tasked with working to review the licensing application for the design and installation of a deep geological disposal site by 2015. The 2006 law sets the facility's licensing for 2025. The parliament demanded that disposal be reversible (recoverable) for at least 100 years. This condition is expected to be stipulated by law by the time of facility licensing. A public debate is scheduled to be held before the license is granted.

To summarize the above-mentioned background, it is as follows. In 1991, the Radioactive Waste Management Act (commonly known as the Bataille Act) studied three options for high-level radioactive waste: separation and transmutation, long-term storage, and geological disposal, for 15 years. Andra was responsible for considering geological disposal and conducted research in three clay regions and one granite region. The results of the three options were evaluated in 2005, and in 2006, deep geological disposal was chosen by law. On its website, Andra presents the view of the French Nuclear Regulatory Authority (ASN) that "not all high-level radioactive waste can be disposed of by separation and transmutation, long-term storage is not a permanent solution, deep geological disposal is the only solution" (Andra 2023d).

The Bure site is a consolidation of two adjacent regions out of the three clay regions and one granite region that Andra studied according to the Bataille Act, with a clay layer from 160 million years ago spread underground. Construction of the underground laboratory began in 2000, and by 2004, the borehole had reached the center of the clay layer about 500 m underground and about 160 m thick. Initially, the area around the laboratory, covering 250 square kilometers, was considered, and later detailed investigations were conducted within a range of 30 square kilometers. The clay layer is water-resistant and excellent in retaining radioactive substances. By 2010, thermal conductivity and other properties had been measured, and numerous core samples had been collected. Currently, full-scale transport tunnels and disposal holes are being created, and tests are being conducted on the placement of waste, the sealing of disposal holes, and the possibility of retrieving waste placed in disposal holes (Andra 2023e).

There are two shafts to go from the surface to the underground laboratory, and they are accessed by elevator. The surface facility to accept the waste to be disposed of is located a little distance away, and there are plans to transport the waste to the high-level waste and long-lived intermediate-level radioactive waste disposal site, which is planned to be built adjacent to the underground laboratory at 500 m underground, via a sloping tunnel. A railway to transport the waste is planned to be laid to the surface facility. The operating period of the disposal site (the period until disposal is completed and monitoring begins) is assumed to be 100 years. A video explains the overview of the facility (Andra 2023f). It is also explained that Andra plays a role in improving local life, economy, and employment (Andra 2023g). France is currently adopting reprocessing and plutonium utilization as a policy, but for example, if the energy policy changes in future and the reprocessing of spent fuel is terminated, it is stated that spent fuel can be directly disposed of at Cigeo (Andra 2023h).

The French geological disposal project is steadily progressing step by step, leaving room for future generations to make choices, such as recoverability, over the next 100 years. The concept and approach of long-term projects spanning generations can serve as a reference for Japan. Furthermore, in a government document reported to the International Atomic Energy Agency in the early 2000s, France stated its policy of "reprocess spent fuels to the amount of plutonium used in nuclear power plants". It is not reprocessing of whole spent fuels. In Japan, this is not known well. France is leading the world in the front and back ends of the nuclear fuel cycle, and its policy is robust.

3.2.4 Decommissioning of Nuclear Facilities and Radioactive Waste in Germany

Decommissioning of Nuclear Facilities

When East and West Germany were unified in 1990, there were five light water reactor-type nuclear reactors designed by the former Soviet Union in former East Germany and three more under construction, but all of them failed to meet the safety standards of West Germany and stopped operating or construction.

In Germany, a coalition government announced a phase-out policy for nuclear power generation in 1998. The government rescinded the phase-out policy in 2009, but in 2011, in response to the TEPCO Fukushima accident, it immediately shut down eight nuclear power plants. Until then, about 25% of the electricity was supplied by nuclear power, but in 2020, six nuclear power plants supplied about 10% of the electricity. Three of these were also shut down at the end of 2021. The remaining three were scheduled to stop at the end of 2022, but due to the energy crisis caused by Russia's invasion of Ukraine, the shutdown was postponed to 2023, and all operations were stopped on April 15, 2023. In Germany, the cost of power generation is low, but due to taxes and levies, electricity rates are extremely high (WNA 2023).

Germany is a federal state. The federal nuclear research institutes in Karlsruhe and Jülich have been integrated and transformed into the Karlsruhe Institute of Technology and the Jülich Research Center, respectively, and decommissioning measures for experimental reactors and research reactors for fast reactors and high-temperature gas reactors have been promoted. At the Karlsruhe Institute of Technology, nuclear research is being conducted. The large research reactor of the Technical University of Munich is in operation, and material research is being conducted.

In Germany, power companies have been consolidated into four: RWE, E.ON, Vattenfall Europe, and EnBW, following the liberalization of electricity in 1998. RWE supplies power to western Germany, E.ON to central Germany from south to north, Vattenfall to northeastern Germany, and EnBW to southwestern Germany. Vattenfall Europe is a subsidiary of the Swedish state-owned power company Vattenfall, which entered Germany when RWE and E.ON were forced to abandon their assets in former East Germany during the liberalization of electricity. EnBW is located in southwestern Germany and is owned by the state of Baden-Württemberg and local governments.

Nuclear power plants in Germany can be broadly divided into the following three categories.

1. Nuclear power plants owned by private power companies.
2. Former East German nuclear power plants that were closed due to the reunification of East and West Germany (located in Greifswald on the Baltic Sea).
3. Experimental reactors and prototype reactors built in the early stages of nuclear research and development.

The entities responsible for the decommissioning measures of the category 1 are the power companies, the category 2 is EWN GmbH (Energiewerke Nord GmbH), and the category 3 is the Federal Ministry of Education and Research (BMBF). Since the only shareholder of EWN GmbH is the Federal Ministry of Finance, the decommissioning measures for the category 2 and 3 are carried out under the responsibility of the federal government.

The federal government states that the ultimate goal of decommissioning measures is to release the nuclear facilities where the measures are carried out from their regulations. The final basic goals of the decommissioning measures are either site release by dismantling and removal, or indefinite use conversion of the structures. The methods of decommissioning include "immediate dismantling" and "safe storage". "Immediate dismantling" is to dismantle all or part of the nuclear facilities and release them from nuclear regulations. "Safe storage" is to put the facilities in a state where almost no maintenance is required for a long period of time, and to delay the timing of dismantling and release from regulations. In the latter case, it is required to pass on the knowledge of the operating experience of the facility's employees to the employees who carry out the decommissioning measures. It is considered preferable to choose immediate dismantling rather than safe storage as a decommissioning measure, but the choice is left to each organization (BASE 2023, 2021).

Decommissioning (dismantling) in Germany can be carried out by obtaining permission in stages. For example, first the turbine building, then the equipment inside the reactor building, then the equipment inside the reactor containment vessel outside the biological shield, then the removal of the reactor pressure vessel and biological shield, then the dismantling and removal of all components remaining in the radiation control area, decontamination of concrete structures, termination of nuclear facility licensing, and finally the demolition of the building and site recovery. The removal of reactor fuel, the treatment and disposal of radioactive waste generated during operation, and the decontamination of the facility can be carried out within the scope of operation permission, without obtaining permission for decommissioning, after the end of operation. However, waste that cannot be normally processed, such as heavy water used in reactors, needs to be carried out after obtaining decommissioning permission.

In the case of commercial nuclear power plants, the cost of decommissioning measures, based on the principle of polluter pays, is borne by the power companies. Power companies are obliged to recover costs from electricity charges and are accumulating costs as reserves. The cost of decommissioning measures for former East German nuclear power plants and research and development facilities is borne by the federal government.

The cost of decommissioning nuclear power plants in Germany is evaluated using a calculation program called STILLICO2 (NRW 2007). This program, based on the experience of decommissioning nuclear power plants in Germany, is also used in neighboring European countries such as Belgium and the Netherlands.

Figure 3.8 shows the centralized interim storage facilities and disposal sites for radioactive waste in Germany. In Germany, the Konrad disposal site is under construction, and as of 2021, there are no operating disposal sites for low-level, intermediate-level, and high-level radioactive waste. Therefore, in the decommissioning of nuclear facilities, it is common to provide an interim storage facility for the generated radioactive waste near the decommissioning site. This is also the case with the decommissioning of the Greifswald nuclear power plant (Rehs 2011). After paying the cost to the interim storage facility and moving the radioactive waste, the responsibility for the disposal of the radioactive waste shifts to the interim storage facility.

In addition to individual interim storage facilities, Germany has three centralized interim storage facilities. The Gorleben facility stores vitrified high-level waste returned from France, where spent fuel was reprocessed. The Ahaus facility is located in a town adjacent to the French border in North Rhine-Westphalia, northwestern Germany, and stores spent fuel from research reactors and power reactors. The Lubmin interim storage facility is adjacent to the decommissioning site of the Greifswald nuclear power plant on the Baltic Sea, and can sort and store all types of radioactive waste (BASE 2024a).

The status of decommissioning of nuclear power plants and nuclear facilities in Germany is published by the Federal Office for the Safety of Nuclear Waste Management. As of January 2024, it is as follows (BASE 2024b).

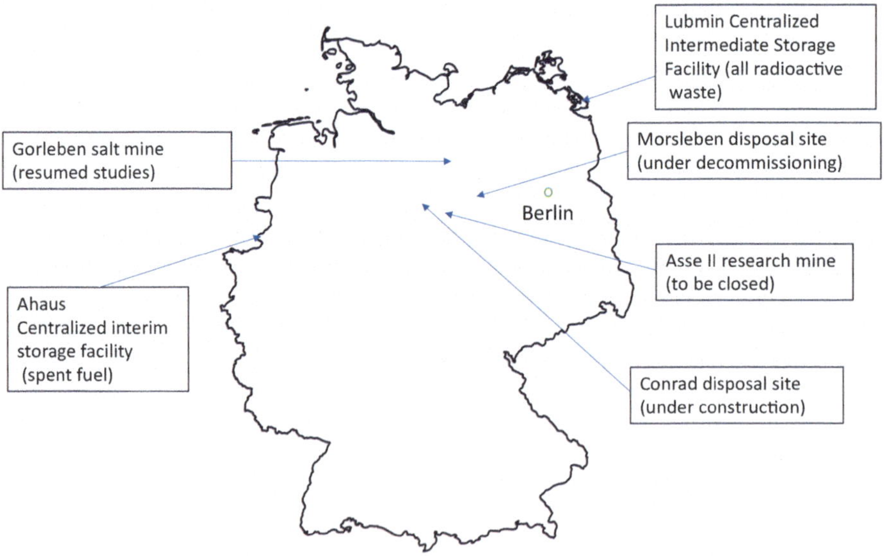

Fig. 3.8 Central interim storage and repository of radioactive waste in Germany

(1) Decommissioned nuclear power plants and prototype reactors: 5 planned for decommissioning application, 28 under decommissioning, 3 completed decommissioning and deregulated.
(2) Decommissioned research reactors: 3 under decommissioning permit application, 6 under decommissioning, 31 completed decommissioning and deregulated.
(3) Nuclear fuel cycle facilities (excluding interim storage facilities and final disposal sites): 2 under decommissioning, 9 completed decommissioning and deregulated.

In reprocessing and fast reactors, the reprocessing pilot plant WAK (with an annual reprocessing capacity of 35 tons, stopped in 1990) and the fast experimental reactor KNK-II (with an electrical output of 17,000 kilowatts, stopped in 1991) at the federal Karlsruhe Research Institute (now Karlsruhe Institute of Technology) are under decommissioning. The German fast prototype reactor SNR300 was built but never started operation.

Germany is a federal state, so the responsibility for nuclear-related permits, including decommissioning measures, lies with the states. The Federal Ministry for the Environment, Nature Conservation and Nuclear Safety (BMU) creates nuclear regulatory documents and supervises the nuclear administration carried out by the states under federal delegation. The BMU has advisory committees such as the Nuclear Safety Committee and the Radiation Protection Committee. Under the BMU is the Federal Office for Radiation Protection (BfS). The authorities of each state carry out licensing and supervision. The Technical Inspection Association (TÜB) and the Nuclear Safety Association (GRS) review licensing documents and create safety certificates on behalf of the states. They do not have decision-making authority.

Radioactive Waste Disposal

In Germany, the federal government creates a *national waste management plan and updates it every three years, advancing the management and disposal of radioactive waste*. The underlying list includes estimates of the amount of radioactive waste generated until 2080.

In Germany, the implementation of radioactive waste disposal is under the jurisdiction of the Federal Ministry for the Environment, Nature Conservation and Nuclear Safety (BMU), and the licensing is carried out by its subordinate agency, the Federal Office for the Safety of Nuclear Waste Management (BASE). The implementing body for disposal is the Federal Company for Radioactive Waste Disposal (BGE). The BGE is a 100% government-owned agency supervised by the BMU.

In Germany, it was decided in the 1960s to create a radioactive waste disposal site underground. For this purpose, abandoned mines are used as disposal sites. Germany's radioactive waste is distinguished as either heat-generating or non-heat-generating. The Konrad disposal site is planned as a disposal site for radioactive waste where heat generation can be ignored. The heat generated by the decay of radioactive substances in the waste, which does not raise the temperature of the waste body containing that radioactive waste by an average of more than 3 degrees, is used as a guideline for non-heat-generating waste. This corresponds to the natural temperature variation of 3 degrees in the strata 100 m underground. This heat generation is equivalent to the waste body having a heat generation of 200 watts per cubic meter.

Non-heat-generating radioactive waste includes pipes from nuclear power plants, ion exchange resins, filters for exhaust and wastewater, protective clothing, decontaminants, waste generated in laboratories, sealed radiation sources, sludge, and suspended matter. These are low-level and intermediate-level radioactive waste.

Heat-generating radioactive waste is waste with a high concentration of radioactive substances and a large amount of decay heat. This includes concentrated liquids containing fission products, metals produced when fuel assemblies are cut for reprocessing used fuel, residues generated during reprocessing, radioactive waste generated when reprocessing was commissioned to the UK and France, and unprocessed used fuel. These are high-level and intermediate-level radioactive wastes. Used fuel itself is not currently considered waste and is being stored temporarily. The transportation of used fuel for reprocessing in the UK and France ended in 2005. (BASE 2022).

The disposal site for radioactive waste in Germany is shown in Fig. 3.8. The Konrad disposal site is an old iron mine that was in operation from 1965 to 1975, and a disposal site for non-heat-generating radioactive waste is being constructed using the old mine shaft. The Morsleben disposal site is a potassium and rock salt mine over 100 years old, where non-heat-generating intermediate-level radioactive waste was disposed of from 1971 to 1998. Disposal has ended, and unused cavities have been backfilled. It is currently in the process of being decommissioned as a disposal site. The Asse II mine is a potassium and rock salt mine from 100 years ago, used by the Radiation & Environment Association (now the Helmholtz Research Center in Munich) for disposal research from 1965 to 1995. From 1967 to 1978, drums containing radioactive waste were brought in. These are being recovered and are scheduled to be decommissioned (closed) (BASE 2017).

Gorleben is located near the former East German border, and the underground rock salt dome became a candidate site for a disposal site for high-level radioactive waste (heat-generating radioactive waste) in 1977. However, subsequent investigations found that the clay layer above the rock salt layer was unstable and there was a risk of the rock salt dome being infiltrated by groundwater, making it unsuitable as a disposal site. The aforementioned centralized interim storage facility for high-level waste in Gorleben is separate from this, and the two are about 2 km apart.

In Germany, the search for new high-level radioactive waste disposal sites has been underway since the 2010s. Since 2017, a federal government-owned company (BGE) has been in charge of this, with the goal of finding candidate sites by 2031. This company also operates and manages the final disposal sites of Konrad and Morsleben and the old mines of Asse II and Gorleben. Site selection is carried out in the following steps.

(1) Selection of Suitable Areas: First, areas unsuitable for disposal are eliminated based on six criteria. Among those that meet the criteria, areas that are more suitable and those that are not are further divided from a geological science

perspective, and a report is published. Then, a preliminary safety investigation is conducted. Based on these results, BGE proposes areas for surface investigation to the Federal Office for the Safety of Nuclear Waste Management (BASE).

(2) Surface Investigation.

In the selected areas, BGE conducts a surface investigation to deepen the preliminary investigation. BGE analyzes the socio-economic potential of the area. These are used in the proposal for the site where the underground investigation will be conducted. BASE evaluates the proposal and decides on the site for the underground investigation. A review by the federal court is also conducted.

(C) Underground Investigation.

BGE conducts an underground investigation, creates a comprehensive safety analysis report, and submits it to BASE. BASE conducts an environmental assessment.

Based on these results, BASE proposes and decides on a disposal site for heat-generating radioactive waste (BGE 2024).

3.2.5 Disposal Measures and Radioactive Waste of Nuclear Facilities in Sweden

Disposal Measures for Nuclear Power Plants in Sweden

After the first oil crisis, Sweden built nuclear power plants in the 1970s due to the high cost of thermal power generation and concerns about stable supply. Previously, 12 nuclear power plants (9 boiling water light water reactors, 3 pressurized water light water reactors) were in operation, supplying about 80% of the electricity through hydroelectric and nuclear power. The locations of Sweden's nuclear power plants and radioactive waste disposal sites/interim storage facilities are shown in Fig. 3.9. The 1979 Three Mile Island nuclear power plant accident in the United States had an impact, and a national referendum on the pros and cons of nuclear power generation was held in 1980. Subsequently, influenced by the Chernobyl accident, the Barsebäck nuclear power plant, located 50 km across the strait from the capital of neighboring Denmark, Copenhagen, ceased operations in 1999 and 2005. On the other hand, about 1.6 million kilowatts of output enhancement (increased power generation output due to replacement with more efficient power generation turbines, etc.) has been made at operating power plants. After the TEPCO Fukushima accident in 2011, two units of the Oskarshamn power plant and then two units of the Ringhals power plant ceased operations by 2020 due to management decisions considering the cost of enhanced safety measures and the downturn in electricity demand, and the low cost of hydropower generation. All of these power plants are in the process of being decommissioned. Currently, three units at Forsmark, one unit at Oskarshamn, and two units at Ringhals, a total of six nuclear power plants, are in operation. All of these started operations in the 1980s. They supply about 30% of Sweden's electricity. The owners of the Forsmark and Ringhals nuclear power plants are Vattenfall (a state-owned power company owned by the Swedish government), and the Oskarshamn

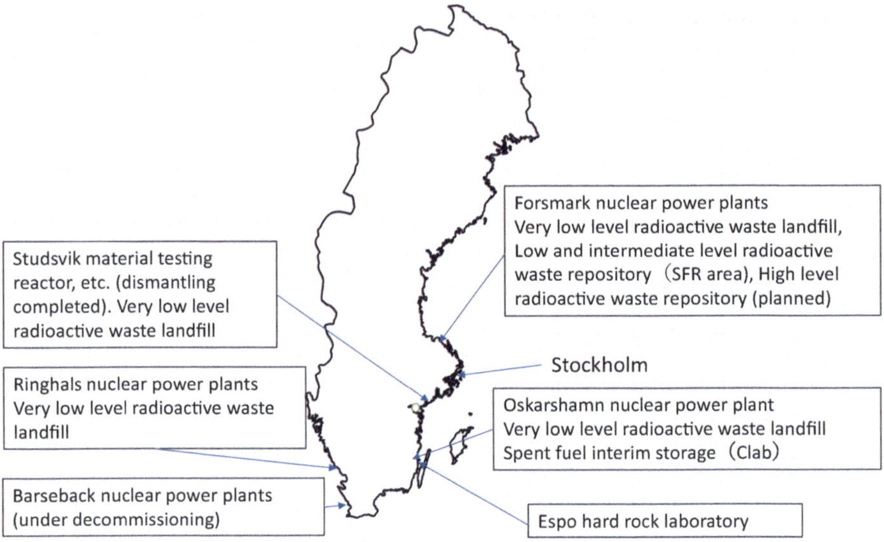

Fig. 3.9 Nuclear power plants, radioactive waste disposal sites and interim storage facilities in Sweden

power plant is owned by Uniper SE (formerly E.ON Sweden, a private company) and Fortum Sweden (a company owned by the Finnish government) each owning just over 50% and just under 50% of the shares, respectively.

In other major decommissioning measures, the decommissioning of the AGESTA pressurized heavy water reactor, which had been supplying electricity and heat for heating to Stockholm since 1974, began in 2020. The dismantling of the R2 material test reactor of Studsvik, a former national nuclear research institute, began in 2015, and in 2020, it is expected that an application will be made for the release of the remaining buildings and underground structures from nuclear regulation. The decommissioning measures for the Randstadt uranium mine and refining facility were carried out from 2015 to 2017, and in 2019 the site was released from nuclear regulation (MOE-GOS 2020).

Disposal of Radioactive Waste in Sweden

The administration of radioactive waste in Sweden is under the jurisdiction of the Ministry of the Environment. The Ministry of the Environment is responsible for nuclear safety (including security), radiation protection, and nuclear damage compensation. The direct responsibility for the operation and decommissioning of nuclear power plants and nuclear facilities, and the treatment and disposal of the resulting spent fuel and radioactive waste, lies with each company and institution. The four nuclear power companies have a responsibility to establish a management system and disposal site necessary for the disposal of radioactive waste and spent fuel generated by the operation and decommissioning of nuclear power plants (NEA 2020a). SKB (Swedish Nuclear Fuel and Waste Management Co) is a management company for spent fuel and radioactive waste created jointly by nuclear power companies. It is responsible for the transportation and disposal of radioactive waste, as well as the necessary research for disposal. Each nuclear power company is responsible for the landfill of very low-level radioactive waste within the power plant site.

Studsvik is a company that manages waste generated in the medical, research, and non-nuclear power sectors. Its disposal can be carried out at the disposal site operated by SKB.

AB SVAFO is a company jointly established by nuclear power companies, responsible for the decommissioning measures of old research reactors and development facilities, and the management of historical waste (legacy waste). Legacy waste can be disposed of at the disposal site operated by SKB.

Regulations are made by the parliament, and the administration is in charge of the Ministry of Environment. The Swedish Radiation Safety Authority (SSM) is responsible for nuclear safety and regulation. The National Council for Nuclear Waste is an independent advisory committee under the jurisdiction of the Ministry of Environment. It conducts research on decommissioning measures and radioactive waste and advises the government and the Ministry of Environment.

The classification of radioactive waste in Sweden is not defined by law, but it is divided into the following four types according to the difference in disposal methods:

1. Very low-level radioactive waste,
2. Short-lived low and intermediate-level radioactive waste,
3. Long-lived low and intermediate-level radioactive waste,
4. Spent fuel.

Very low-level radioactive waste is waste generated from the decommissioning and operation of nuclear power plants and the process of fuel processing for reactors. It is disposed of in shallow ground within the facility area of nuclear power plants, or if it is below the standard, it is disposed of at regular waste disposal sites or recycled. There are landfill (shallow ground burial) disposal sites at Forsmark, Oskarshamn, Ringhals nuclear power plants, and Studsvik, which was formerly a national nuclear research institute. The amount of disposal is an order of magnitude larger at each nuclear power plant.

Short-lived low and intermediate-level radioactive waste is generated from decommissioning and operation, as well as sealed radiation sources and industrial waste. It is disposed of at the disposal site for short-lived radioactive waste (SFR). The SFR is built and operating underground (under the seabed) near the Forsmark nuclear power plant.

Long-lived low and intermediate-level radioactive waste includes structural materials inside the reactor vessel of nuclear power plants, transuranic waste, and high-dose sealed sources. It is planned to be disposed of at the disposal site for long-lived waste (SFL). The SKB is expected to apply for a disposal site in the late 2030s, depending on how long the nuclear power plants will operate.

Used fuel is managed at a centralized interim storage facility called Clab, built near the Oskarshamn power plant. Construction began in 1980 and it has been in operation since 1985. The disposal site for used fuel is planned to be built near the Forsmark nuclear power plant in Östhammar. Long-term site activities have been carried out, and as of October 2020, the understanding of the local Östhammar municipality for the site has been obtained (SKB 2020). In Sweden, after 40 years of site activities, it is expected that the used fuel disposal site (SFK) will be in operation, following Finland.

In relation to geological disposal, there is the Äspö Hard Rock Laboratory near the Oskarshamn power plant. At a depth of 460 m, research is being conducted on the behavior of barriers (storage containers, buffer materials, backfill materials, rocks) for disposing of used fuel. Research on the disposal of low and intermediate-level radioactive waste with long half-lives is also being conducted. The above-ground part includes a high-level radioactive waste storage container laboratory and a multipurpose laboratory (bentonite laboratory). Bentonite is a general term for clay mainly composed of a mineral called montmorillonite, also known as swelling clay. It is widely used in the industrial and construction fields as a waterproof material and binder. In geological disposal, it is installed to surround the outside of the storage container for high-level radioactive waste, and is used to prevent the intrusion of groundwater for a long period.

As mentioned later, in Sweden, disposal and recycling at general disposal sites for waste that is not radioactive waste generated by decommissioning measures, etc. (clearance items) is progressing. For example, in 2004, about 500 tons of metal scrap (items below 500 becquerels/kg) were melted and reused (NEA 2020a, p. 11). Studsvik has a large metal melting furnace and accepts the melting of foreign waste.

3.2.6 Decommissioning of Nuclear Facilities and Radioactive Waste in Finland

Finland has few energy resources in its own country, imports coal from Russia and Poland, and electricity from Sweden, and is promoting nuclear power generation. Finland's nuclear power plants and radioactive waste disposal sites are shown in Fig. 3.10.

Finland has two nuclear power plants each in Loviisa and Olkiluoto, supplying about 30% of the electricity. Loviisa has two pressurized water light water reactors called VVER type made in the former Soviet Union, and Olkiluoto has two boiling water light water reactors made in Sweden, which have been in operation since the late 1970s. The pressurized water light water reactor in Loviisa is a modified version of the former Soviet Union's VVER type, with improvements made for Finland, such as the addition of a reactor containment vessel from the Western pressurized water light water reactor. Both power plants have extremely high operating rates, reaching 95% in the last 10 years. Power output has also been increased, with the Olkiluoto power plant, which initially had an electrical output of 660,000 kilowatts, now operating at 880,000 kilowatts, and the Loviisa power plant, which initially had an output of

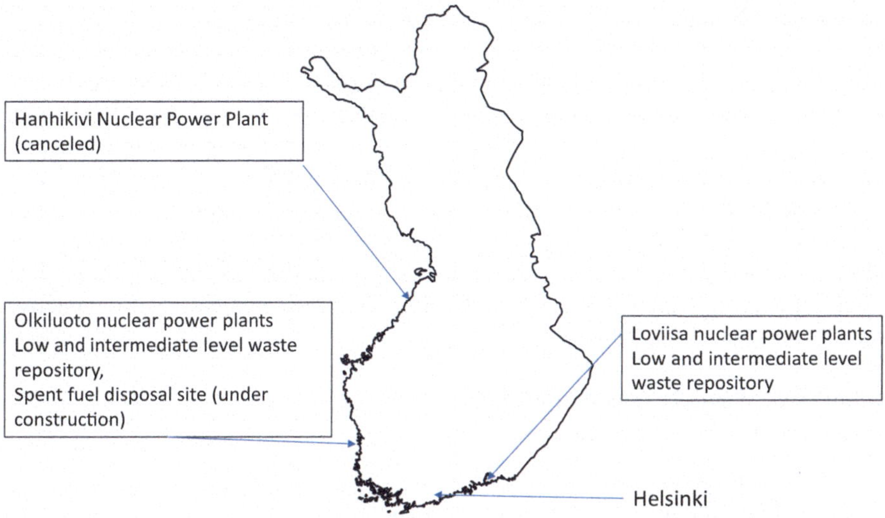

Fig. 3.10 Nuclear power plants and radioactive waste disposal sites in Finland

420,000 kilowatts, now generating about 510,000 kilowatts. Construction of Unit 3 is completed at Olkiluoto (WNA 2024).

The companies operating nuclear power plants are Fortum Power and Heat (FPH) for the Loviisa power plant and Teollisuuden Voima (TVO) for Olkiluoto, both of which are responsible for the disposal of used fuel and radioactive waste resulting from power generation. TVO is a private company that sells the electricity generated at Olkiluoto to its shareholder companies at wholesale prices. Fortum owns numerous hydroelectric power plants and conducts transmission business mainly in Finland and its surrounding countries. In Finland, Fennovoima has applied for permission to build the Hanhikivi nuclear power plant in Pyhäjoki in the central west of Finland using a Russian pressurized water reactor (STUK 2020). After Russia's invasion of Ukraine, Fennovoima canceled its order for the power plant from Rosatom in May 2022.

There are no nuclear power plants in Finland that are undergoing decommissioning. Used fuel is stored on the power plant site until the start of geological disposal. The used fuel from the Loviisa reactor was taken over by Russia until 1996, but is now stored on the power plant site. The small research reactor (TRIGA type) made in the USA, which started operation in 1961, was operated at VTT, a research center in the suburbs of Helsinki, Finland, but was shut down in 2015, applied for decommissioning in 2021, and plans to decontaminate the reactor building and release it from nuclear regulation in 2023. The TRIGA type research reactor is made in the USA and used in many countries around the world. The used fuel from it, not limited to Finland's, is taken over by the USA (VTT 2024).

The agencies and organizations related to radioactive waste and decommissioning in Finland are the government's Ministry of Economic Affairs and Employment, MEAE. The responsibility for long-term radioactive waste management lies with the

Radiation and Nuclear Safety Authority (STUK), a department under the Ministry of Social Affairs and Health. Nuclear power companies are responsible for the interim storage of spent fuel, low and intermediate-level radioactive waste, and decommissioning measures. Posiva, a company jointly established by Fortum and TVO, which operate nuclear power plants, is responsible for the preparation and implementation of the geological disposal of spent fuel. The national research institute, VTT, is responsible for the decommissioning of its research reactors and hot cells (facilities for handling nuclear fuel materials and radioactive substances). STUK operates an interim storage facility for waste generated from radiation use other than nuclear power generation, which is attached to the low- and intermediate-level radioactive waste disposal site in Olkiluoto.

Radioactive Waste Disposal in Finland

Radioactive waste is classified into.

1. high-level radioactive waste,
2. low and intermediate-level radioactive waste,
3. very low-level radioactive waste, and
4. radioactive waste generated from medical, industrial, and research activities,
5. waste below the clearance level.

Depending on the source, there is radioactive waste associated with the use of nuclear energy and radioactive waste associated with the use of radiation.

Finland has prohibited the export and import of spent fuel, such as taking spent fuel to foreign countries for reprocessing, and the export of radioactive waste generated from nuclear power generation to overseas, by law in 1994. High-level radioactive waste is spent fuel from nuclear power plants, and Posiva is advancing the construction of a geological disposal site near the Olkiluoto power plant, where it will be disposed of after completion.

The low and intermediate-level radioactive waste disposal site was built about 100 m underground near the Olkiluoto and Loviisa power plants. It has been in operation since 1992 and 1998, and waste is disposed of in cavities (silos) and tunnels underground. The Olkiluoto disposal site also accepts low and intermediate-level radioactive waste generated from medical, industrial, and research activities. It is noteworthy that the disposal sites of nuclear power companies also accept radioactive waste other than that from nuclear power generation.

There is no disposal site for very low-level radioactive waste yet, but plans are underway. In 2021, TVO submitted an environmental impact assessment report to the responsible ministry, MEAE, for the construction of a disposal site near the Olkiluoto power plant. MEAE has stated that it meets the environmental impact assessment criteria. The disposal site is planned to be built near the surface, but attaching it to the existing low and intermediate-level disposal site is also considered as an alternative. Public hearings will be held on the plan. The environmental impact assessment report and MEAE's conclusion are available on the Finnish government's homepage (MEAE 2024).

In Finland, there is a "Principle Decision" (Decision-in There is a legal system called the "Principle" which is also applied in the Atomic Energy Act. In the licensing of nuclear power plant construction, used fuel storage facilities, and disposal facilities, it is necessary to go through the following three processes, the first of which is the "Principle Decision". The Radiation and Nuclear Safety Center (STUK) conducts safety assessments at each stage, and the Ministry of Employment and Economy (MEAE) is responsible for the administrative decisions (NEA 2020b, p. 5).

(1) Principle Decision: The government makes the licensing decision after obtaining the approval of the local municipality and the parliament in advance.
(2) Construction permission by the government.
(3) Operating permission by the government.

In Finland, this system required consideration of the decommissioning measures and radioactive waste generated when planning the construction of a nuclear power plant in the past. The Principle Decision legal system is also applied in the licensing of used fuel storage and disposal facilities.

This process also determined the disposal site in the selection of the geological disposal site. First, the stage and target period for the selection of the disposal site were set in 1983. Then, site confirmation surveys were conducted from 1983 to 1985. Geological information was collected, literature surveys were conducted to avoid large fracture zones and select stable bedrock blocks, and many target areas consisting of 100 square kilometers were selected from all over Finland. Furthermore, based on literature surveys on the size of the bedrock, transportation of used fuel, population density, etc., the target areas were narrowed down, and about 100 areas of about 10 square kilometers were selected. From 1986 to 1992, preliminary site characteristic surveys were conducted in five areas by drilling, etc., and detailed site characteristic surveys were conducted in four areas considered more suitable from 1993 to 2000. Posiva Corporation compiled a report on long-term safety, selected Olkiluoto in the municipality of Eurajoki, and applied based on the Principle Decision procedure in 1999. These were examined by STUK (Radiation and Nuclear Safety Authority), and the opinion was submitted to the government in 2000. Based on this opinion, the Eurajoki municipality voted in the parliament and expressed acceptance of the disposal site. In response to this, the government made a Principle Decision in 2001, which was approved by the parliament in 2001. As a result, Olkiluoto (Eurajoki municipality) has been decided as the final disposal site. Construction of the underground characteristic survey facility (Onkalo) has been started since 2004. In Finland, activities that involve expenses that influence judgments in the government and parliament (such as large-scale underground excavation) are prohibited before the Principle Decision.

Posiva Corporation obtained permission for the construction of the geological disposal site in 2015, and construction of the geological disposal site and the facility (Onkalo) for storing fuel assemblies in containers began the following year. After obtaining operating permission, disposal is scheduled to start in 2024. The operating period is 45 years, and a regular safety review is planned every 15 years (RWMC 2021, pp. 132–147 of STUK 2020, pp. 14–15 of NDA 2013). In a public opinion survey in Finland, for the first time in 2019, the public's trust in geological disposal exceeded the skeptical view (36%) at 37%. Since the 1980s, trust has been gradually increasing, excluding the years of the Chernobyl accident and the TEPCO accident. In other words, the plan for the geological disposal site was advanced during a time when there were many skeptical citizens.

3.2.7 *Recycling of Non-radioactive Waste (Clearance System)*

The majority of waste generated by the decommissioning of nuclear power plants (about 97%) is not radioactive waste. Many of these have a radioactive substance concentration lower than the concentration of low-level radioactive waste defined by law, or the same as that of substances in nature. The "clearance system" is a system that allows for the reuse or disposal of radioactive waste generated by the operation and decommissioning of nuclear power plants, which has a low concentration of radioactive substances and has little impact on human health, with the approval/confirmation of the country, and is implemented in various countries. The European Community and the International Atomic Energy Agency have summarized their thoughts. The Nuclear Energy Agency of the Organisation for Economic Co-operation and Development (OECD/NEA) has summarized the current status of the clearance system in member countries (NEA 2008).

Clearance is defined by the International Atomic Energy Agency as "not applying regulations for radiological protection to radioactive substances or radioactive objects if they follow a predetermined procedure". The concept of clearance is closely related to the concept of exemption, but strictly speaking, clearance is the process of deregulation, and exemption is the process of deciding that there is no need for restrictions by regulation from the beginning, and there is a slight difference. For example, the criteria for clearance vary depending on how the waste is used or disposed of afterward, or whether the decommissioned building is in an area where public access is restricted.

The concept of clearance corresponds to the concept of "triviality of exposure", for example in the following cases.

(1) When the individual exposure risk caused by the act or source is considered "trivial".
(2) When the collective radiation dose caused by the act or source is sufficiently low in general situations.
(3) When the possibility of the act or source causing exposure exceeding the dose limit is sufficiently low.

When creating the criteria for clearance, each country generally adopts the standard that the public exposure dose is less than 10 microsieverts/year. This is also the case in Japan.

Clearance includes "unconditional clearance/unrestricted use" and "conditional clearance/restricted use".

Clearance means that if the criteria are met, it is not subject to the restrictions of (nuclear and radiation) regulations, and waste that meets the clearance criteria can be treated as regular waste. For example, metals can be reused, recycled, sold, or transferred to a third party, and can be used for any purpose without being considered as radioactive substances. This is called "unconditional clearance". "Conditional clearance" or "clearance for specific purposes" is, for example, when metals are not directly reused by melting, when only the demolition of buildings is carried out without reuse, or when they are disposed of in regular disposal sites. Even in these cases, they are not subject to regulation after clearance has been implemented.

The procedures for calculating the clearance criteria for metals and buildings have been compiled by international agencies. Depending on whether it is conditional or not, the criteria values change. The main radioactive substances generated in nuclear power generation and nuclear fuel reprocessing facilities are.

1. weak beta-emitting nuclides (H-3, C-14, Ni-63),
2. strong beta/gamma-emitting nuclides (Co-60, Cs-137),
3. strong beta-emitting nuclides (Sr-90),
4. alpha-emitting nuclides (U-235, Am-241, Pu-239).

Clearance criteria values (becquerels per gram) for these nuclides have been established for unconditional clearance, scrap metals for reuse/recycling, building rubble, buildings for reuse, and c building demolition only, (Table 3.1 of NEA 2008).

Materials such as concrete in buildings contain uranium and thorium, which are naturally occurring radioactive substances, and radioactive substances generated by their radioactive decay, and K-40 (a nuclide that was generated in the Big Bang at the time of cosmic creation and remains because of its long half-life), but there is no need to consider these naturally occurring radioactive substances in the measurements for clearance determination. In the determination of clearance, only radioactive substances generated in the process of using nuclear energy and radiation need to be considered.

The term clearance is from the perspective of decommissioning measures and radioactive waste treatment and disposal, but when viewed from the perspective of reuse and recycling, it can be understood that clearance is within the framework of a circular society. At the beginning of this chapter, we discussed general waste. The reuse of general waste is a means to realize a circular society, and the proportion of reuse has been increasing year by year. *The clearance system is a means to contribute to a circular society in the nuclear field.*

If concrete and metal generated by the decommissioning of nuclear power plants are not reused after clearance, it is necessary to cut down mountains to extract rocks, soil, and ores, consume energy to process them, and create new cement, gravel, concrete, and metal. The amount of waste generated by decommissioning is far greater than the waste generated by operation. Decommissioning in the nuclear field is expected to increase in future. There is a demand to increase the proportion of recycling and reuse of waste generated by decommissioning.

There are, for example, the following methods to aim for recycling in decommissioning:

(1) Items contaminated with radioactive substances on the surface are decontaminated, go through unconditional clearance, and are then reused or melted.
(2) Items that generate radioactive substances inside and have radioactivity below the standard are reused after conditional clearance.
(3) Items with radioactivity above the standard are reused in an environment regulated by nuclear and radiation.

(1) applies to metal pipes, power turbines, and other equipment used outside the vicinity of the reactor (precisely outside the range where metal and concrete are radiated by neutrons generated in the reactor), and pipes used in nuclear fuel cycle facilities. Buildings can also be reused by removing surface contamination from the walls. If there are cracks in the wall and contamination has seeped in, that part also needs to be decontaminated or the concrete removed to meet the standard. When demolishing a building, the used rebar can be reused after unconditional clearance.

(2) and (3) are, for example, reused as iron radiation shields used in large accelerator facilities.

In the West, the recycling and reuse of primary waste generated by the decommissioning of nuclear power plants and the like have a history of nearly half a century. In the United States, there are private companies that conduct decommissioning as a business and process for recycling and reusing waste generated by decommissioning. In Europe, a company in Sweden, which was a national nuclear research institute half a century ago, is conducting tests on reactor fuel and materials, and has made clearance and reuse of waste generated by decommissioning one of its main jobs. Both companies also process and recycle waste from outside their own countries.

Furthermore, while the United States does not have a legalized clearance system, standards for releasing nuclear power plant sites (grounds) have been legalized through decommissioning measures. Three types of termination approvals are available: unrestricted release, restricted release, and alternative standards. The standard for unrestricted site release is less than 250 microsieverts/year in total effective dose equivalent due to residual radioactivity. The standard for restricted release is less than 1000 microsieverts/year when the unrestricted release standard cannot be met despite efforts, and institutional control of the site is imposed even after the decommissioning measures are completed. Alternative standards are prepared for cases where the release standards cannot be met even with restrictions on-site use. In all cases, it is required to follow the concept of ALARA in addition to the dose equivalent value. As the United States is a federal system, there are power plants where lower release standards than 250 microsieverts/year have been applied depending on the state.

Let's discuss a little more specifically what kind of recycling and reuse is being done (NEA 2017). Metals are relatively expensive and easy to decontaminate among the substances generated in decommissioning, so recycling is progressing. Iron is used not only in power plant structures but also in building reinforcements, and is the most abundant. Copper is used in power cables and is expensive, so it is being recycled. Lead has chemical toxicity, and aluminum, when mixed with other substances and discarded, reacts chemically with the contained moisture to generate hydrogen gas, making it unsuitable for disposal as radioactive waste, so recycling is required. When treated as waste, there are restrictions such as limiting aluminum to less than 10 kg per drum can containing radioactive waste. Iron is recycled after decontaminating surface contamination and meeting the unconditional clearance. Metals are melted in facilities within the controlled area, then ingotized and cleared. The fission product cesium 137 volatilizes by melting, so its contamination can be removed. Some of the radioactive substances contained in the metal may transfer to the slag (melted waste) when the metal is melted. Metals generated in decommissioning may also be melted in civilian facilities mixed with other (non-radioactive) metals. In that case, the melting facility is regularly inspected, and it is required to record the handling of the resulting metal and the review of the process to its final destination (conditional clearance).

Concrete waste is most commonly generated during decommissioning. In the demolition of buildings, surface contamination is removed in advance and the concentration of radioactivity is measured. After this process, the concrete waste generated by demolition can be treated in the same way as concrete waste generated by the demolition of ordinary buildings. There are also cases where the concrete is cut into small blocks for demolition. Furthermore, there are cases where the concrete blocks are crushed. The concrete waste obtained in this way is used for site leveling and landfill, and is used for site recovery. When reusing concrete on site, it is permissible to have a higher concentration of radioactive substances than unconditional clearance because public exposure can be avoided.

The turbine of a generator is a large structure made of stainless steel. There have been cases where surface contamination was removed and handed over to a civilian recycling contractor. There have also been cases where the heat exchangers (steam generators) of nuclear power plants were transported to the country of the company that undertakes the decommissioning and clearance as they are, decontaminated, and melted. The radioactive waste generated in the process of decontamination is returned to the commissioned country and disposed of as low-level radioactive waste. There have been cases where lead generated by demolition was melted on site and reused to make radiation shielding blocks. If it can be reused within the nuclear facility, it can be reused without clearance. There are also cases where some of the buildings of the decommissioned power plant, such as the turbine building, are left for future use or used as warehouses without being demolished.

3.2.8 Decommissioning of Nuclear Facilities and Radioactive Waste in Japan

Decommissioning of Nuclear Power Reactors

The first nuclear power generation in Japan was carried out by the Japan Atomic Energy Research Institute (now the Japan Atomic Energy Agency) with Japan Power Demonstration Reactor (JPDR) in 1963. In 1966, the Tokai Power Plant of the Japan Atomic Power Corporation started the first commercial power generation. Subsequently, due to the impact of the two oil shocks, many nuclear power plants were built throughout Japan over a period of about 30 years from the 1970s. In Japan, before the liberalization of electricity, power was supplied by 10 power companies in each region. Of these, nine power companies, excluding Okinawa Electric Power Company, have built and operated nuclear power plants. The Japan Atomic Power Corporation is a company specializing in nuclear power generation, funded by power companies in Japan, and all of these are private companies.

The JPDR ceased operation in 1976 and was used for field testing aimed at developing decommissioning technology for power reactors. The main body began to be dismantled in 1986 and was completed in 1996. Radioactive waste was stored in shielded containers, and field tests were conducted to bury concrete waste with extremely low radiation levels in shallow ground within the research institute. Most of the waste generated by the dismantling of the building was non-radioactive concrete, which was crushed and used to fill in depressions caused by the dismantling of the JPDR, and as the foundation for roads and structures within the institute (Japan Atomic Energy Agency 2024).

The Tokai Power Station (Tokai Unit 1) is a natural uranium fueled and carbon dioxide cooled reactor, known as the Calder Hall type, developed and commercialized in the UK. It began commercial operation in 1966 and ended operation in 1998. Decommissioning measures began in 2001, including the removal of heat exchangers. Clearance items and non-radioactive waste were also removed. In 2015, permission was obtained from the Nuclear Regulation Authority to bury the lowest concentration L3 waste among low-level radioactive waste in a shallow ground trench facility within the power station, and burial work has been carried out over a 10-year plan. The management period for the trench burial facility is 50 years after the end of burial. The concentration of radioactive substances in groundwater is regularly measured. The graphite used in the reactor of the Tokai Power Station is sealed in the reactor vessel until a disposal method and site are determined (Japan Atomic Power Company 2024). This is the same method as in the UK.

The Japan Atomic Energy Agency's "Fugen" is a prototype power reactor (a reactor to confirm whether a new type of reactor can be used for power generation) with heavy water moderated boiling light water cooling. Because it has a high efficiency of using plutonium fuel, it was called a new type of conversion reactor. It was built on the Tsuruga Peninsula in Fukui Prefecture, started generating power in 1978, and was operated until 2003. It is currently being decommissioned.

Units 1 and 2 of the Chubu Electric Power Company Hamaoka Power Station ceased operation in 2009 and have transitioned to decommissioning. The removal of fuel, decontamination of radioactive substances attached to equipment and pipes where reactor water circulates, dismantling and removal of facilities around the reactor area, dismantling and removal of the reactor area, and dismantling and removal of buildings are being carried out in parallel with the treatment and disposal of waste (Chubu Electric Power Company 2013).

The above refers to power reactors that transitioned to decommissioning before the TEPCO accident in 2011. In Japan, before the TEPCO accident, there were three commercial nuclear power plants, including the Tokai Power Plant, that had ended their commercial operations. However, after the TEPCO accident, many nuclear

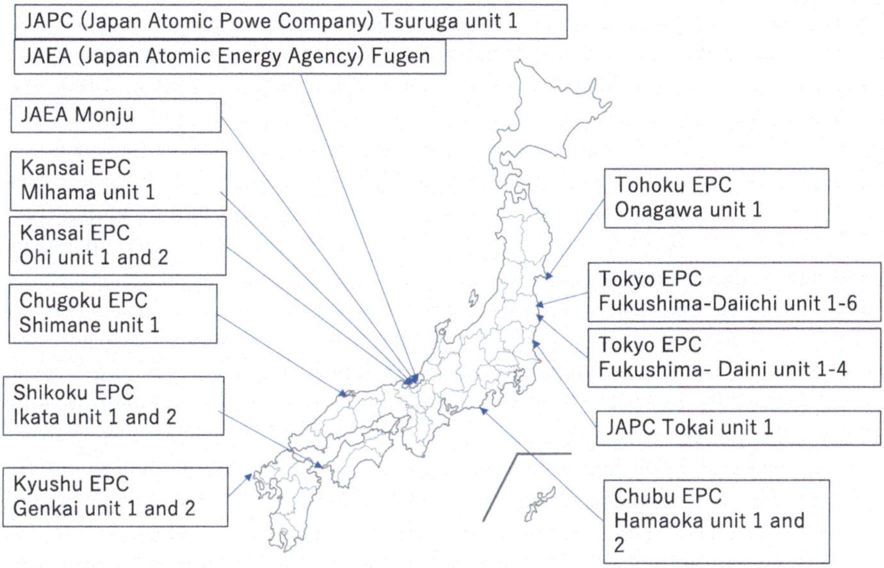

Fig. 3.11 Nuclear power plants under decommissioning in Japan

power plants, especially those with smaller power outputs, abandoned their operations and transitioned to decommissioning due to management decisions on whether they could recover the cost of facility investments to comply with new regulations. The nuclear power plants in Japan that are in the process of decommissioning are shown in Fig. 3.11.

Regarding safety regulations, in order to carry out decommissioning, it is necessary to submit a decommissioning plan to the Nuclear Regulation Authority (NRA) and receive approval. Approval is also required when changing the decommissioning plan. Confirmation is also required when decommissioning is completed. After receiving confirmation of completion, the reactor is exempted from the application of the Reactor Regulation Law (Nuclear Regulation Authority 2024a). The website of the Nuclear Regulation Authority publishes the status of decommissioning not only for power reactors but also for research reactors, reprocessing businesses, and disposal businesses. By law, the installers of reactors are required to create and publish a policy for implementing decommissioning when they start operating the facility.

Decommissioning of Nuclear Research and Development Facilities

Decommissioning is also being carried out for nuclear facilities used for research and development, such as research reactors and nuclear fuel handling facilities. Japan's first research reactor, JRR-1, was operated from 1957 to 1969. After it was shut down, nuclear fuel was removed and it was sealed and is now on display as a memorial exhibition hall within the Japan Atomic Energy Agency in Tokai Village, Ibaraki Prefecture. The nuclear-powered ship "Mutsu" was built with Japanese technology, launched in 1969, and achieved a nuclear fission chain reaction at sea in 1974. However, due to the detection of radiation leakage during the power increase test, it was modified and conducted a power increase test and sea trial in 1990, and completed experimental navigation in 1992. The following year, the spent fuel was removed, and in 1995, the reactor compartment was removed in one piece and stored in a storage building. Along with the reactor control room and steering room, it is open to the public as the Mutsu Science and Technology Museum in Mutsu City, Aomori Prefecture. As of March 2011, research reactors that had stopped operation, in addition to these, include two at private universities, one at a national university, and two at private companies, for a total of six.

The Japan Atomic Energy Agency (JAEA) is a central nuclear research and development institution in Japan, formed in 2005 through the merger of the Japan Atomic Energy Research Institute and Japan Nuclear Cycle Development Institute (JNC, formerly PNC, Power Reactor and Nuclear Fuel Development Corporation). In 2018, the JAEA published a backend roadmap, outlining a long-term policy for the next 70 years on decommissioning measures, waste disposal, and management of nuclear fuel materials (Japan Atomic Energy Agency 2018a, b). For the first 10 years, specific plans are presented in the "Facility Medium to Long-term Plan". During this period, while ensuring the immediate safety of the facilities through new regulatory standards, seismic measures, aging countermeasures, and risk reduction measures, backend measures (decommissioning, waste disposal, and nuclear fuel management) are being advanced (Japan Atomic Energy Agency 2022).

The first phase involves advancing the decommissioning of "Monju", "Fugen", and "Tokai Reprocessing Plant (TRP)", as well as primarily advancing the decommissioning of relatively small-scale facilities with less waste generation. From the second phase in 2029 onwards, full-scale decommissioning measures will be implemented. The entire period for backend measures is 70 years, and the estimated cost, including facility dismantling costs and waste disposal costs, is approximately 1.9 trillion yen. The most expensive is the decommissioning of the Tokai Reprocessing Plant (TRP), which also has a long period of 70 years. The sodium-cooled fast reactor "Monju" and the advanced thermal reactor "Fugen" have larger facility dismantling costs than waste disposal costs. The Japan Atomic Energy Agency has established an organization to manage the entire backend and an organization specialized in decommissioning management of "Monju" and "Fugen", and has started work. In research reactors, JRR4 and JMTR (Material Testing Reactor) have given up on continued operation and have transitioned to decommissioning, but consideration is being given to constructing a new research reactor in the Tsuruga area of Fukui Prefecture.

Processing and Disposal of Radioactive Waste

Japan's radioactive waste is broadly divided into high-level radioactive waste and low-level radioactive waste. Low-level radioactive waste is divided into waste for intermediate-depth disposal, pit disposal, and trench disposal, depending on the disposal method, and may be referred to as L1, L2, and L3 waste, respectively. These correspond to intermediate-level radioactive waste, low-level radioactive waste, and very low-level radioactive waste. Intermediate-depth disposal is sometimes called surplus depth disposal and is a disposal at a depth of 50 m or more underground. In addition to the above, low-level radioactive waste includes uranium waste and radioactive waste containing transuranic elements (TRU waste). These are generated at uranium enrichment and fuel processing facilities and reprocessing facilities, respectively. Disposal of uranium waste is being considered according to the radioactivity concentration of the uranium contained. TRU waste with high radioactivity concentration is considered to be waste equivalent to geological disposal. High-level radioactive waste is a glass solidified body generated by reprocessing and is disposed of in deep geological layers. Used fuel is not considered radioactive waste in Japan, but it is a target for geological disposal in some countries. In the burial regulations of the law (Nuclear Reactor Regulation Law), the burial of high-level radioactive waste is called Type 1 waste burial, and the burial of low-level radioactive waste is called Type 2 waste burial.

In the classification by the source of waste generation, there are those associated with nuclear power generation, those associated with research and development, and those associated with medical care, etc. Radioactive isotopes (radioisotopes) are used in radiation use, but in the law, their use is regulated by a law separate from the Nuclear Reactor Regulation Law, the law concerning the regulation of radioactive isotopes (commonly known as the RI Law). Radioactive waste generated by the use of radioactive isotopes has been made possible to be finally disposed of as waste under the Nuclear Reactor Regulation Law due to the amendment of the RI Law in 2017. The use of radioactive isotopes for medical purposes is regulated by the Medical Law and the Veterinary Law, and is under the jurisdiction of the Ministry of Health, Labour and Welfare. Since there is no law concerning final disposal in the Medical Law, storage disposal continues. For waste generated by the use of radioactive isotopes in the medical field, the development of relevant laws and regulations for rationalizing treatment and disposal is being sought and considered (Ministry of Health, Labour and Welfare, 2021). Most of the radioactive isotopes used for medical purposes have a short half-life, and their radioactivity decays during use and storage. RI waste is collected and accepted for disposal by the Japan Radio Isotope Association (JRIA).

Radioactive waste operates under the principle of generator responsibility, requiring the business that generated the waste to be responsible for its disposal. In the case of nuclear power generation, the main body responsible for the disposal of low-level radioactive waste is the electric utility (power company). The pit disposal is carried out by Japan Nuclear Fuel Ltd. (JNFL), which was established with funding from power companies, at the Low-Level Radioactive Waste Disposal Center in Rokkasho Village, Aomori Prefecture, since 1992. For intermediate-depth disposal (surplus depth disposal), a test cavity has been created about 70 m underground, and tests are being conducted.

The Nuclear Waste Management Organization of Japan (NUMO) is responsible for the implementation of high-level radioactive waste disposal, and preparations are being made for site selection. In 2017, they published a scientific characteristic map, color-coding the undesirable and desirable areas and their requirements on a map of Japan. They plan to conduct literature surveys based on applications from local governments or proposals from the national government, listen to the opinions of the local community, select a district, and conduct preliminary and detailed surveys. It is legally required to listen to the opinions of local governments.

The main body responsible for the disposal of low-level radioactive waste generated by research institutions, universities, medical institutions, private companies, etc. (referred to as research facility waste) is the Japan Atomic Energy Agency, which accounts for the majority of waste generation and has technical knowledge about disposal. This was established in a legal amendment in 2008. The development of trench disposal facilities and pit disposal facilities for waste disposal is being promoted along with waste treatment and waste body production (Japan Atomic Energy Agency 2016).

Figure 3.12 shows Japan's nuclear fuel cycle facilities. In Mutsu City, Aomori Prefecture, Recyclable-Fuel Storage Co., Ltd., funded by the power companies, is constructing an interim storage facility for spent nuclear fuel from nuclear power plants. Note that there are also interim storage facilities built on the sites of power companies' nuclear power plants. In Rokkasho Village, Aomori Prefecture, Japan Nuclear Fuel Ltd.'s Low-Level Radioactive Waste Disposal Center (L2 waste pit disposal site), a uranium enrichment plant, and a high-level radioactive waste (glassified waste generated from overseas reprocessing of spent fuel) storage management center are in operation. The reprocessing plant has been built and is undergoing testing for commercial operation. A mixed oxide (MOX) fuel processing plant is being built to use the plutonium obtained at the reprocessing plant in nuclear power plants.

In Tokai Village, Ibaraki Prefecture, Japan Atomic Power Company operates a facility (Trench Disposal Site for L3 Waste) where very low-level radioactive waste generated from the decommissioning measures of the Tokai Power Station (Tokai Unit 1) is buried. The very low-level radioactive waste disposal facility of the Atomic

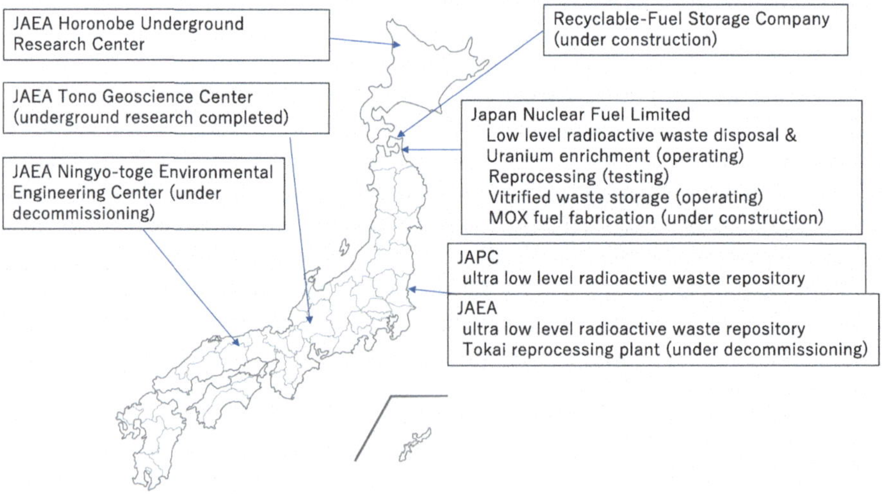

Fig. 3.12 Nuclear fuel cycle facilities in Japan. *Note* Uranium conversion and fuel fabrication plants locate in Ibaragi, Kanagawa and Osaka prefectures

Energy Science Institute of the Japan Atomic Energy Agency (JAEA) is disposing of waste generated from the decommissioning measures of JPDR and is under surveillance. At the Nuclear Fuel Cycle Research Institute of JAEA, preparations for the decommissioning of various facilities, including the Tokai Reprocessing Facility, are underway. The decommissioning of the Tokai Reprocessing Facility is the most costly and time-consuming among the decommissioning measures of JAEA.

The Ningyo Pass Environmental Technology Center of JAEA is located at Ningyo Pass on the border of Okayama and Tottori Prefectures, where the first uranium outcrop deposit was discovered in Japan in 1955. Decommissioning measures were taken for facilities that conducted technology development for uranium exploration, mining, refining, conversion, and uranium enrichment.

At the Horonobe Underground Research Center of JAEA in Horonobe Town, Hokkaido, research and development for geological disposal of high-level radioactive waste is being conducted by digging a shaft deep underground and creating a tunnel. An agreement has been made with the local government not to make it a disposal site. JAEA also conducted research on geological disposal at the Tono Geoscience Center in Gifu Prefecture, but this has ended underground research and is backfilling the tunnel in accordance with an agreement with the local community. This is granite, and the geology is different from the sedimentary rock of Horonobe, and the knowledge gained is useful. The Tono Geoscience Center began when a uranium deposit was discovered in the 1960s and a prospecting office was set up in 1965. The Horonobe Research Center was built in the 2000s. In addition to these, there are business establishments of private companies that manufacture nuclear fuel for nuclear power generation in Tokai Village, Ibaraki Prefecture, Yokosuka City, Kanagawa Prefecture, and Kumatori Town, Osaka Prefecture.

There is an international treaty on the safety of spent fuel management and radioactive waste management. The ratifying countries submit reports to the International Atomic Energy Agency every three years. This allows us to know the status of decommissioning measures and the generation and treatment of radioactive waste in each country. In the appendix, the inventory (stock quantity) of spent fuel and radioactive waste is reported. The inventory of spent fuel and radioactive waste from Japan's nuclear power plants and research and development facilities can also be known from this report (NRA 2017). Furthermore, the descriptions in this Japanese report are heavily focused on explaining laws and regulations, making it difficult to grasp the overall picture compared to reports from other countries. I wonder if the authors of this report truly understand the knowledge and information behind these laws and regulations.

As shown in this example, Japanese administrators are experts in procedures (laws), but it looks that they may lack understanding of the knowledge behind them. After the TEPCO Fukushima accident, the author call to "establish a responsible nuclear regulation system," and within these calls, there were opinions stating "nurture experts with deep knowledge" (Oka 2011). Japanese administrators are excellent, responsible, and diligent individuals, but due to frequent changes in their assignments, it is difficult for them to investigate and understand the basis of specific fields and acquire deep knowledge. This role lies with nuclear experts (researchers and engineers). They need to serve as technical support teams to assist administrators. However, in Japan, not only in the field of radioactive waste and decommissioning, there are hardly any systematically organized and easy-to-understand reports. Without a knowledge base composed of human resources and information, it may be difficult to robustly utilize nuclear power.

Clearance System

The clearance system is a system that confirms whether the concentration of radioactive substances in materials used in nuclear power and radiation applications, such as when recycling or disposing of these materials, is below a certain standard (clearance level). If confirmed, these substances and materials can be handled as if they are not subject to regulations such as nuclear reactors, nuclear fuel materials, and radiation. This system is also in operation in Japan, and metals and fragmented concrete generated from decommissioning nuclear power plants or replacing equipment (such as power turbines) have been confirmed. (Nuclear Regulation Authority 2024b). As decommissioning of nuclear power plants and other facilities becomes more prevalent in future, reuse through clearance will become increasingly important as a contribution to a circular society in the nuclear field.

Decommissioning of the Tokyo Electric Power Company's Fukushima Daiichi Nuclear Power Plant

At the Tokyo Electric Power Company's Fukushima Daiichi Power Plant, three reactors experienced a core meltdown. The adjacent Unit 4 was also damaged when hydrogen generated by the accident flowed in from the adjacent Unit 3 via the piping to the common exhaust tower. The decommissioning of these reactors is being carried out with the cooperation of the government. The government has established a "Decommissioning and Contaminated Water Countermeasures Ministerial Meeting," where progress reports and confirmations are made, and government responses are decided. A medium- to long-term roadmap has been created, and progress management is being conducted. The main decommissioning tasks include "contaminated water countermeasures," "fuel removal from spent fuel pools," "fuel debris removal," and "waste countermeasures".

The contaminated water is produced by mixing water containing radioactive substances used to cool the melted nuclear reactor fuel with groundwater flowing from the mountain side to the sea side of the power plant, and rainwater entering the damaged buildings. This contaminated water, with the exception of tritium, has most of its radioactive substances removed by a purification device and is stored in tanks (Tokyo Electric Power Company 2024a). Since 2023, this purified water (treated water) has been diluted to reduce the concentration of tritium and has been discharged into the sea. The results of environmental monitoring are reported from the homepage of Fukushima Prefecture (Fukushima Revitalization Information Portal Site).

Used fuel removed from the reactor due to fuel exchange operations conducted before the accident is left in the spent fuel pool on the upper part of the reactor building. Since the building and crane were damaged in the accident, new cranes and workrooms have been set up, and the removal work from the pool is being carried out. The spent fuel of Units 3 and 4 has already been removed. Work is being carried out to remove the fuel from Units 1 and 2.

The solidified material cooled from melted fuel and metal is called fuel debris. Investigations have been conducted to determine where the fuel debris of Units 1, 2, and 3 is located. It is estimated that there is a lot at the bottom of the reactor pressure vessel for Unit 2, and a lot at the bottom of the reactor containment vessel for Units 1 and 3. Various investigations are being carried out to remove the fuel debris.

The waste generated by the decommissioning work is stored within the power plant premises. There are plans to build incineration facilities and volume reduction treatment facilities to reduce the amount of waste by incinerating combustibles and reducing the volume of metals and concrete, and to construct waste storage facilities.

Improvements have been made to the working environment of the decommissioning work within the premises. Dust prevention and decontamination have made it possible to work in most areas of the premises in regular clothes and dust masks (Tokyo Electric Power Company 2024b). The J-Village, which spans Naraha Town and Hirono Town, has been loved as a national soccer training center, but after the accident, it was used as a relay base for the Fukushima Daiichi nuclear accident response managed by the government, with dormitories and parking lots built. Workers changed into protective clothing here and headed to the site 20 km away. In 2018, it ended its role as a base and has been reopened as a training center. *Even in the decommissioning measures of the accident reactors of TEPCO's Fukushima Daiichi, a response considering the balance of risks and benefits is necessary in future.*

Restoration of Fukushima's Environment (Decontamination and Management of Removed Soil)

Decontamination to remove radioactive substances released into the environment due to the TEPCO accident has been carried out. Due to decontamination and the natural decay of radioactivity, as well as the movement of radioactive substances by wind and rain, the spatial radiation dose rate in Fukushima Prefecture has significantly decreased compared to immediately after the accident. Decontamination was carried out directly by the national government (Ministry of the Environment) in areas with high contamination, and by each municipality in other areas. The area-wide decontamination in Fukushima Prefecture, excluding the difficult-to-return areas, was completed in March 2018. The removed soil that had been stored at temporary storage sites and on-site has been transported to interim storage facilities and has greatly decreased. The land used as temporary storage sites has been returned to the owners in a condition that does not hinder subsequent use (Fukushima Prefecture 2021).

The interim storage facility is located within a 14 square kilometer area around the TEPCO Fukushima Daiichi Nuclear Power Plant site, taking into account the topography and land acquisition status. There are soil storage facilities, incineration facilities, and waste storage facilities. The soil storage facility stores soil on a waterproof sheet, and the concentration of radioactive substances in the leachate is monitored. The stored soil and other materials are promised to be finally disposed of outside the prefecture within 30 years after the start of interim storage. A demonstration project to reuse soil with low concentrations of radioactive substances has begun (Ministry of the Environment 2024a).

Debris from buildings contaminated with radioactive substances released in the TEPCO accident, cleanup waste, agricultural and forestry waste, etc., with not very high concentrations of radioactive substances, can be disposed of in controlled disposal sites along with local household waste. These are incinerated and landfilled separately from decontaminated soil (Ministry of the Environment 2018). Note that the regulation of decontaminated soil and these radioactive wastes is not under the Reactor Regulation Law, but under the Special Measures Law on Dealing with Environmental Pollution by Radioactive Substances Released by the TEPCO Accident.

Within the areas that have been designated as difficult-to-return zones where residence will be restricted in future, areas where evacuation has been lifted and residence has become possible due to decontamination (specified reconstruction and regeneration base areas) have been established, and the return of residents who wish to do so has become possible (Ministry of the Environment 2024b).

Reconstruction of Fukushima

Fukushima Prefecture operates a portal site titled "Fukushima Reconstruction Station," disseminating the progress of reconstruction to residents and people both domestically and internationally. They regularly publish a booklet titled "Progress of Reconstruction and Regeneration". They list the steady progress of decommissioning, responsible handling of treated water, significant reduction in spatial dose rate, development of infrastructure such as roads, establishment of large-scale development bases for field robots and hydrogen energy research fields, and branding of Fukushima products as aspects of ongoing reconstruction. However, as of November 2023, they also mention the existence of over 25,000 evacuees, the final disposal of removed soil outside the prefecture within 30 years after the start of intermediate storage, and the price difference between Fukushima's agricultural, forestry and fishery products and those of the rest of the country as challenges in the process of reconstruction. As of May 2012, the number of evacuees inside and outside the prefecture was about 160,000. It decreased about 26,000 as of February 2024.

The Fukushima Innovation Coast Concept is a national project aimed at rebuilding the industrial base of the Hamadori region of Fukushima Prefecture (the Pacific side of Fukushima Prefecture), which was lost due to the TEPCO accident. It is committed to developing specific areas such as decommissioning, robots/drones, energy/environment/recycling, agriculture, forestry and fisheries, medical-related, and aerospace, as well as industrial accumulation, human resource development, expansion of exchange population, information dissemination, and infrastructure development for living environment. The second phase of reconstruction and creation is set for the five years from April 2021, and the reconstruction plan and regeneration plan have started (Fukushima Prefecture 2023, 2024). As one of the nuclear-related parties, I can only apologize for the hardships suffered by the people of Fukushima

Prefecture, but I look forward to future development. The Nuclear White Paper of Japan Atomic Energy Commission published in July 2021 features a special section titled "Ten Years Since the Tokyo Electric Power Company's Fukushima Daiichi Nuclear Power Plant Accident" (Japan Atomic Energy Commission 2020). The Nuclear White Paper announced in 2019 features a special section on "Decommissioning Measures and Management of Nuclear Facilities" (Japan Atomic Energy Commission 2018).

3.3 Summary and Discussion

The recycling of resources and disposal of waste are the core concepts of a circular society. In Japan, general waste disposal and industrial waste disposal used to be major social issues, as exemplified by the "Suginami Garbage War," but they have now become established in society, and waste disposal is an important task for realizing a circular society. Factories, buildings, equipment, and pipes that have served their purpose are dismantled and treated as industrial waste, and are then reused and disposed of. Even nuclear facilities need to be decommissioned and their waste disposed of once they have served their purpose.

Radioactive waste is a form of industrial waste. In the decommissioning of nuclear facilities, a large amount of industrial waste and a small amount of radioactive waste are generated. This chapter describes radioactive waste in relation to general waste. The decommissioning of nuclear power plants and nuclear research and development facilities needs to be considered in conjunction with the recycling and disposal of the waste generated, so each country's approach to decommissioning and waste disposal is described together.

Radioactive waste is, considered a major unresolved issue in nuclear power generation in Japan, but in the West, disposal is progressing. Labeling radioactive waste as something special and negative contradicts the concept of a circular society. In the decommissioning of nuclear power plants and nuclear research and development facilities, a large amount of non-radioactive waste (industrial waste) and a small amount of radioactive waste are generated. In the West, rules have been established for the interface between low-level radioactive waste disposal and industrial waste disposal, and the recycling and disposal of waste generated by decommissioning are progressing.

In the United States, the decommissioning of nuclear power plants is progressing. There are companies that specialize in decommissioning nuclear power plants. Rules and disposal sites have been prepared so that private companies can handle decommissioning and radioactive waste disposal. Industrial waste is under the jurisdiction of the Environmental Protection Agency (EPA), and the actual administration is handled by the state or county, with disposal sites provided. Low-level radioactive waste has four disposal sites nationwide, where it is transported and disposed of. Spent fuel is stored in storage containers. The United States has many nuclear facilities for the development of nuclear weapons. The issues have been recognized, the overall picture of decommissioning and environmental recovery has been grasped, the necessary expenses have been estimated, and decommissioning has been completed at many sites over the past 30 years. The U.S. Government Accountability Office has made recommendations in its reports, which are referred to in Congress and reflected in the budget and policy.

The decommissioning of nuclear facilities in *France, the UK, and Germany is managed by responsible agencies that regularly calculate the amount and cost of radioactive waste generated by decommissioning and manage the progress of the plan.* The Board of Audit constantly monitors and reports on the costs and issues of decommissioning measures, encouraging the parliament to promote improvements. I have heard from European experts that the key to decommissioning nuclear facilities lies in calculations (such as the type and amount of waste). In these countries, calculations are made every three years, reflected in the decommissioning plan and costs, and the progress is evaluated by a third-party organization. In Japan, a plan for decommissioning research and development facilities has been made, but what about after that?

In Europe, despite difficulties, progress is being made in the treatment and disposal of high-level radioactive waste, such as spent fuel. Finland decided on a disposal site candidate in 2001, construction of the disposal site is underway, and final disposal is planned to start in 2024. Finland disposes of spent fuel directly without reprocessing. France is also making steady progress. France's disposal is not final, but it could become final. It takes 100 years from the start of operation to the closure of the disposal site. The current generation does not decide whether to close the disposal site. France plans to dispose of glassified bodies containing high-level radioactive waste generated by reprocessing spent fuel, but this disposal site can also directly dispose of spent fuel. The United States is struggling because the government decided on the disposal site from the top down. In 2020, a disposal site candidate was decided in Sweden.

In Finland, for the first time in 2019, public trust in geological disposal exceeded skepticism (37% vs. 36%). In Finland, the plan for a geological disposal site was advanced during a time when there were many skeptical citizens. In Japan, activities are being carried out to raise public awareness about high-level radioactive waste disposal sites. It is not necessary to directly obtain the consensus of the entire population about the location of the radioactive waste disposal site. It is the rule to proceed within a parliamentary democracy. Was it necessary to directly obtain the agreement of all local government residents for the location of the waste incineration plant? In Japan, too, the location of geological disposal sites is being promoted. Building trust with the local community takes time. It is necessary to have a careful dialogue and build trust. We need to learn from the lessons of the United States, where the government decided on the location of the high-level radioactive waste disposal site from the top down, causing confusion.

To add to the discussion on high-level radioactive waste treatment and disposal, *Japan needs to first expand the storage capacity of spent fuel.* This does not mean that spent fuel does not need to be disposed of forever, but spent fuel can also be considered a resource containing plutonium. The technology to place it in a storage container and fix it on the ground is established. Japan Atomic Energy Commission stated in its "Basic Concept of Plutonium Use" issued in 2018 that it is necessary to expand the storage capacity of spent fuel.

Japan is lagging behind Europe and America not only in high-level but also in low-level radioactive waste disposal. In order to proceed with the decommissioning measures of nuclear power plants and nuclear research and development facilities, Japan needs to dispose of low-level radioactive waste in collaboration with industrial waste disposal. Japan Atomic Energy Commission has publicly stated its view on the "concept of low-level radioactive waste disposal" at the end of 2021.

In the relationship between society and radioactive waste, why have many Japanese people come to believe that radioactive waste is an insurmountable problem? Is there not a similarity to how nuclear power generation, despite its safety record, has come to be perceived as extremely dangerous compared to other power generation methods? Safety is psychologically a matter of risk. The more we talk about the safety of radioactive waste disposal to the public, the more it may have been perceived as dangerous.

It is necessary to compile the results of research and development on geological disposal into a detailed technical report. However, the Nuclear Regulation Authority is the one who listens and judges on behalf of the public when it comes to safety. Discussions about safety and detailed technical matters should be conducted with the regulatory side, and detailed reports are necessary for that purpose. On the websites of the geological disposal agencies in the UK and France, there are no technical safety explanations, such as the characteristics of the overpack (metal container) of the glass solidified body. The UK's geological disposal agency explains what geology and strata are.

A safety guide for the general public should be created and made available so that locals and the public can find it when they search. The explanation of the health effects of radiation, which is the basis of the safety of nuclear power generation and radioactive waste, has been greatly enriched by the Ministry of the Environment after the TEPCO accident. In order to reach the public, efforts such as linking so that it appears at the top of the search when the public searches for it are necessary.

Almost all researchers in radioactive waste and reactor safety are technical experts and are researching technology. In reactor safety, they are researching technology to prevent core meltdown and reduce radioactive material release in the event of an accident. However, did they not embarrass themselves in the TEPCO Fukushima accident by overconfidently stating that "the nuclear reactor containment vessel will not break"? Wasn't the emergency response, which is a measure after radioactive materials have been released into the environment, neglected? Have they learned from the lessons of emergency responses to disasters outside the nuclear field? In radioactive waste disposal, while technical research is important, the problem cannot be solved by technology alone. Just as there are related fields of disaster prevention and emergency response in general disasters in reactor safety, there are related fields of general waste in radioactive waste. *General waste disposal and radioactive waste disposal are interconnected fields.* It's good to have confidence in the technical safety of radioactive waste, but it's not advisable to think that local understanding of radioactive waste disposal can be achieved through technical explanations alone. When technical safety explanations are communicated to the public, they are perceived as risk explanations, which could potentially backfire. Dialogue-based understanding and trust-building with locals is necessary. For high-level radioactive waste disposal, wouldn't it be good to prepare various levels of documents about European experiences, strata, geology, and underground environments, so that they can be found by both locals and the public? Detailed technical reports on the safety of geological disposal should serve as part of this.

The treatment and disposal of radioactive waste and the decommissioning of nuclear facilities in the West are carried out considering the balance of risk (cost) and benefit, and disposal systems and sites are established and implemented. *In Japan too, the decommissioning and treatment and disposal should be carried out based on the concept of risk-benefit balance.*

Although this book does not cover the safety management of radioactive waste and spent fuel in China and South Korea, each country, in accordance with international treaties, submits a report to the International Atomic Energy Agency every three years (China 2020; NSSC 2020).

Exercise

1. Research and summarize the construction of an incineration plant in Suginami Ward, Tokyo in the 1960s and its subsequent history.
2. Summarize the history and current situation of industrial waste treatment and disposal in Japan.
3. Summarize the licensing procedures for decommissioning nuclear power plants in the United States.
4. Summarize the treatment and disposal of waste and radioactive waste generated by the decommissioning of nuclear power plants in the United States.
5. Summarize the location, construction, and subsequent history of Yucca Mountain, a high-level radioactive waste disposal site in the United States.
6. Summarize the decommissioning of nuclear reactors and reprocessing facilities for nuclear development at Hanford in Washington State, USA.
7. Summarize the treatment and disposal of general waste in the United States.
8. Summarize the decommissioning and waste disposal of the UK's Magnox reactor, a graphite-moderated gas-cooled reactor.
9. Summarize the decommissioning of Sellafield in the UK.
10. Summarize the decommissioning at Dounreay in the UK.
11. Discuss the UK's radioactive waste and its disposal sites.
12. Summarize the past history, lessons learned, current status, and plans for the selection of geological disposal sites in the UK.
13. Compare the features of the items and content on the websites of the UK's Radioactive Waste Management (RWM) and Japan's Nuclear Waste Management Organization (NUWMO). Investigate how the progress of each organization's work is audited.

14. Describe the outline of the "Law on the Management and Research of Radioactive Waste" enacted in France in 1991 (commonly known as the Bataille Law), and the research and evaluation results of high-level radioactive waste disposal conducted in accordance with it.
15. Discuss the decommissioning measures of reprocessing facilities in France, not only in La Hague but also in Marcoule.
16. Discuss the disposal and disposal sites of radioactive waste in France.
17. Discuss the disposal and disposal sites of radioactive waste in Germany.
18. Discuss the disposal and disposal sites of low-level radioactive waste in Sweden.
19. Summarize the site selection activities and history of geological disposal sites in Sweden.
20. Summarize the site selection activities and current status of geological disposal sites in Finland.
21. Summarize the decommissioning measures for Japan's JPDR (Japan Power Demonstration Reactor).
22. Summarize the decommissioning measures for Units 1 and 2 of the Chubu Electric Power Hamaoka Nuclear Power Station.
23. Compare the current status of radioactive waste treatment and disposal in Japan with that in the United States, the UK, and France, and summarize the issues.
24. Summarize the environmental restoration and reconstruction activities in areas contaminated with radioactive substances due to the accident at the Tokyo Electric Power Company's Fukushima Daiichi Nuclear Power Station.
25. Investigate and summarize the past policies and main activities related to the disposal of high-level radioactive waste in Japan.
26. Summarize the key points and logic of the "Comprehensive Technical Report: Realizing Safe Geological Disposal in Our Country NUMO-TR-20–01, February 2021—Building a Safety Case for Appropriate Site Selection" created by the Nuclear Waste Management Organization of Japan.
27. Summarize the overall picture and issues of the decommissioning measures of the facilities of the Japan Atomic Energy Agency.
28. Summarize the overall picture and issues of the decommissioning measures of nuclear power plants in Japan.

References

3KQ (2016) Public dialogue on geological disposal and working with communities: report by 3KQ, March. https://assets.publishing.service.gov.uk/government/uploads/system/uploads/attachment_data/file/603176/GDF_public_dialogue_final_report.pdf. Accessed 30 June 2024

Andra (2023a) Waste classification, Andra https://international.andra.fr/radioactive-waste-france/waste-classification. Accessed 30 June 2024

Andra (2023b) Operational facilities, Andra https://international.andra.fr/operational-facilities. Accessed 30 June 2024

Andra (2023c) Solutions for low-level, long-lived waste (LL-LLW), Andra. https://international.andra.fr/projects/solutions-low-level-long-lived-waste-ll-llw. Accessed 30 June 2024

Andra (2023d). More than 25 years' research, Andra https://international.andra.fr/projects/cigeo/more-25-years-research. Accessed 30 June 2024

Andra (2023e) 30 years of scientific expertise at the service of the Cigéo project, Andra. https://international.andra.fr/30-years-scientific-expertise-service-cigeo-project. Accessed 30 June 2024

Andra (2023f) Project siting and facilities overview, Andra https://international.andra.fr/projects/cigeo/cigeos-facilities-and-operation/project-siting-and-facilities-overview. Accessed 30 June 2024

Andra (2023g) A project anchored in its region, Andra, https://international.andra.fr/projects/cigeo/project-anchored-its-region. Accessed 30 June 2024

Andra (2023h) Protection from the most hazardous radioactive waste, Cigeo can adapt to different energy policy scenarios. https://international.andra.fr/projects/cigeo/protection-most-hazardous-radioactive-waste/protection-most-hazardous-radioactive-waste. Accessed 30 June 2024

ASN (2009) ASN policy concerning the decommissioning and delicensing of basic nuclear installations, Revision 0.v3 April 2009. https://www.french-nuclear-safety.fr/content/download/54408/file/ASN%20policy%20concerning%20the%20decommissioning.pdf. Accessed 30 June 2024

ASN (2021) National plan on management of radioactive materials and waste (PNGMDR), ASN. http://www.french-nuclear-safety.fr/References/National-plan-on-management-of-radioactive-materials-and-waste-PNGMDR. Accessed 30 June 2024

BASE (2017). Introducing the concept of final disposal, Federal Office for the safety of nuclear waste management, BASE https://www.ccomptes.fr/sites/default/files/EzPublish/thematic_public_report_costs_nuclear_%20power_sector_012012.pdf. Accessed 30 June 2024

BASE (2021). Guide to the decommissioning, the safe enclosure and the dismantling of facilities or parts thereof as defined in § 7 of the Atomic Energy Act, 16 September. https://www.base.bund.de/SharedDocs/Downloads/BASE/EN/hns/3-bmub/3-73-Decommissioning-Guide.pdf?__blob=publicationFile&v=2. Accessed 30 June 2024

BASE (2022) Waste types, Federal Office for the safety of nuclear waste management, BASE. https://www.base.bund.de/EN/nwm/waste/types/types.html. Accessed 30 June 2024

BASE (2023) Decommissioning of nuclear facilities, Federal Office for the safety of nuclear waste management (BASE). https://www.base.bund.de/EN/ns/decommissioning/decommissioning/decommissioning_node.html. Accessed 30 June 2024

BASE (2024a) Central interim storage facilities, Federal Office for the safety of nuclear waste management, BASE. https://www.base.bund.de/EN/nwm/interim-storage/central/central.html. Accessed 30 June 2024

BASE (2024b). Nuclear facilities in Germany, Part II: nuclear facilities. In: Decommissioning, Federal Office for the safety of nuclear waste management, BASE. https://www.base.bund.de/SharedDocs/Downloads/BASE/EN/reports/kt/nuclear-facilities-decommissioning.pdf?__blob=publicationFile&v=26. Accessed 30 June 2024

BEIS (2018) Implementing geological disposal-working with communities, an updated framework for the long-term management of higher activity radioactive waste. Department for Business, Energy & Industrial Strategy, December. https://assets.publishing.service.gov.uk/government/uploads/system/uploads/attachment_data/file/766643/Implementing_Geological_Disposal_-_Working_with_Communities.pdf. Accessed 30 June 2024

BGE (2024) Site selection procedure, Bundesgesellschaft für Endlagerung, BGE. https://www.bge.de/en/repository-search/site-selection-procedure/. Accessed 30 June 2024

CEA (2016) Dismantling nuclear facilities at the CEA, June 28. https://www.cea.fr/english/Pages/research-areas/nuclear-energy/cea-research-on-dismantling-nuclear-facilities-at-the-cea.aspx. Accessed 30 June 2024

China (2020) The People's Republic of China, fifth national report for the joint convention on the safety of spent fuel management and on the safety of radioactive waste management prepared for the seventh review meeting, September. Beijing, China. https://www.iaea.org/sites/default/files/china-7rm_english.pdf. Accessed 30 June 2024

Chubu Electric Power Company (2013) About the decommissioning of units 1 and 2 of the Hamaoka Nuclear Power Station, Chubu Electric Power Co., Ltd., June 25 (in Japanese). https://www.meti.go.jp/shingikai/enecho/denryoku_gas/denki_ryokin/hairo_kaikei/pdf/01_001_06_00.pdf. Accessed 30 June 2024

Cour des comptes (2012) The cost of nuclear power, Thematic public report, January. Cour des comptes. https://www.ccomptes.fr/sites/default/files/EzPublish/thematic_public_report_costs_nuclear_%20power_sector_012012.pdf. Accessed 30 June 2024

DECC (2014) Implementing geological disposal. Department of Energy & Climate Change. https://assets.publishing.service.gov.uk/government/uploads/system/uploads/attachment_data/file/332890/GDF_White_Paper_FINAL.pdf. Accessed 30 June 2024

DEFRA (2023) UK statistics on waste, DEFRA. Government Statistical Service, June 28. https://www.gov.uk/government/statistics/uk-waste-data/uk-statistics-on-waste. Accessed 30 June 2024

DOE (2014) Evolution and history of the Department of Energy and the Office of Environmental Management. DOE-EM, March 5. https://www.energy.gov/sites/default/files/2014/04/f15/Evolution_History_DOE_042314.pdf. Accessed 30 June 2024

DOE (2021) EM strategic vision: 2021–2031. DOE, Office of Environmental Management. https://www.energy.gov/sites/default/files/2021-04/EM-Strategic-Vision-2021-2031.pdf. Accessed 30 June 2024

DOE (2023a). Cleanup sites. DOE, Office of Environmental Management. https://www.energy.gov/em/cleanup-sites. Accessed 30 June 2024

DOE (2023b) LM sites. DOE, Office of Legacy Management. https://www.energy.gov/lm/office-legacy-management. Accessed 30 June 2024

Dounreay (2021) Your guide to Scotland's centre of excellence in nuclear decommissioning. https://assets.publishing.service.gov.uk/government/uploads/system/uploads/attachment_data/file/985222/Dounreay_2021.pdf. Accessed 30 June 2024

Environment Agency (2010) Criteria for setting limits on the discharge of radioactive waste from nuclear sites. Environmental Permitting Regulations (England and Wales), Environment Agency, UK. https://assets.publishing.service.gov.uk/government/uploads/system/uploads/attachment_data/file/296486/geho0612buqp-e-e.pdf. Accessed 30 June 2024

EPA (2020) Advancing sustainable materials management: 2018 fact sheet. https://www.epa.gov/sites/default/files/2020-11/documents/2018_ff_fact_sheet.pdf. Accessed 30 June 2024

EPA (2023a) National overview: facts and figures on materials, wastes and recycling. https://www.epa.gov/facts-and-figures-about-materials-waste-and-recycling/national-overview-facts-and-figures-materials. Accessed 30 June 2024

EPA (2023b) Waste management hierarchy and homeland security incidents. https://www.epa.gov/homeland-security-waste/waste-management-hierarchy-and-homeland-security-incidents. Accessed 30 June 2024

EPA (2023c) U.S. state and local waste and materials characterization reports. https://www.epa.gov/facts-and-figures-about-materials-waste-and-recycling/us-state-and-local-waste-and-materials. Accessed 30 June 2024

EPA (2023d) Sustainable management of construction and demolition materials. https://www.epa.gov/smm/sustainable-management-construction-and-demolition-materials. Accessed 30 June 2024

EPA (2023e) Resources for businesses, states, and local governments. https://www.epa.gov/recycle/resources-businesses-states-and-local-governments. Accessed 30 June 2024

EPA (2023f) Guidance about planning for natural disaster debris. https://www.epa.gov/homeland-security-waste/guidance-about-planning-natural-disaster-debris. Accessed 30 June 2024

EPA (2023g) Learn the basics of hazardous waste. https://www.epa.gov/hw/learn-basics-hazardous-waste. Accessed 30 June 2024

Equitis Consulting (2016) Working with communities geological disposal, literature review—geological disposal. Commissioned by DECC to support the Initial Actions workstream—April 2016. Equitis Consulting. https://assets.publishing.service.gov.uk/government/uploads/system/

uploads/attachment_data/file/637265/gdf-working-with-communities-literature-review.pdf. Accessed 30 June 2024

Ervin PF (2008) Characterisation progress at the windscale pile reactors, challenges and results. In: WM2008 Conference, February. Phenix, AZ, Abstract #8463. https://archivedproceedings.econference.io/wmsym/2008/pdfs/8463.pdf. Accessed 30 June 2024

Forwood M (2008) The legacy of reprocessing in the United Kingdom, a research report of the international panel on fissile materials, April. https://fissilematerials.org/library/rr05.pdf. Accessed 30 June 2024

Fukushima Prefecture (2021) Efforts for decontamination in Fukushima Prefecture, Fukushima Prefecture living environment department decontamination measures division, June (in Japanese). https://www.pref.fukushima.lg.jp/uploaded/attachment/455020.pdf. Accessed 30 June 2024

Fukushima Prefecture (2023) Progress of reconstruction and regeneration, new Fukushima reconstruction and regeneration headquarters, March 25 (in Japanese). https://www.pref.fukushima.lg.jp/uploaded/attachment/637212.pdf. Accessed 30 June 2024

Fukushima Prefecture (2024) Fukushima revitalization portal site. https://www.pref.fukushima.lg.jp/site/portal/ (in Japanese). https://www.pref.fukushima.lg.jp/site/portal-english/ (in English). Accessed 30 June 2024

GAO (2023) Nuclear waste disposal. U.S. Government Accountability Office. https://www.gao.gov/nuclear-waste-disposal. Accessed 30 June 2024

Gov.UK (2018) Scope and exemptions from the radioactive substances legislation in England, Wales and Northern Ireland, guidance document, August. https://assets.publishing.service.gov.uk/media/5b643627e5274a76f73e20ce/RSL_Guidance_update_BEIS_format_v5_180803.pdf. Accessed 30 June 2024

Gov.UK (2021a) Waste: detailed information. Gov.UK. https://www.gov.uk/topic/environmental-management/waste. Accessed 30 June 2024

Gov.UK (2021b) Radioactive substances regulation (RSR) for non-nuclear sites. Gov.UK. https://www.gov.uk/government/collections/radioactive-substances-regulation-for-non-nuclear-sites. Accessed 30 June 2024

Harwell (2023) Harwell, science and innovation. https://www.harwellcampus.com/. Accessed 30 June 2024

Holt G (1999) Radioactive graphite management at UK Magnox Nuclear Power Stations, IAEA-NGWM/CD-01, Technical Committee Meeting on Nuclear Graphite Waste Management, Manchester (United Kingdom), 18–20 Oct. https://inis.iaea.org/collection/NCLCollectionStore/_Public/32/039/32039338.pdf?r=1. Accessed 30 June 2024

IAEA (2010) Progress in radioactive graphite waste management. IAEA-TECDOC-1647. https://www-pub.iaea.org/MTCD/Publications/PDF/te_1647_web.pdf. Accessed 30 June 2024

Japan Atomic Energy Agency (2016) About the disposal of waste from research facilities. Japan Atomic Energy Agency, Backend Research and Development Division, January 1 (in Japanese). https://www.jaea.go.jp/04/maisetsu/project/pdf/20180322kyougikai16_shiryou16-4-2.pdf. Accessed 30 June 2024

Japan Atomic Energy Agency (2018a) Backend roadmap, December 26 (in Japanese). https://www.jaea.go.jp/about_JAEA/backend_roadmap/s02.pdf. Accessed 30 June 2024

Japan Atomic Energy Agency (2018b) About the backend roadmap, Yasushi Taguchi, 4th Management Advisory Council Materials 4-4, March 28 (in Japanese). https://www.jaea.go.jp/about_JAEA/advisory/3-no04/s4-4.pdf. Accessed 30 June 2024

Japan Atomic Energy Agency (2022) Medium and long-term facility plan, April 1 (in Japanese). https://www.jaea.go.jp/about_JAEA/facilities_plan/. Accessed 30 June 2024

Japan Atomic Energy Agency (2024) Decommissioning of JPDR, Japan Atomic Energy Agency (in Japanese). https://www.jaea.go.jp/04/ntokai/decommissioning/01/decommissioning_01_01.html. Accessed 30 June 2024

Japan Atomic Energy Commission (2018) White paper on nuclear energy 2018. https://www.aec.go.jp/kettei/hakusho/2018/ (in Japanese). https://www.aec.go.jp/kettei/hakusho/2018/pdf/gaiyo_e.pdf (summary in English). Accessed 30 June 2024

Japan Atomic Energy Commission (2020) White paper on nuclear energy 2020. https://www.aec.go.jp/kettei/hakusho/2020/ (in Japanese) https://www.aec.go.jp/kettei/hakusho/2020/pdf/gaiyo_e.pdf (summary in English). Accessed 30 June 2024

Japan Atomic Power Company (2024) Decommissioning measures for Tokai Power Station. Japan Atomic Power Company (in Japanese). http://www.japc.co.jp/tokai/haishi/tokai_haishi.html. Accessed 30, June 2024

LLW Repository LTD (2017a) Past, present and future. LLW Repository Ltd. https://webarchive.nationalarchives.gov.uk/ukgwa/20170412103134/http://llwrsite.com/our-company/past-present-and-future/. Accessed 30 June 2024

LLW Repository LTD (2017b) The 2011 environmental safety case, non-technical summary. LLW Repository Ltd. https://webarchive.nationalarchives.gov.uk/ukgwa/20170412103307/http://llwrsite.com/national-repository/esc-permit-approval/. Accessed 30 June 2024

LLW Repository LTD (2018) LLWR Plan 2018–2023. LLW Repository Ltd. https://assets.publishing.service.gov.uk/government/uploads/system/uploads/attachment_data/file/757075/LLWR_Plan_2018-2023__double_page_.pdf. Accessed 30 June 2024

MEAE (2024) Environmental impact assessment on the final disposal of very low-level nuclear waste in the Olkiluoto nuclear power plant area. Ministry of Economic Affairs and Employment of Finland. https://tem.fi/en/vllw-eia. Accessed 30 June 2024

Ministry of Health, Labour and Welfare (2021) Issues concerning the disposal of radioactive contaminants for medical use, study group on proper management of medical radiation. Ministry of Health, Labour and Welfare, June 24 (in Japanese). https://www.niph.go.jp/h-crisis/wp-content/uploads/2021/06/20210624100347_content_10801000_000796682.pdf. Accessed 30 June 2024

Ministry of the Environment (2018) Landfill disposal business of specific waste, information site for landfill of specific waste. Ministry of the Environment, July (in Japanese). http://shiteihaiki.env.go.jp/pamphlet/pdf/landfill_disposal_pamphlet_var6.pdf. Accessed 30 June 2024

Ministry of the Environment (2020) Reiwa 2nd edition environmental, circular society, biodiversity white paper, Part 2 chapter 3 circular society (in Japanese). https://www.env.go.jp/policy/hakusyo/r02/pdf/2_3.pdf. Accessed 30 June 2024

Ministry of the Environment (2024a) Management of removed soil, decontamination information site. Ministry of the Environment (in Japanese). http://josen.env.go.jp/soil/index.html. Accessed 30 June 2024

Ministry of the Environment (2024b) Decontamination information site. Ministry of the Environment (in Japanese). http://josen.env.go.jp/. Accessed 30 June 2024

MOE-GOS (2020) Sweden's seventh national report under the joint convention on the safety of spent fuel management and on the safety of radioactive waste management. Ministry of Environment, Government of Sweden, October 26. https://www.regeringen.se/contentassets/ace68a7f79ec46bea20dffc76bfbeb85/swedens-seventh-national-report-under-the-joint-convention-on-the-safety-of-spent-fuel-management-and-on-the-safety-of-radioactive-waste-management-ds-2020-21/. Accessed 30 June 2024

NAO (2018) The nuclear decommissioning authority: progress with reducing risk at Sellafield, June. National Audit Office. https://www.nao.org.uk/wp-content/uploads/2018/06/The-Nuclear-Decommissioning-Authority-progress-with-reducing-risk-at-Sellafield.pdf. Accessed 30 June 2024

NAO (2020a) Progress report: terminating the Magnox contract, October. National Audit Office. https://www.nao.org.uk/wp-content/uploads/2020/09/Progress-report-Terminating-the-Magnox-contract.pdf. Accessed 30 June 2024

NAO (2020b) Managing infrastructure projects on nuclear-regulated sites, January 2020. National Audit Office. https://www.nao.org.uk/wp-content/uploads/2020/01/Managing-infrastructure-projects-on-nuclear-regulated-sites.pdf. Accessed 30 June 2024

NDA (2013) Geological disposal overview of international siting processes, September. Nuclear Decommissioning Authority. https://assets.publishing.service.gov.uk/government/uploads/system/uploads/attachment_data/file/456820/Overview_of_international_siting_processes_September_2013.pdf. Accessed 30 June 2024

NDA (2017) Radioactive wastes in the UK: context and methodology report, March, BEIS, Nuclear Decommissioning Authority, p 13. https://ukinventory.nda.gov.uk/wp-content/uploads/2014/01/2016UKRWMI-Context-and-methodology.pdf. Accessed 30 June 2024

NDA (2019a) Mission progress report, nuclear decommissioning authority, Published July. https://assets.publishing.service.gov.uk/media/5d1df0f240f0b609f298d9b2/Mission_Progress_Report_July_2019.pdf. Accessed 30 June 2024

NDA (2019b) Nuclear provision: the cost of cleaning up Britain's historic nuclear sites, July 4. https://www.gov.uk/government/publications/nuclear-provision-explaining-the-cost-of-cleaning-up-britains-nuclear-legacy/nuclear-provision-explaining-the-cost-of-cleaning-up-britains-nuclear-legacy. Accessed 30 June 2024

NDA (2020a) Draft strategy for consultation, August. Nuclear Decommissioning Authority. https://www.gov.uk/government/consultations/nuclear-decommissioning-authority-nda-draft-strategy-for-consultation. Accessed 30 June 2024

NDA (2020b) NDA annual report and accounts 2019 to 2020, July. https://www.gov.uk/government/publications/nuclear-decommissioning-authority-annual-report-and-accounts-2019-to-2020/nda-annual-report-and-accounts-2019-to-2020. Accessed 30 June 2024

NDA (2020c) Case study: timing of the Magnox reactor decommissioning strategy, draft strategy. Published August 2020 for consultation, NDA, p 33. https://assets.publishing.service.gov.uk/media/5f4632a8d3bf7f699b486d06/NDA_Draft_Strategy_4_for_public_consultation.pdf. Accessed 30 June 2024

NDA (2020d) Draft strategy published August 2020 for consultation. Nuclear Decommissioning Authority, p 29 Case Study: Ponds and Silos at Sellafield. https://assets.publishing.service.gov.uk/media/5f4632a8d3bf7f699b486d06/NDA_Draft_Strategy_4_for_public_consultation.pdf. Accessed 30 June 2024

NDA (2021) Nuclear decommissioning authority strategy effective from March 18. https://www.gov.uk/government/publications/nuclear-decommissioning-authority-strategy-effective-from-march-2021. Accessed 30 June 2024

NDA (2023a) Corporate information, about us. Nuclear Decommissioning Authority. https://www.gov.uk/government/organisations/nuclear-decommissioning-authority/about. Accessed 30 June 2024

NDA (2023b) What are the main waste categories, about radioactive waste. Nuclear Decommissioning Authority. https://ukinventory.nda.gov.uk/about-radioactive-waste/what-is-radioactivity/what-are-the-main-waste-categories/. Accessed 30 June 2024

NEA (2008) Release of radioactive materials and buildings from regulatory control, OECD/NEA No. 6403. ISBN 978-92-64-99061-6. https://www.oecd-nea.org/upload/docs/application/pdf/2019-12/nea6403-regulatory.pdf. Accessed 30 June 2024

NEA (2017) Recycling and reuse of materials arising from the decommissioning of nuclear facilities, OECD/NEA No. 7310. https://www.oecd-nea.org/jcms/pl_15012/recycling-and-reuse-of-materials-arising-from-the-decommissioning-of-nuclear-facilities?details=true. Accessed 30 June 2024

NEA (2018) Nuclear reactor dismantling demonstrator. OECD/NEA NI 2050 R&D Cooperate program proposal, EDF-DP2D, February. https://www.oecd-nea.org/upload/docs/application/pdf/2020-07/04pieraccinini2050decommissioning_.pdf. Accessed 30 June 2024

NEA (2020a) Radioactive waste management and decommissioning in Sweden. OECD/NEA. https://www.oecd-nea.org/upload/docs/application/pdf/2020-07/sweden_report_web.pdf. Accessed 30 June 2024

NEA (2020b). Radioactive waste management and decommissioning in Finland, July. https://www.oecd-nea.org/upload/docs/application/pdf/2020-07/finland_report_web.pdf. Accessed 30 June 2024

NRA (2017) National report of Japan for the sixth review meeting, joint convention on the safety of spent fuel management and on the safety of radioactive waste management, October. Nuclear Regulation Authority/Agency. https://www.nra.go.jp/data/000232629.pdf (in English). https://www.nsr.go.jp/data/000232312.pdf (in Japanese). Accessed 30 June 2024

Nuclear Regulation Authority (2024a) Overview of safety regulations at the decommissioning stage (in Japanese). https://www.nsr.go.jp/activity/regulation/reactor/haishi/haishi0.html. Accessed 30 June 2024

Nuclear Regulation Authority (2024b). Overview of the clearance system. Nuclear Regulation Authority (in Japanese). https://www.nsr.go.jp/activity/regulation/nuclearfuel/haiki4.html. Accessed 30 June 2024

NRC (2019) Backgrounder on decommissioning nuclear power plants. Office of Public Affairs. USNRC, August. https://www.nrc.gov/docs/ML0403/ML040340625.pdf. Accessed 30 June 2024

NRC (2023) Sites undergoing decommissioning (by location or name) USNRC. https://www.nrc.gov/info-finder/decommissioning/index.html. Accessed 30 June 2024

NRW (2007). Comparison among different decommissioning funds methodologies for nuclear installations, Country Report Germany, Service Contract TREN/05/NUCL/S07.55436, Wuppertal, North Rhine-Westphalia Science Center 2007. https://epub.wupperinst.org/frontdoor/deliver/index/docId/2604/file/2604_EUDecommFunds_DE.pdf. Accessed 30 June 2024

NSSC (2020). Korean seventh national report under the joint convention on the safety of spent fuel management and on the safety of radioactive waste management, October, Nuclear Security and Safety Commission, Republic of Korea. https://www.iaea.org/sites/default/files/korean-rep-of-7rm.pdf. Accessed 30 June 2024

Nuclear Waste Services (2020) Site evaluation, how we will evaluate sites in England. Nuclear Waste Services, GDF. https://assets.publishing.service.gov.uk/government/uploads/system/uploads/attachment_data/file/866407/RWM_Site_Evaluation_for_England_2020.pdf. Accessed 30 June 2024

Oka Y (2011) Building a "system" for responsible nuclear regulation, Waseda Online, East Japan Earthquake Special Feature, Yomiuri Online November 14, 2011 (in Japanese)

Oka Y (2019a) Reprocessing plant and decommissioning measures in La Hague, France, Atomic Energy Commission Email Magazine, Issue 282, December 6 (in Japanese). https://www.aec.go.jp/mailmagagine/backnumber/2019-0282.html. Accessed 30 June 2024

Oka Y (2019b) Decommissioning measures and energy policy of nuclear research and development facilities in France, Atomic Energy Commission Email Magazine, Issue 262, January 25 (in Japanese). https://www.aec.go.jp/mailmagagine/backnumber/2019-0262.html. Accessed 30 June 2024

Oka Y (2020a) Decommissioning measures and waste disposal of nuclear facilities in the UK, Atomic Energy Commission Email Magazine, January 24, Issue (in Japanese). https://www.aec.go.jp/mailmagagine/backnumber/2020-0285.html. Accessed 30 June 2024

Oka Y (2020b) Geological disposal and information provision in the UK, Atomic Energy Commission Email Magazine, March 6, Issue (in Japanese). https://www.aec.go.jp/mailmagagine/backnumber/2020-0288.html. Accessed 30 June 2024

ONR (2015) Basic principles of radioactive waste management. Office of Nuclear Regulations, February. https://www.onr.org.uk/media/4yznxreg/basic-principles.pdf. Accessed 30 June 2024

ONR (2021) On-site disposal of solid radioactive waste on nuclear licensed sites, Joint Regulators' Statement of Common Understanding. https://www.onr.org.uk/documents/2021/joint-regulators-statement.pdf. Accessed 30 June 2024

ONR (2023a) Harwell. Office for Nuclear Regulation. https://www.onr.org.uk/our-work/what-we-regulate/sellafield-decommissioning-fuel-and-waste/sites/harwell/. Accessed 30 June 2024

ONR (2023b) Decommissioning, fuel and waste (DFW). Office for Nuclear Regulation UK. https://www.onr.org.uk/decomissioning-overview.htm. Accessed 30 June 2024

Orano (2023a) Orano la Hague's presentation. The world leader in recycling used nuclear fuels, ORANO. https://www.orano.group/en/nuclear-expertise/from-exploration-to-recycling/world-leader-in-recycling-used-nuclear-fuels. Accessed 30 June 2024

Orano (2023b) Recycling used nuclear fuel—Orano la Hague—English. https://www.youtube.com/watch?v=V0UJSlKIy8g. Accessed 30 June 2024

RWMC (2021) About the disposal of high-level radioactive waste in various countries, 2021 edn. Radioactive Waste Management Funding and Management Center. (in Japanese). https://www2.rwmc.or.jp/publications:hlwkj2021. Accessed 30 June 2024

Rehs B (2011) Decommissioning in Germany: Greifswald NPPs. In: 12th Meeting of the Working Party on Decommissioning and Dismantling (WPDD), 15–17th November. Paris, OECD/NEA. https://www.oecd-nea.org/rwm/wpdd/10/documents/11____Decomm_Story_Greifswald_____Rehs.pdf. Accessed 30 June 2024

RWM (2023a) Introduction to geological disposal. Radioactive waste management. https://assets.publishing.service.gov.uk/government/uploads/system/uploads/attachment_data/file/835825/RWM_Prospectus_EnglandV1_a.pdf. Accessed 30 June 2024

RWM (2023b). Community guidance, how we will work with communities in England. Radioactive Waste Management. https://assets.publishing.service.gov.uk/government/uploads/system/uploads/attachment_data/file/835826/RWM_CommunityGuidance_ENGLAND_a.pdf. Accessed 30 June 2024

RWMC (2022) About the facilities and sites related to radioactive waste in various countries, 2022 edn. Radioactive Waste Management Funding and Management Center (in Japanese). https://www2.rwmc.or.jp/publications:rwmhg2022. Accessed 30 June 2024

Schneider M, Marignac Y (2008) Spent fuel processing in France, a research report of the international panel on fissile materials, April. https://fissilematerials.org/library/rr04.pdf. Accessed 30 June 2024

SKB (2020) Historic decision—Östhammar municipality agrees to the final repository. SKB, October. https://skb.com/nyhet/historic-decision-osthammar-municipality-agrees-to-the-final-repository/. Accessed 30 June 2024

STUK (2020) Joint convention on the safety of spent fuel management and on the safety of radioactive waste management. In: 7th Finnish National Report as referred to in Article 32 of the Convention, STUK, October. ISBN 978-952-309-488-8. https://www.julkari.fi/bitstream/handle/10024/140601/stuk-b259.pdf?sequence=1&isAllowed=y. Accessed 30 June 2024

Suginami Study Club (2014) The 'Tokyo Garbage War' over the Suginami incineration plant (in Japanese). https://www.suginamigaku.org/2014/10/oldnew-seisokojo.html. Accessed 30 June 2024

Thompson PJ et al (2002). Radiation protection during the early stages of site decommissioning at the Ukaea's Dounreay site. https://inis.iaea.org/collection/NCLCollectionStore/_Public/37/115/37115737.pdf. Accessed 30 June 2024

Tokyo Electric Power Company (2024a) Status of contaminated water measures (in Japanese). https://www.tepco.co.jp/decommission/progress/watermanagement/. Accessed 30 June 2024

Tokyo Electric Power Company (2024b) Status of decommissioning work (in Japanese). https://www.tepco.co.jp/decommission/progress/. Accessed 30 June 2024

Tokyo 23 Wards Cleaning Division Union (2023) List of incineration plant facilities in the 23 wards of Tokyo. https://www.union.tokyo23-seisou.lg.jp/kojo/index.html (in Japanese). https://www.union.tokyo23-seisou.lg.jp.e.de.hp.transer.com/kojo/index.html (in English). Accessed 30 June 2024

UK Parliament (2020) The nuclear decommissioning authority's management of the Magnox contract. UK Parliament, Committees. https://committees.parliament.uk/work/597/the-nuclear-decommissioning-authoritys-management-of-the-magnox-contract/publications/. Accessed 30 June 2024

URSUS Consulting (2016) Evaluation of public dialogue on community involvement in siting a geological disposal facility, April. URSUS Consulting. https://assets.publishing.service.gov.uk/

government/uploads/system/uploads/attachment_data/file/608653/ursus_evaluation_report_09.04.2016_v2.pdf. Accessed 30 June 2024
VTT (2024) Decommissioning of FiR 1 nuclear reactor. VTT. https://www.vttresearch.com/en/ourservices/decommissioning-fir-1-nuclear-reactor. Accessed 30 June 2024
Waters et al. (2018) National geological screening: Northern England region. British Geological Survey, 6–8. https://nora.nerc.ac.uk/id/eprint/527049/1/CR17103N.pdf. Accessed 30 June 2024
WNA (2016) Nuclear development in the United Kingdom. Power reactor decommissioning. World Nuclear Association. https://www.world-nuclear.org/information-library/country-profiles/countries-t-z/appendices/nuclear-development-in-the-united-kingdom.aspx. Accessed 30 June 2024
WNA (2023a) Nuclear power in Germany. World Nuclear Association. https://world-nuclear.org/information-library/country-profiles/countries-g-n/germany.aspx. Accessed 30 June 2024
WNA (2024) Nuclear power in Finland, June. World Nuclear Association. https://world-nuclear.org/information-library/country-profiles/countries-a-f/finland. Accessed 30 June 2024
Yamada F (2019) Progress and regional dialogue on radioactive waste disposal projects in the UK. Nuclear Environment Center Topics, No. 130, September (in Japanese). https://www.rwmc.or.jp/library/file/topicsNo.131_2.pdf. Accessed 30 June 2024

Chapter 4
Legal Framework for Nuclear Use

This chapter first discusses nuclear disarmament and non-proliferation and safeguard measures, then introduces the nuclear damage compensation system, outlining the systems in the United States and Japan. It then discusses laws and organizations related to nuclear use and safety regulations in the United States, France, and Japan, and introduces the structure and concept of nuclear law. It discusses the U.S. Nuclear Regulatory Commission as an example of administrative agency accountability. Finally, it introduces the safety review of nuclear power plants in Japan.

4.1 Non-proliferation and Safeguard Measures

Dynamite, which has been used in road and dam construction and has benefited society, has also been used in wars as a bomb. Nuclear technology can generate enormous energy from a small amount of material. It is humans who decide how to use this technology, and it is necessary to prevent nuclear technology from being used in wars or terrorism. Atomic bombs (bombs that use nuclear fission reactions) and hydrogen bombs (bombs that use nuclear fusion reactions, which are detonated using nuclear fission reactions) are weapons of mass destruction, and their proliferation must be prevented. Among nuclear technologies, uranium enrichment technology, which produces highly enriched uranium used in these nuclear weapons, and reprocessing technology, which extracts plutonium from spent fuel, are particularly strictly managed to prevent proliferation.

Y. Oka, *Nuclear Power and Society*, Nuclear Science and Technology,
https://doi.org/10.1007/978-981-97-8951-1_4

4.1.1 Nuclear Disarmament

The Treaty on the Non-Proliferation of Nuclear Weapons (NPT) came into effect in 1970, and currently about 190 countries are signatories to this treaty. Japan ratified it in 1976. This treaty defines the five countries that manufactured and detonated nuclear weapons before 1966—the United States, Russia, the United Kingdom, France, and China—as "nuclear-weapon states", and prevents the spread of nuclear weapons to countries other than these. It also stipulates that the treaty states have an obligation to negotiate nuclear disarmament in good faith, and that the peaceful use of nuclear energy is an "inalienable right" of the treaty states (MOFA 2021).

The number of states party to the NPT has increased since 1990. South Africa, which had possessed nuclear weapons, abandoned them and joined as a "non-nuclear-weapon state" in 1991. In 1992, China and France joined as "nuclear-weapon states", and Belarus, Ukraine, and Kazakhstan transferred their nuclear weapons to Russia and joined as "non-nuclear-weapon states" by 1994. Argentina and Brazil joined in 1995 and 1998, respectively. The treaty was indefinitely extended in 1995. India, Pakistan, Israel, and South Sudan are not members of the treaty.

In the 1990s, issues of suspected nuclear weapons development arose in Iraq and North Korea, both of which are treaty members. India and Pakistan conducted nuclear tests in 1998. North Korea announced its withdrawal from the treaty in 1993 and 2003, and has conducted several nuclear tests since 2006.

The other treaty, the *Treaty on the Prohibition of Nuclear Weapons* was drafted in 1996, and came into effect in January 2021 after the number of ratifying countries exceeded 50 in October 2020. The Treaty on the Prohibition of Nuclear Weapons completely and comprehensively prohibits nuclear weapons for all signatory states upon joining. While the Treaty on the Non-Proliferation of Nuclear Weapons focuses on the transfer, manufacture, and acquisition of nuclear weapons, the Treaty on the Prohibition of Nuclear Weapons prohibits the use of nuclear weapons itself. Japan has not joined the Treaty on the Prohibition of Nuclear Weapons. This is because the Japanese government believes it is necessary to maintain the deterrent power of the United States, which possesses nuclear weapons, under the Japan-US alliance, as it is difficult to deter opponents like North Korea, who hint at the use of nuclear weapons, with conventional weapons alone. This can be confirmed on the Ministry of Foreign Affairs' website.

The Treaty on the Prohibition of Nuclear Weapons is more of a humanitarian treaty than a disarmament treaty. In the field of security, the international community is not yet a society that can be managed solely by high morality, but a world where sovereign states compete in a survival of the fittest. The current security environment is not only not mature enough to abandon nuclear weapons, but there are also countries that are openly advancing nuclear development and increasing the number of nuclear warheads. In such a situation, the Treaty on the Prohibition of Nuclear Weapons is too detached from reality. The Treaty on the Non-Proliferation of Nuclear Weapons (NPT) maintains its vector toward nuclear abolition and forms the foundation of international security, and countries, including Japan, that are members of the NPT must uphold the legitimacy of the NPT (Sano 2022a, b).

In addition, Japan is working through the United Nations to pass resolutions for the abolition of nuclear weapons, the early entry into force of the Comprehensive Nuclear-Test-Ban Treaty (CTBT), and the start of negotiations for the Fissile Material Cut-off Treaty (FMCT). Japan ratified the Comprehensive Nuclear-Test-Ban Treaty in 1997, and many countries have signed it, but the treaty has not yet entered into force because the ratification by the required countries has not been completed (MOFA 2023a).

4.1.2 Nuclear Non-proliferation

Nuclear non-proliferation refers to the prevention of the spread of nuclear weapons, not increasing the number of countries possessing nuclear weapons, and countries that possess nuclear weapons reducing their number of nuclear weapons. The former is called the prevention of horizontal proliferation, and the latter is called the prevention of vertical proliferation.

Figure 4.1 shows the legal framework for nuclear non-proliferation. United Nations Security Council (UNSC) resolutions are approved when nine out of the fifteen members of the Security Council vote in favor and none of the permanent members oppose (no veto is exercised). For example, there are resolutions on counter-terrorism (SCR1373, 2001) and the prevention of the proliferation of weapons of mass destruction (SCR1540, 2004).

At the international level, there are multilateral treaties such as the Treaty on the Non-Proliferation of Nuclear Weapons. Agreements include technical details based on treaties. In the field of nuclear non-proliferation, the International Atomic Energy Agency and each country have concluded safeguards agreements. *Safeguards* are an inspection system to ensure that the use of nuclear materials and other nuclear equipment, such as uranium and plutonium, is limited to peaceful purposes, that they

Fig. 4.1 Nuclear non-proliferation legal framework

are not diverted to nuclear explosive devices or other military purposes, that there are no undeclared nuclear materials, and that no nuclear activities are being conducted (ISCN 2024).

The International Atomic Energy Agency (IAEA) conducts verification, including inspections, under the Safeguards Agreement to ensure that nuclear power is not diverted for military purposes such as the production of nuclear weapons in non-nuclear-weapon states. Non-nuclear-weapon states under the Treaty on the Non-Proliferation of Nuclear Weapons (NPT) are obligated to conclude a Safeguards Agreement (Full Scope Safeguards Agreement) with the IAEA. In Japan, the agreement came into effect in 1977, and domestic nuclear facilities are subject to IAEA inspections. Many IAEA member states do not operate nuclear power plants. Countries that do not possess nuclear materials domestically, or possess only small amounts, and do not own nuclear facilities or permit their construction, conclude a Small Quantities Protocol with the IAEA, and their inspection burdens are exempted or reduced.

The threat of nuclear terrorism includes radiation damage caused by the theft of nuclear weapons, the acquisition of nuclear materials for the purpose of manufacturing nuclear explosive devices, the malicious use of radiation sources including "dirty bombs", and sabotage and destruction of nuclear facilities. "Nuclear security" refers to "prevention, detection, and response to theft, sabotage, unauthorized access, illegal transfer, or other malicious acts involving nuclear material, other radioactive substances, or their associated facilities". Measures to prevent the theft of nuclear materials at each stage, such as sabotage and destruction of nuclear facilities, transportation and storage of nuclear materials, and use in nuclear facilities, are mandated for nuclear operators. Specifically, these include setting up protected areas and managing access. The national Nuclear Regulation Authority regularly verifies the effectiveness of the protective measures taken by nuclear operators. The transportation of nuclear fuel materials is also under the jurisdiction of the ministry stipulated by law. In 2005, the "International Convention for the Suppression of Acts of Nuclear Terrorism" (hereinafter referred to as the "Nuclear Terrorism Prevention Convention") was adopted by the United Nations General Assembly and came into effect in 2007.

In response to India's nuclear test in 1974, the Nuclear Suppliers Group (NSG), which includes countries capable of supplying nuclear-related materials and equipment, was established, with about 50 countries, including Japan, participating. NSG participating countries select guidelines that define the export conditions for items and related technologies designed or manufactured for use in nuclear activities, and conduct export control based on these guidelines (MOFA 2023b).

Japan is the only non-nuclear-weapon state that has been promoting the use of civilian plutonium by reprocessing spent fuel from nuclear power plants. Although Japan is an excellent student in nuclear material management (safeguard measures), high transparency is required for the peaceful use of plutonium. For this reason, Japan Atomic Energy Commission established the "Basic Concept of Plutonium Use in Our Country" in 2018. The main points are to "authorize reprocessing only to the extent necessary for pluthermal (use in power reactors of nuclear fuel containing reprocessed plutonium)", "promote steady reduction of overseas-held plutonium by encouraging cooperation among electric power companies", and "steadily implement measures toward the expansion of spent fuel storage" (Nuclear White Paper 2019).

France, a nuclear-weapon state, has been promoting the use of civilian plutonium in nuclear power plants by reprocessing spent fuel from nuclear power plants and processing it into nuclear fuel. France reported its spent fuel management policy, called the Equal flow principle, "To avoid accumulation of unused separated plutonium, plutonium is reprocessed only when it is used and only in the amount used", at the 2003 IAEA "Convention on the Safety of Spent Fuel Management and Radioactive Waste Management" Contracting Parties Meeting (Review Meeting) (ASN 2003, p. 9).

Not limited to France, the contracting parties to this treaty, including Japan, regularly submit reports on the management of spent fuel and radioactive waste to the International Atomic Energy Agency (IAEA). From these materials, we can understand the situation of spent fuel and radioactive waste management in each country.

When I was discussing the concept of plutonium use at the Atomic Energy Commission, there were not a few people in Japan's nuclear industry who advocated "whole reprocessing (reprocessing whole spent fuels)". However, producing more separated plutonium than the amount used can be a major business and political risk if there is no prospect of consumption. Most of the high-enriched uranium and spent fuel used in Japan's research reactors were made in the United States, so the United States took them back. Therefore, the treatment and disposal of these materials did not pose a business risk for Japanese universities with research reactors. The high-enriched uranium fuel of the research reactor of the University of Tokyo, which I had been in charge of operating and managing for many years, was also able to be returned to the United States, which was a great relief. The return took place after I retired from Japan Atomic Energy Commission.

Plutonium made in Japan is not made in the United States, so there is no mechanism for the United States to take back Japan's separated plutonium. This is even more the case for electric power companies, which are private companies. Furthermore, from a technical point of view, the composition of separated plutonium deteriorates due to the decay of plutonium-241. Therefore, separated plutonium needs to be consumed as fuel within 3 or 4 years after reprocessing. Separated plutonium is a "perishable" item. Old perishables cannot be used in light water reactors unless they are dissolved and processed again.

The call for "whole reprocessing" was loud in Japan, but those who were shouting it did not realize the risk of accumulating large amounts of separated plutonium in order to promote the nuclear fuel cycle. A similar example is that those involved in nuclear power who loudly advocate for verbal risk communication to the public, do not realize that they are giving the impression that nuclear power is dangerous. Moreover, a common feature of those who advocate for total reprocessing and risk communication is that they are not involved in local dialogue activities or reprocessing projects. They may think they are cheerleaders, but nuclear power stakeholders should not strangle their own use of nuclear power.

The Japanese nuclear power industry is a collectivist one with vague boundaries of stakeholders. The consolidated opinions are not robust and may not reflect past knowledge, including that of the West. The location of responsibility is also vague. Since accountability is not functioning in Japan's administration, there is a risk of repeating the same mistakes. Instead of collectivism, each expert and professional organization should fulfill their respective responsibilities.

4.2 Nuclear Damage Compensation System

4.2.1 Overview

Operators of nuclear power plants have comprehensive liability for accidents they cause, regardless of fault (strict liability). Therefore, in many countries, these companies are insured for third-party liability. As the compensation range can extend across borders, national compensation systems are often supplemented by international treaties. For damages beyond the liability covered by insurance, it is generally considered that the state is the ultimate bearer of responsibility, as with other industrial accidents.

Strict liability means that the victim does not need to prove the perpetrator's fault. Comprehensive liability means that the company operating the reactor takes on all claims for accident damage. Even if the scope of the accident extends across borders, the courts of the country where the accident occurred have jurisdiction over damage claims. This prevents lawsuits from being filed in a country that is convenient for the plaintiff. The fact that there is a cap on the amount of compensation the company that caused the accident can pay often causes controversy. With a cap, the company can transfer responsibility to either the state or a collective of power companies, or both. There is an opinion that the state, by permitting the construction and operation of nuclear power plants, implicitly takes on the responsibility that arises from this, which is the same as with other infrastructure. On the other hand, there is also an opinion questioning whether the public would accept the company responsible for the accident shirking its responsibility.

Insurance against nuclear damage is different from insurance for general industries in that there is a pool of insurance (a group of insurance companies) for this purpose domestically and there is an international mutual insurance mechanism. The insurance pool began in the UK in 1956.

In the international framework, there are the Vienna Convention of the International Atomic Energy Agency (effective 1977) and the Paris Convention of the Organisation for Economic Co-operation and Development (OECD) (effective 1968, strengthened by the Brussels Supplementary Convention in 1963). The latter is mainly joined by European countries, and the former is mainly joined by countries other than these. These two conventions are linked by a joint protocol adopted in 1988, which stipulates comprehensive liability of power companies (exemption of liability of reactor manufacturers), no-fault liability, a cap on liability compensation, a 10-year claim period from the accident, jurisdiction in the courts of the accident country, non-discrimination of victims by nationality, and that damage is limited to property, health, and life, and does not include environmental damage or protective measures or economic damage. Subsequently, coordination of the two conventions was sought by amending the Vienna Convention and the Paris Convention, and in 2004 an amendment was made to allow countries with unlimited liability to participate in the Paris Convention. The amended Vienna Convention came into effect in 2003, but the amended Paris Convention and the 2004 amended Brussels Supplementary Convention have not come into effect.

The CSC (Convention on Supplementary Compensation for Nuclear Damage) was adopted by the IAEA in 1997. It is a treaty to build a global liability system by supplementing measures under domestic law on compensation for nuclear damage, from the perspective of expanding the amount of compensation. It supplements measures under domestic law that have principles that do not contradict the Vienna Convention and the Paris Convention (as stipulated in the annex), from the perspective of expanding the amount of compensation, and is an agreement to build a global liability system. Even countries that have not concluded either the Vienna Convention or the Paris Convention can become CSC target countries. The United States and Canada participate in the CSC, and Japan, which had not participated in either the Vienna Convention or the Paris Convention, joined the CSC in 2015. The CSC came into effect in 2015. India also joined the CSC in 2016. It is expected that the CSC will become the mainstream in future. China and South Korea have not yet joined any of the conventions.

4.2.2 US Nuclear Damage Compensation System

In the United States, there is a nuclear damage compensation law called the Price-Anderson Act. This law was established in 1957 and is the world's first comprehensive nuclear damage compensation law. The amount of compensation, etc., is reviewed periodically, and an extension of 20 years was made in 2005. It applies to all nuclear-related facilities, including not only nuclear power plants but also research reactors, uranium enrichment plants, radioactive waste disposal sites, etc., although the amount of compensation varies.

Firstly, as a primary compensation measure, nuclear power operators are enrolled in private insurance of 450 million dollars per reactor (as of 2017). Secondly, as a secondary compensation measure, there is an insurance pool covered by the insurance premiums paid by all nuclear power operators, with a total of about 13 billion dollars prepared per accident as the liability amount of the nuclear power operators. So far, the amount paid out as compensation, including the 70 million dollars for the Three Mile Island nuclear accident in 1979, is about 150 million dollars.

The U.S. nuclear damage compensation system stipulates the financial responsibility of nuclear power operators, not their legal responsibility. On the other hand, international treaties are about legal responsibility, and the U.S. did not participate in international treaties on nuclear damage compensation. The CSC has provisions in its annex that take into account the U.S. system, enabling the U.S. to join, and the U.S. ratified the CSC in 2008 (WNA 2021).

4.2.3 Japan's Nuclear Damage Compensation System

The law on nuclear damage compensation in Japan was enacted in 1961, the dawn of nuclear power use in Japan. At that time, U.S.-made nuclear power plants were being introduced worldwide, and to prevent U.S. reactor manufacturers from bearing product liability, the U.S. asked export partner countries to enact laws, so not only Japan but also other countries established compensation systems. Therefore, the basic principles of the compensation system are almost universally common, and so is Japan's nuclear damage compensation system.

The nuclear damage compensation system generally aims to "protect victims" and "promote the sound development of the nuclear power industry". The sound development of the nuclear power industry means, for example, limiting the scope of compensation liability to companies that are operating nuclear power generation, thereby enabling private companies such as reactor manufacturers and construction companies to participate in the nuclear power business. The essence of the nuclear damage compensation system is the concentration of responsibility on nuclear power operators, no-fault liability, mandatory damage compensation measures, limitation of the amount of compensation so that the economic responsibility of nuclear power operators does not become infinite, and state compensation in case the operator cannot fulfill its compensation liability. In Japan, Germany, and Switzerland, there is no set limit for compensation liability. Figure 4.2 shows Japan's nuclear damage compensation system (MEXT 2024; JAERO 2023; JAIF 2024).

Nuclear power operators are obligated to enter into nuclear damage liability insurance contracts with private insurance companies and nuclear damage compensation contracts with the government. Nuclear power operators are subject to strict liability and concentration of responsibility. Nuclear power operators not only refer to organizations that operate nuclear power plants and reactors, but also those that process and reprocess nuclear fuel, use nuclear fuel materials, and store spent fuel. This includes not only operators of nuclear reactors, but also those who process nuclear fuel. National research institutions and universities that operate research reactors are also considered nuclear power operators, and the nuclear damage compensation system applies to them. The amount of insurance varies depending on the output of the reactor and the type of business. The compensation amount is capped at 120 billion yen. There is no limit to the amount of damage, and it is unlimited liability. However, nuclear power operators are exempt from responsibility for abnormally large natural disasters and social unrest (such as wars). If nuclear damage exceeding the compensation amount occurs, the government will assist the operator in compensating for the damage when necessary. The Nuclear Damage Compensation Act has an application

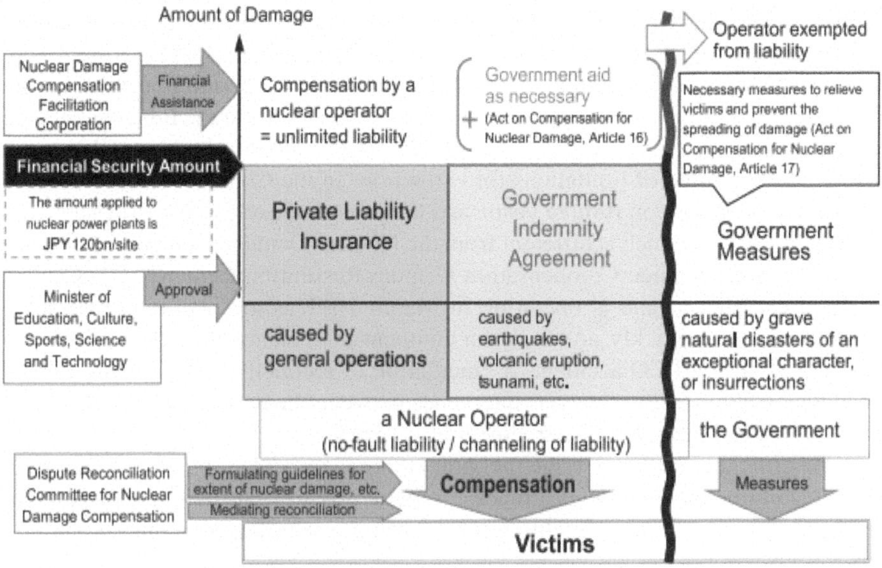

Fig. 4.2 Japan's compensation system for nuclear damage. *Source* OECD/NEA No.7089, 2012

deadline, so the content will be reviewed by the deadline. Recently, it was revised in 2018. Japan Atomic Energy Commission has set up a special committee to review the nuclear compensation system and compiled a report in 2018. You can learn about the issues and results of the review of Japan's nuclear damage compensation system (JAEC 2018).

The TEPCO Fukushima accident that occurred in March 2011 is considered not to fall under the category of abnormally large natural disasters based on the parliamentary responses at the time the Nuclear Damage Compensation Law was enacted in Japan, and compensation procedures are being carried out on the premise that TEPCO has primary liability for damage compensation. The amount of damage is far greater than 120 billion yen, so the Nuclear Damage Compensation Support Organization Law was created in August 2011, and the government, through the Nuclear Damage Compensation Support Organization, provides support for damage compensation to TEPCO, and a system has been built in which electric power companies that operate nuclear power generation respond to payments and the like through mutual aid. Subsequently, the Nuclear Compensation Organization also undertook the decommissioning-related work of the TEPCO Fukushima Daiichi Nuclear Power Station, and its name was changed to the Nuclear Damage Compensation and Decommissioning Facilitation Corporation (NDF) in 2014. TEPCO uses the funds received from the NDF to compensate victims and repay the funds received from the profits of future corporate management. It should be noted that the government is supposed to fill in for accidents due to disclaimers and compensation after the 10-year compensation period from the accident (when damage occurred). The Nuclear Damage

Compensation Special Law enacted in 2013 stipulates that the statute of limitations for extinction is 10 years from the time the damage and the perpetrator were known (when it became possible to exercise rights), and the period of exclusion (the period until the right to claim damages is lost) is 20 years from the time the damage occurred (when the right to claim arose). TEPCO has announced that it will respond flexibly to the statute of limitations for extinction. In the Civil Code, the statute of limitations for extinction is three years, and the period of exclusion is 20 years from the time of the tort, which is different from the Nuclear Damage Compensation Law.

The Nuclear Damage Compensation Dispute Resolution Center (ADR Center) has been established and is operating in March 2011 as an institution to resolve disputes smoothly, quickly, and fairly for compensation claims from those who were damaged by the TEPCO accident. A "mediation of reconciliation" is being carried out in which a third party (lawyer) intervenes between the parties and strives for the approval of the dispute by the agreement of the parties.

In the criticality accident at the fuel processing facility (JCO) in Tokai Village in 1999, the amount paid by the insurance company was 1 billion yen, and the amount paid by the parent company, Sumitomo Metal Mining, was the remaining 13.5 billion yen.

4.3 Laws and Organizations Related to the Use of Nuclear Power and Safety Regulations

4.3.1 Laws and Organizations Related to Nuclear Power in the United States

In the United States, about 100 nuclear power plants and more than 30 experimental research reactors are in operation. 2. In the United States, nuclear power plants are operated by private power companies, except for those owned by the Tennessee Valley Authority. Many research reactors are owned by universities.

In the United States, the Atomic Energy Commission (AEC) was established by the Atomic Energy Act of 1946. The use of nuclear power for civilian purposes began with the amendment of the Atomic Energy Act in 1954. The AEC was also responsible for safety regulations, but with the Energy Reorganization Act of 1974, the Nuclear Regulatory Commission (NRC) was created, and the AEC's other duties were taken over by the Energy Research and Development Administration (ERDA). The Department of Energy Organization Act abolished the ERDA and assigned the AEC's non-regulatory duties to the Department of Energy (DOE), which continues to this day (NEA 2016). It is noteworthy that the establishment of the Nuclear Regulatory Commission in the United States took place before the Three Mile Island (TMI) nuclear power plant accident in 1979. The recognition of the need for independence of regulatory agencies and its implementation took place in the United States before the TMI accident. When using radioactive materials for civilian purposes, it is necessary to obtain a license from the NRC. The United States is a federal country. States regulate nuclear activities that the federal government does not involve in, under conditions that do not conflict with federal law. The safety regulation of facilities under the jurisdiction of the Department of Energy, such as national research institutes, is carried out by the Department of Energy itself.

The Price-Anderson Act, which deals with nuclear damage compensation, was established with the amendment of the Atomic Energy Act in 1954. In 1966, the Price-Anderson Act was amended to clarify no-fault liability. In 1978, the Atomic Energy Act was amended and the Nuclear Non-Proliferation Act was established. The NRC has established standards for controlling the export of nuclear materials and technology in the United States. Measures to strengthen nuclear non-proliferation and to ensure a stable supply of nuclear fuel to countries that comply with nuclear non-proliferation policies are stipulated. In terms of radioactive waste, the Uranium Mill Tailings Radiation Control Act of 1978, the Radioactive Waste Policy Act of 1982, and the Low-Level Radioactive Waste Policy Amendments Act of 1985 are stipulated. These laws are under the jurisdiction of the NRC. Uranium mill tailings are the remaining pieces of ore that are produced when uranium ore is sorted. This regulates the disposal, long-term stabilization, and management of these tailings. The response to low-level radioactive waste was also stipulated. The Radioactive Waste Policy Act of 1982 stipulates the establishment of a high-level radioactive waste disposal site at Yucca Mountain in Nevada. The Low-Level Radioactive Waste Policy Amendments Act of 1985 stipulates that states are granted the authority to dispose of low-level radioactive waste generated within the state. In the United States, the disposal of private low-level radioactive waste is under the jurisdiction of the states, not the federal government.

The Energy Policy Act of 1992 revised the nuclear reactor licensing process in response to the liberalization of the electricity market. Reactor design standards were standardized (Design Certification, DC), and the two-step process of construction permit and operation permit was combined into a construction and operation permit. Combined Operating License (COL) can be issued before construction. The Energy Policy Act of 2005 stipulates federal loan guarantees, tax deductions, and compensation for delays in licensing procedures to promote the construction of new nuclear power plants, which are difficult to attract investment due to their huge construction costs.

In the United States, since there are states that have caused major power outages after electricity liberalization, the progress of electricity liberalization has stopped, and there are now liberalized and non-liberalized states. The ease of obtaining federal government guarantees for new nuclear power plant construction differs between non-liberalized and liberalized states, with non-liberalized states having better conditions. In 2005, the Price-Anderson Act was extended for 20 years and the insurance premium was revised. In 2007, with the ratification of the Convention on Supplementary Compensation for Nuclear Damage (CSC) by the United States, its implementation law was enacted.

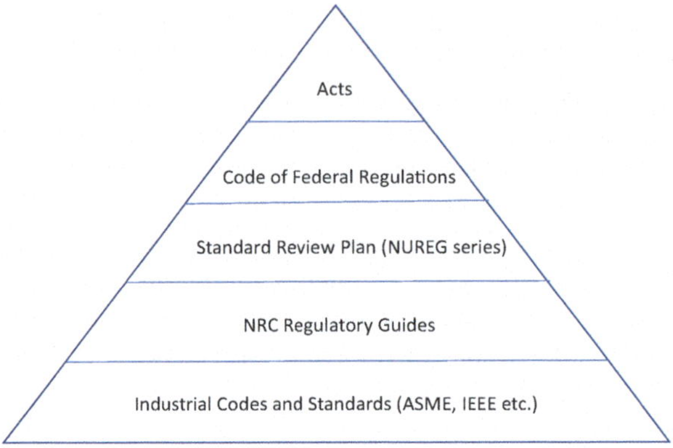

Fig. 4.3 US nuclear safety legal system

The legal system for the safety of nuclear power plants in the United States, as shown in Fig. 4.3, consists of laws (Acts), federal regulations (Code of Federal Regulation), Standard Review Plans, Regulatory Guides, and private standards (Industrial Codes and Standards).

Among the federal regulations related to nuclear power generation are 10CFR50 related to licensing, 10CFR52 related to early site permits, Design Certification (DC), and Combined Operating License (COL), and 10CFR54 related to operating license renewal.

The Standard Review Plan is a guidebook for the reviewers of the Nuclear Regulatory Commission's Reactor Regulation Office when conducting safety reviews of nuclear power plant construction and operation applications. This ensures the quality and uniformity of the review, provides a basis for safety evaluation, and helps improve the disclosure of regulatory information and understanding of the review process.

Regulatory guides provide guidance to applicants on the techniques and necessary data used by reviewers when evaluating specific parts of NRC rules, issues, etc., and are classified into 10 categories such as power reactors, research and test reactors, and fuel and nuclear material facilities.

Furthermore, there are technical standards related to the manufacture and performance maintenance of nuclear power plant equipment, such as the American Society of Mechanical Engineers (ASME) technical standards for nuclear reactor pressure vessels and piping, and the Institute of Electrical and Electronics Engineers (IEEE) technical standards for electrical and control systems.

Nuclear power plants that have obtained operating licenses are subject to regular inspections by the NRC over the period during which power generation operations are conducted. A senior inspector is stationed at each nuclear power plant site to monitor operations. The NRC's regional offices may also conduct special inspections. In this case, inspectors may be dispatched from the NRC headquarters. The inspection results are documented and made public. Furthermore, from the perspective of the nuclear power plant side proactively seeking improvement, under the slogan of *risk-based performance-based regulation*, there is the Reactor Oversight Process. The process is carried out based on performance indicators. When a special event (such as an accident) occurs during operation, the NRC forms a team to investigate. The nuclear power plant regularly, or when a special event occurs, prepares a report and submits it to the NRC. The operating period of a nuclear power plant is initially authorized for 40 years, and then it can be renewed every 20 years with prior application and approval. In international treaties, the United States ratified the Nuclear Safety Convention, which was established in 1994, in 1999.

I will discuss the institutional framework (organization) of nuclear power use in the United States. The federal government's regulatory organization (agency) is the Nuclear Regulatory Commission (NRC). The NRC was established in 1974 and is responsible for regulating nuclear materials and nuclear facilities. The NRC has five commissioners (top executives) appointed by the President. The commissioners clearly state whether they are on the side of the Republican or Democratic Party, the two major political parties in the United States, and are chosen so as not to be too biased toward either party. The term is five years, and holding other jobs is prohibited. The NRC has bureaus such as the Reactor Regulation Bureau, Nuclear Material Safety and Security Measures Bureau, Nuclear Regulatory Research Bureau, Nuclear Security Incident Response, and regional offices. Regional offices are located in four regions across the country and are responsible for inspecting and licensing nuclear power plants in the region, as well as liaising with states. The NRC is mandated by the Comprehensive Budget Adjustment Act of 1990 to obtain 90% of its budget from licensing fees.

Federal agencies have advisory committees established under the Federal Advisory Committee Act, which provide balanced perspectives and advice from technical and policy perspectives. The NRC has an Advisory Committee on Reactor Safety (ACRS).

The Department of Energy (DOE) was established in 1977 and is responsible for the research, development, and demonstration of civilian nuclear power previously carried out by the AEC and then the ERDA. In addition, the DOE, along with the Ministry of Defense (DOD), is responsible for the military use of nuclear energy. The DOE approves contracts for special nuclear materials and uranium enrichment services to foreign countries, manages the export of nuclear equipment, reactors, and nuclear materials, and also handles the return (repatriation) of U.S.-origin nuclear materials previously sent to foreign countries. While the DOE is responsible for all energy, including renewable energy and fossil fuels, in relation to nuclear power, it has the Nuclear Energy Bureau, which is responsible for nuclear energy policy, the Environmental Management Bureau, which is responsible for the environmental recovery of sites such as national laboratories that developed nuclear weapons during World War II, the Loan Programs Office, which is responsible for loans (borrowings and guarantees) for energy projects, the Legacy Management Bureau, which is responsible for the stabilization and restoration of abandoned uranium mines etc. for nuclear weapons development, and the Science Bureau, which is responsible for research and development in accelerators and elementary particle science.

Under the Department of Energy, there are the Idaho, Argonne, Oak Ridge, Livermore, Sandia, Los Alamos, Pacific Northwest, Brookhaven, and Lawrence Berkeley National Laboratories. During World War II, these national laboratories were involved in the development of nuclear weapons and their materials, and after the war, they conducted not only military research but also the development of nuclear reactors and nuclear-related experimental research. Various experimental reactors were built and tested at the Idaho National Laboratory's nuclear reactor test site. The Idaho National Laboratory was once known as the Argonne West Laboratory. The Argonne National Laboratory, located in the suburbs of Chicago, conducted research using experimental nuclear reactors and accelerators. Today, a wide range of scientific and engineering research is being conducted. The Lawrence Livermore National Laboratory is located a little inland from San Francisco in Livermore, where research on nuclear weapons using lasers and nuclear security is being conducted. The Sandia National Laboratories, located in New Mexico, conduct tests on electronic and large-scale equipment, including military use. With the budget of the NRC, research on the safety of nuclear reactors and core meltdown accidents is conducted and knowledge is compiled. The Los Alamos National Laboratory, also in New Mexico, developed atomic bombs during World War II. Today, it conducts both military research and general scientific and engineering research. The Pacific Northwest National Laboratory is located inland in Washington state, where plutonium for nuclear weapons was produced. Today, it conducts research in chemistry and earth science. The Brookhaven National Laboratory is located on Long Island in southeastern New York, where advanced scientific research, such as particle physics, is conducted. The Lawrence Berkeley National Laboratory is located in Berkeley, facing the San Francisco Bay, and is adjacent to the University of California, Berkeley. It was created by Lawrence, who invented the cyclotron accelerator, and now conducts non-military research in a wide range of scientific fields. These national laboratories are operated by private companies that have management and operation contracts with the Department of Energy. Many large facilities, such as accelerators, at the national laboratories are available for joint use.

The Federal Energy Regulatory Commission (FERC) is an independent commission composed of five commissioners. Its role is to help consumers receive economically efficient, safe, reliable, and secure energy services at reasonable prices through regulation, market mechanisms, and cooperation. It is responsible for the safety and inspection of transmission lines and natural gas pipelines that cross state lines, the extension of operating periods for hydroelectric power plants, and the promotion of competition in wholesale electricity.

In the United States, the disposal of high-level radioactive waste and spent fuel emitted from private nuclear power plants is the responsibility of the Department of Energy. The US Nuclear Waste Technical Review Board) is an independent federal government agency that conducts technical and scientific peer reviews on the treatment and disposal of high-level radioactive waste carried out by the DOE. The results and recommendations are submitted to the federal congress and the Secretary of Energy.

The Department of State (DOS) is equivalent to the foreign ministries of other countries. It implements foreign policy and manages the conclusion of international treaties and international cooperation. Based on the Nuclear Non-Proliferation Act of 1978, the Department of State checks for consistency with U.S. laws and policies.

Within the Department of State is the Bureau of International Security and Nonproliferation (ISN), which implements policies and their execution and proposals related to nuclear non-proliferation and nuclear exports to other countries and international organizations. The ISN is working to prevent the proliferation of weapons of mass destruction, their means of transport, and new conventional weapons. It imposes sanctions on foreign states, companies, and individuals engaged in nuclear proliferation activities, such as Syria and North Korea.

The Environmental Protection Agency (EPA) is responsible for the environment in the United States. The EPA sets limits on radiation exposure and standards for the release of radioactive substances into the environment. The NRC applies and implements the standards set by the EPA for facilities where the NRC grants licenses. In the United States, emergency response is the responsibility of the state government. The EPA supports state emergency responses in the event of oil spills, environmental pollution by hazardous substances, and major disasters. For nuclear and radiation disasters, it has created Protective Action Guides (PAGs).

The Department of Labor (DOL) is in charge of worker welfare and safety. In the case of whistleblowing about violations of nuclear-related laws and regulations, the NRC can respond to discrimination by licensees (nuclear operators) based on the Atomic Energy Act, but the Department of Labor can personally relieve the whistleblowing employee. Relief for whistleblowers also applies when they have given unfavorable testimony in Congress or have been forced to violate laws and regulations. It can also provide relief for internal whistleblowing by contract employees related to the Department of Energy.

The Food and Drug Administration of the Department of Health and Human Services (DHHS) is in charge of the safety and effectiveness of the manufacture and sale of radioactive pharmaceuticals, biological products, and medical devices. The DHHS has a department that conducts radiation therapy research.

The Department of Interior (DOI) is in charge of conserving federally owned public lands and natural resources. The Department of Interior has the U.S. Geological Survey, which cooperates in the geological disposal of radioactive waste carried out by the Department of Energy from a geoscientific technical perspective. While many other countries' interior ministries are in charge of maintaining public order, police, border security, and disaster prevention, the U.S. Department of Interior does not have such duties.

The Department of Transportation (DOT) is in charge of regulations related to the transportation of radioactive materials in cooperation with the NRC. This is under the jurisdiction of DOT's Pipeline and Hazardous Materials Safety Division.

The Federal Emergency Management Agency (FEMA) is in charge of off-site responses during nuclear disasters, and collaborates with the NRC and DOE to coordinate emergency responses at the federal, state, and local levels.

The Tennessee Valley Authority is a federally owned corporation that executes plans for resource conservation, development, and land use in the Tennessee River basin. In addition to hydroelectric power, it operates commercial nuclear power plants and sells electricity.

The White House includes the Office of Management and Budget (OMB) and the Office of Science and Technology Policy (OSTP). The former organizes the budget reflecting the President's vision, creates budget instructions to be sent to Congress, and manages the activities of each administrative agency by executing the established budget. The OSTP advises the President and the White House on the impact of science and technology on domestic and international issues.

Quasi-public institutions in the United States include the American National Standards Institute (ANSI), the National Academy of Science (NAS), the National Council on Radiation Protection and Management (NCRP), and the National Nuclear Data Center.

The American National Standards Institute (ANSI) is an organization for standardizing industrial standards within the United States. Its purpose is to improve the quality of American industry and life through voluntary consensus standards and conformity assessment systems. ANSI itself does not create standards, but certifies and approves standards developed by standard development organizations. The certified standards become American National Standards (ANSI standards). While the Japanese Industrial Standards (JIS) are established by the government, ANSI is an independent non-profit organization in the United States. ANSI is a founding member of the International Organization for Standardization (ISO) and participates as the U.S. representative in the International Accreditation Forum. Although it is a domestic standards organization in the United States, ANSI standards often precede ISO standards.

The National Academy of Science (NAS) is a private, non-profit organization composed of leading scientists in the United States. It was established by the federal congress in 1863. Along with the National Academy of Engineering, established in 1964, and the National Academy of Medicine, established in 1970, it forms the National Academies of Sciences, Engineering, and Medicine, conducting independent, objective analysis and providing advice to the nation, and undertaking activities to solve complex problems.

The National Council on Radiation Protection and Management (NCRP) is a non-profit private organization recognized by the federal government that provides independent scientific analysis, information, and recommendations to support radiation protection, based on the consensus of scientists. The NRC and EPA use the NCRP's recommendations as the basis for their respective radiation protection activities. The NCRP works closely with the International Commission on Radiation Protection (ICRP).

The National Nuclear Data Center (NNDC) is a center that collects, evaluates, and disseminates nuclear physics data for nuclear technology applications. The center's mission is to provide current, accurate, and widely accessible data to the nuclear community. The Center is located at Brookhaven National Laboratory and collects, evaluates, and publishes nuclear data such as the structure of atomic nuclei and the nuclear reaction rate (also known as neutron cross-section) between atomic nuclei and neutrons. Nuclear data is fundamental data necessary for exploring science and applications involving atomic nuclei. The periodic table is one type of nuclear data.

The above is about the U.S. Atomic Energy Act and its related organizations.

U.S. universities, both public and private, have nuclear engineering departments at major state universities and MIT where nuclear research and education are conducted. Many universities also have low-power research reactors. Universities conduct research by obtaining competitive funds and other resources from federal government agencies and national research institutes. University faculty, environmental groups, and consumer groups express their opinions in testimonies in Congress and at public hearings held by the NRC.

The main private organizations related to nuclear power include the Nuclear Energy Institute (NEI), which is operated with funding from nuclear power operators and reactor manufacturing companies, and represents the nuclear industry in expressing opinions to the U.S. Congress. The Electric Power Research Institute (EPRI) is a private research institution that proposes industry-responsive research projects on a non-profit basis, collects budgets, and implements them. Joint research is also conducted with the EPRI, NRC, and national research institutes under the DOE, taking into account conflicts of interest.

The U.S. power reactor manufacturing companies include GE (General Electric) and WH (Westinghouse). GE was founded by Edison. WH is a historic company founded by Westinghouse, who invented AC power transmission. In the U.S., venture companies are proposing the development of nuclear power plants using small reactors. Currently, GE's nuclear business has formed a joint venture with Hitachi's nuclear business, becoming GE Hitachi Nuclear Energy in the U.S. In Japan, it is Hitachi GE Nuclear Energy.

Westinghouse developed the pressurized water light water reactor and conducted business worldwide, but sold its nuclear division to the British Nuclear Fuels Public Corporation (BNFL) in 1998 and closed its business in 1999. The nuclear division acquired by BNFL was bought by Toshiba in 2006, but it went bankrupt due to delays in the construction of nuclear power plants in the U.S., and Toshiba sold it to a Canadian investment company in 2018. It is now operating as Westinghouse Electric Company (Westinghouse Electric Corporation, WE Corporation).

In addition to what has been mentioned, in the United States, companies are active in specific areas such as decommissioning measures and radioactive waste. For example, nuclear power companies can sell decommissioned nuclear power plants to decommissioning specialist companies for decommissioning measures. There are also companies that take over tasks such as licensing from power companies and dealing with the Nuclear Regulatory Commission. Such things are possible under U.S. laws and regulations. Furthermore, many consulting firms are active mainly in Washington DC. The United States has high labor mobility, and mid-career hiring is commonplace. Under this U.S. employment system, there are people who change jobs between companies, consulting firms, and government agencies. The U.S. government and its laws and regulations operate under a strict audit and evaluation system, incorporating various opinions and perspectives, with the view that "the people to be explained to are the citizens, and the policies are for the citizens".

A concrete example, not limited to the nuclear field, is that U.S. agencies explain policies briefly and concisely on their websites in the names of bureau chiefs and section chiefs. This greatly enhances the transparency of policy and administration and serves as the foundation for democracy. The abilities of bureau chiefs and section chiefs also become clear to the public through this. In the United States, the top echelons of administrative agencies are appointed by the administration, so it is also referred to as the basis for evaluation when appointing bureau chiefs and section chiefs. In the United States, checks on policy and administration are carried out not only by members of Congress but also as activities of congressional hearings and the Government Accountability Office (GAO). Reports created by various organizations and experts are made and published, and are referred to as the basis. Although the United States also has a system for administrative agencies to listen to the opinions of experts, unlike Japan, meetings effectively controlled by administrative agencies do not aggregate opinions on policies and execute the results as policies. Also, on the websites of Japanese government agencies, there is no concise and accurate explanation of policies like in the United States.

U.S. Government Accountability Office: GAO

The U.S. Government Accountability Office (GAO) serves as the Board of Audit or the Accounting Audit Office. "Accountability" is translated as "explanation responsibility" in Japan, but it is not accurate because it can be misunderstood as something that can be settled by explanation, so it is better to use "responsibility", "responsibility for results", or "accountability". However, responsibility here means accountability for the results of past decisions and actions, not responsibility for future decisions. The use of accountability in the name is in the United States, while in the UK, for example, the National Audit Office (NAO) uses audit. Both GAO and NAO do not state good or bad, but their role is clarifying things using the power of investigation.

The GAO is an agency affiliated with the federal Congress. Since the 1960s, it has shifted its focus from traditional accounting audits, such as "compliance" audits that inspect whether government spending violates laws and regulations, and "economy and efficiency" audits that investigate whether waste and inefficiency are occurring, to "program evaluation" that conducts policy analysis and project evaluation. In Japan, there is often a discussion about how to objectively create achievement indicators, but the GAO hardly uses such indicators, aiming to "report conclusions and analyses objectively and logically derived from the survey results and provide useful information to policy makers". Furthermore, the GAO focuses on "identifying the causes of problems, not on the performance of the project or the accountability of the person in charge". Some of the GAO's survey items are set by the GAO itself, but many are made at the request of the chairman of the congressional committee. The congressional committees conduct hearings in the committee, referring to the GAO's reports, and decide on policies and budgets.

Attention should be paid to the fact that the accountability of the U.S. government is ensured by meeting the following three conditions.

1. There is information about the decisions and actions taken by the responsible person or organization (the role of the agency),
2. There are people or organizations who receive or discover this information and review, investigate, and report on it (the role of the GAO),

3. There are actions to correct defects based on this information (the role of the federal Congress and the budget committee).

The GAO also produces various reports on nuclear and radiation matters, which are contributing to the improvement of the robustness of nuclear energy policy. In relation to the Department of Energy (DOE), for example, it has issued several reports in the 1970s in response to the ballooning construction budget and delays in the plan for the Clinch River Fast Breeder Reactor (a prototype reactor for fast breeder reactors like Japan's "Monju", a prototype reactor is a reactor for demonstrating new nuclear reactor technology). The GAO's 1979 report was not necessarily negative about the continuation of the construction plan, but the federal Congress, referring to this report and the hearings in Congress, asked the industry for additional support for the construction budget, and as a result, it was not obtained and the construction plan was canceled in 1983. This is not mentioned in the GAO's report, but is mentioned in the following report collected by the U.S. Nuclear Regulatory Commission (NRC) (Breeder Reactor Cooperation 1985).

The Department of Energy, reactor manufacturers, national research institutes, and many nuclear power stakeholders who have been promoting the plan hoped for the continuation of the construction plan. However, since it is a prototype reactor for commercialization, the industry, including power companies that will eventually use fast breeder reactors, was asked for additional construction costs for the prototype reactor. This decision was made because they could not obtain it. Another point to note is that the NRC has been given the role of ensuring transparency of nuclear-related information, and it is publishing the above-mentioned report on its website. Not limited to this report, the NRC plays a role in collecting and publishing documents, so you can read important historical materials related to the use of nuclear power in the United States.

Understanding past experiences is essential for a robust nuclear policy. It is also noteworthy that the company (Breeder Reactor Corporation) established by 753 power companies in the United States, including public power companies that invested in the development of fast breeder reactors, compiled this report. According to this report, the total cost used until the cancelation of the Clinch River Fast Reactor Project was $1.7 billion (including the cost of manufactured equipment of $790 million), of which the burden on power companies was $135 million. The $135 million spent by US power companies is two orders of magnitude less than the total cost (costs from 1974 to 1997, mainly borne by Electricité de France) of €12 billion (2010 euros, 2012 French Cour des Comptes report "Cost of Nuclear Power Generation" page 46) required for the construction and operation of France's commercial fast reactor Superphenix. It should be noted that the burden on US power companies, and hence US consumers, has been reduced due to the robustness of the GAO report and US policies. It is not accurate to hear opinions that President Carter and European opposition movements stopped these fast breeder reactors.

Furthermore, the first to point out the possibility of a power reactor (breeder reactor: also called a dream reactor) that can produce more nuclear fuel material than it consumes while generating power was the Italian genius physicist Enrico

Fermi. Fermi is a genius among geniuses who not only realized the principle of nuclear power generation, the nuclear fission chain reaction, for the first time in the world, but also predicted the theory of beta decay and the existence of neutrinos, and won the Nobel Prize in Physics. However, it is clear that there are challenges that cannot be overcome by just extending the scientific possibilities, in the process from scientific possibilities to commercial use. This is why a robust nuclear policy is required.

The GAO has been creating and publishing various reports related to the U.S. Department of Energy (DOE) itself and its areas of responsibility, such as nuclear and accelerator science, radioactive waste, and others. The DOE was established in 1977, and its budget priorities have shifted from research and development of nuclear energy, such as the Clinch River Breeder Reactor Project in the early days, to nuclear weapons development, and then to decommissioning and waste disposal of nuclear development facilities in the late 1980s. The DOE's jurisdiction is broad, including military and nuclear security, nuclear development, accelerator radiation science research, and basic scientific research. The DOE oversees many national laboratories that developed nuclear technology during World War II, and their operation is contracted out to private companies. Ninety-four percent of the DOE's budget is outsourced through contracts. For these reasons, in the 1990s, there were criticisms of overlapping jobs and roles between national laboratories and inefficiencies in operation, and the DOE's management capabilities were questioned. In response to these criticisms, the Secretary of the DOE initiated reforms in the 1990s, and many review reports were also created. However, safety issues and environmental pollution problems arose at the national laboratories, and the DOE's own reform and governance capabilities were deemed insufficient.

The GAO, over a five-year period from 1995, cited about 100 reports it had created on the DOE and dozens of reports including the DOE's own review reports, and in December 2001, submitted a report titled "The Need for a Fundamental Reevaluation of the Role, Organization, and Accountability of the DOE" to the Federal Energy and Water Development Committee (GAO 2001). The fact that nearly 200 reports were created and referenced in a short period of time to create this report is noteworthy.

Based on the GAO's report and others, the roles and operations of the national laboratories under the DOE's jurisdiction were reviewed. For example, since then, research on nuclear energy has become the role of the Idaho National Laboratory. Meanwhile, unique safety research at other national laboratories continues with budgets such as the NRC's. Since this reform of the Department of Energy, the development of reactors aimed at commercialization is not being carried out primarily by the state, even at the Idaho National Laboratory. The model of commercial reactor development where the state develops and the private sector uses has ended in the United States with the Clinch River Breeder Reactor Project. The Idaho National Laboratory has a reactor for irradiation testing of nuclear reactor materials and fuels, called ATR, and research and development facilities for nuclear reactor fuels and materials, and development and cooperation for commercialization of individual technologies such as accident-resistant fuel irradiation testing are being carried out. In Japan, the Oarai Research Institute of the Japan Atomic Energy Agency had a materials testing reactor, which was used for private fuel material development, but due to aging, it has transitioned to decommissioning measures after the TEPCO accident, and there is no materials testing reactor in Japan.

Independence and Oversight of the GAO and the Nuclear Regulatory Commission

In the United States, as federal Supreme Court justices serve for life, provisions are established to ensure independence for organizations where independence is required. In the case of the GAO, the term of the Comptroller General is long, at 15 years. It is said to function effectively in protecting the GAO from external pressure. The term of a commissioner of the U.S. Nuclear Regulatory Commission (NRC) is five years, but before appointment, the five commissioners must disclose whether they support the Republican or Democratic party, and provisions are in place to ensure that the five do not lean toward supporters of either party.

In the United States, independent administrative agencies like the GAO and NRC are monitored and supervised by the federal Congress. The GAO and NRC are required to create and publish a Strategic Plan detailing their work plans and goals, based on which their work is carried out, and the federal Congress can evaluate their performance and reflect the results. In the United States, mechanisms are in place to prevent independent administrative agencies like the GAO and NRC from behaving arbitrarily.

4.3.2 Nuclear-Related Laws and Organizations in France, the European Union, and Euratom

The main laws related to nuclear safety regulations in France are the Nuclear Transparency Act (TSN law) enacted in 2006, the Radioactive Waste Management Plan law, and their related decrees. The Nuclear Transparency Act established an independent nuclear safety regulatory agency (ASN). Until then, nuclear regulation in France was carried out under the jurisdiction of the Minister of Industry. There had been movements toward enacting laws for transparency and independent regulation since 1998, but they had not been enacted.

The TSN law defines the principle of the operator's primary responsibility for nuclear safety, defines four types of nuclear and radiation-related facilities as Basic Nuclear Installations (BNI), namely, nuclear reactors, facilities related to nuclear fuel and radioactive waste, facilities containing radioactive or fissionable materials, and particle accelerators, and stipulates that permission is required for their establishment, modification, or abolition, or transition to the monitoring stage after operation ends. It also stipulates the establishment of a Local Information Committee (CLI) in the area where the basic nuclear facility is located, and the creation of the "High Council for Transparency and Information on Nuclear Safety (HCTISN)" as an organization to supervise the transparency of nuclear safety. The responsibilities of the state and operators regarding information provision to the public are determined,

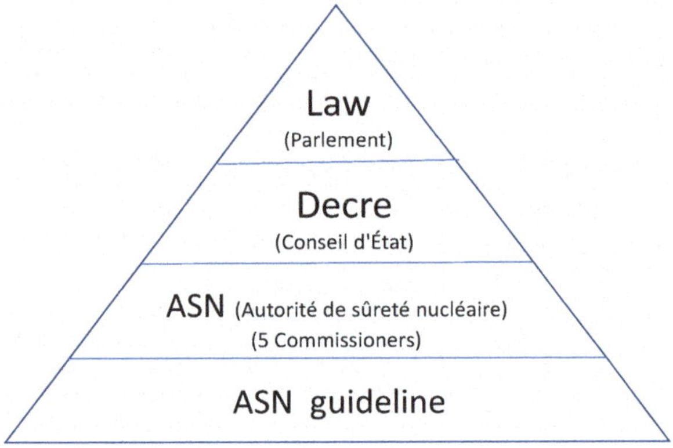

Fig. 4.4 French nuclear safety legal system

and BNI installers are obliged to create and publish reports on nuclear safety. The Radioactive Waste Management Plan law obliges the government to formulate and revise a national plan for the management of radioactive materials and radioactive waste every three years. In addition to these, the Public Health Law is deeply related to nuclear safety regulations.

The legal system related to nuclear safety regulations in France is shown in Fig. 4.4. As shown, laws are under the jurisdiction of the National Assembly, and decrees are administrative acts by the President or Prime Minister based on the French Constitution, issued through cabinet meetings by the State Council, the President, or the Prime Minister. The State Council (Conseil d'Etat) is the highest administrative court and advisory body in France, responsible for drafting laws and administrative regulations, and resolving administrative disputes (administrative litigation). The ASN establishes rules related to nuclear regulation. These have binding force under the law. ASN guidelines are documents that explain how to comply with regulatory rules, but they do not have legal binding force.

In addition to French domestic laws, there are international treaties and the European Atomic Energy Community (Euratom). In international treaties, France has ratified the Nuclear Safety Convention and the Convention on the Management of Spent Fuel and Radioactive Waste. In terms of nuclear damage compensation, it has ratified the Paris Convention and the Brussels Convention and their amendments, and the corresponding domestic laws have been established for each treaty.

Euratom was established in 1957 by the Treaty of Rome, along with the European Economic Community (EEC). In the 1965 Merger Treaty, the operating bodies of the European Coal and Steel Community, which was established in 1951, and the European Economic Community, which was established in 1957, were merged, and it was one of the three pillars of the European Communities. In 1992, the European Union (EU) was established by the Maastricht Treaty, and not only economic and energy cooperation, but also judicial and internal affairs cooperation, and cooperation in common foreign and security policy were carried out. In the 2007 Treaty of Lisbon, the legal personality of the European Community (EC) was inherited by the European Union (EU), but Euratom has not been fully integrated into the

European Union and continues to exist as an independent entity with a separate legal personality within the European Union. Euratom supports nuclear research and development jointly conducted by European countries. Europe participates in the International Thermonuclear Experimental Reactor (ITER) project as the EU. The use and research and development of nuclear power in Europe handled by Euratom is greatly influenced by France.

In addition, the European Union (EU) is a political and economic integration body that carries out economic and monetary union, common foreign and security policy, and police and criminal justice cooperation. For the economic and monetary union, part of national sovereignty is delegated. It forms the world's largest single market implementing a unified trade policy toward non-member countries. In other areas as well, while assuming the authority of member states, it takes a common position as the EU to the maximum extent, and politically speaks with "one voice".

The European Union is neither a federal government nor an intergovernmental organization. The European Union is a union that constitutes a new legal order in international law. EU law is a unique legal system that is implemented in parallel with the laws within the member states of the European Union, directly affecting the legal systems of the member states, and particularly in economic and social policies, it takes precedence over domestic law. Since each European country has its own domestic law and administrative system, the European Union addresses common issues that cannot be addressed by each member state alone, and cross-border issues, in addition to economics, diplomacy and security, and police and criminal fields. Environmental issues are an example of this.

The main institutions of the EU are the "European Council", composed of the EU and the leaders of the member states, the "European Commission", which is responsible for drafting EU law, policy implementation and budget execution, the "EU Court of Justice", which judges the compliance and application of EU law, the "Council of the EU", the main decision-making body composed of ministers of the member state governments, and the "European Parliament", composed of 751 members. When the European Council nominates a candidate for the President of the European Commission, it must take into account the results of the European Parliament elections. The European Parliament can approve or reject the candidate for President. Members of the European Parliament are divided not by their country of origin, but by their party group. In the 2019 election results, the European People's Party (Christian Democratic Party) and the Progressive Alliance of Social Democrats are the two major forces. The former is a center-right, and the latter is a left-wing that wants to advance socialism in accordance with parliamentary democracy. The Green Party, which advocates against nuclear power, accounts for about 10% of the total number of members. The number of members from each country is roughly proportional to the population, but there are slightly more in countries with small populations. The electoral system is determined by each country, and there is no unified one. Also, the EU does not have a constitution. The accountability of the European Union (EU) is mentioned in the European Civil Protection and Humanitarian Aid Operations, but its main job is to streamline the execution of the EU budget, and it does not seem to be playing a role in checking EU policies like the United States. The EU takes up policies that are easy to gain sympathy from the citizens of the member states, such as environmental issues, and has a strong influence on the world, but the European Parliament cannot directly elect the President of the European Commission, so the function of checking policies is weak.

The institutional framework for the use of nuclear energy in France is described below (NEA 2011).

The Nuclear Safety Authority (ASN) is the core organization responsible for nuclear and radiation safety regulations, and is composed of five commissioners. The ASN Secretariat is overseen by the Secretary General and has 11 regional offices.

The President of France fulfills the role of preserving the country's independence, territorial integrity, and treaty compliance in accordance with the constitution. Under the President, there is a Nuclear Policy Council that deals with overall nuclear policy. The President also chairs the National Defense and Security Council. In France, the Prime Minister is in charge of the government's operations. The Prime Minister also chairs cross-ministerial committees on nuclear safety and nuclear emergency response committees, among other nuclear energy-related committees. The Euratom Technical Committee is also under the Prime Minister's jurisdiction. The French Atomic Energy Commission was established by law of the French Atomic Energy Authority (CEA), but it deals with matters beyond those related to the CEA.

Under the French Constitution, the organization of the executive branch, such as ministries, falls within the government's own authority and can theoretically be changed freely. Ministries are established by decrees of the Council of State. In addition to the Ministry of the Interior, Ministry of Justice, Ministry of Defense, and Ministry of Foreign Affairs, there are others such as the Ministry of Finance and Economy, Ministry of Environment and Energy, and Ministry of Higher Education and Research, although their names vary slightly.

Public and quasi-public institutions include the Atomic Energy and Alternative Energies Commission (CEA), Electricité de France (EDF), the National Agency for Radioactive Waste Management (ANDRA, sometimes written as Andra), the Institute for Radiological Protection and Nuclear Safety (IRSN), and the National Institute for Nuclear and Particle Physics (IN2P3).

The CEA is a public institution created in 1945 by General de Gaulle's interim government, and is a government agency responsible for the development and application of civilian and military nuclear power. It was legally established by a decree in 1970. In 2010, it changed its name from the Atomic Energy Commission to the Atomic Energy and Alternative Energies Commission. It has about 20,000 employees. Its civilian research institutes are located near Paris in Saclay and Fontenay-aux-Roses, Grenoble, Cadarache in Provence, and Marcoule and Pierrelatte in the southeast and south of France, respectively, where the Valrho Institute is located. Saclay and Fontenay-aux-Roses conduct basic nuclear research, Grenoble conducts renewable energy research, Cadarache conducts nuclear safety and reactor research, and Valrho conducts research and development on the nuclear fuel cycle. Since its establishment in 1970, the CEA has aimed to industrialize the technologies it has developed, creating entities such as TECHNICATOME (for submarine reactors) and COGEMA (for the nuclear fuel cycle), in which EDF also invested. The names of French nuclear companies have changed over time. TECHNICATOME became AREVA TA and then TECHNICATOME again, while COGEMA became ORANO. The French Atomic Energy Commission, which was established by the 1970 decree that established the CEA, has a close relationship with the CEA. The chair of the Atomic Energy Commission is either the Director of the CEA or someone nominated by the Prime Minister. Furthermore, France has a High Commissioner for Atomic Energy with a three-year term, who plays a role in expressing opinions on nuclear policy to the Director of the CEA and the government.

Electricité de France (EDF) was established in 1946 by the nationalization law of gas and electricity in France, and it monopolized power generation, transmission, distribution, and sales. In 2004, in response to the deregulation of the EU electricity market, privatization was attempted, but even now the French government owns about 84% of the shares. EDF generates almost all of the electricity used in France. EDF's transmission division was also split off, but it is a member of the EDF Group. In the UK, EDF ENERGY operates nuclear power plants and supplies electricity and gas. In the United States, EDF owns just under 50% of the shares in three nuclear power plants operated by Exelon. AREVA, a French reactor manufacturer, fell into a management crisis due to the delay in the construction of a nuclear power plant in Finland in the 2010s, and the French government reorganized the French nuclear industry by strengthening EDF's involvement. The reactor and nuclear fuel manufacturing business became FRAMATOME, a subsidiary of EDF. The nuclear fuel cycle business became ORANO.

The National Radioactive Waste Management Agency (ANDRA) was originally an organization within the CEA, but was established by law in 1991. It is responsible for the long-term management of radioactive waste. Its role has been modified and added by the Radioactive Waste Management Law of 2006. It is supervised by the Ministry of Energy and Environment. ANDRA is responsible for the creation of reports on the inventory status of radioactive substances and radioactive waste every three years, the implementation of geological disposal, the establishment and operation of radioactive waste storage and disposal sites, the collection and transportation of radioactive waste, the creation and disclosure of information, and overseas deployment.

The Institute for Radiological Protection and Nuclear Safety (IRSN) was established in 2001. Its predecessor was a research institute established within the CEA in 1976. Through its research on radiological protection and nuclear safety, IRSN serves as a technical support organization for the ASN, the nuclear safety regulatory authority.

The Institute of Nuclear and Particle Physics (IN2P3) was established in 1971 as a research institute for nuclear and particle physics, and is coordinating research in this field at universities in France and research institutions abroad. In Europe, the European Organization for Nuclear Research (CERN) is located in a region straddling the border between France and Switzerland to the west of Geneva, where research in particle physics is conducted using a circular accelerator (Large Proton Accelerator) and a Large Hadron Accelerator with a total length of 27 km. CERN is operated and used by member countries (22 European countries and Israel), but also accepts use from non-member countries. It was established in 1954.

The above are the laws and related organizations related to nuclear power in France.

In France, an example of the parliament breaking through the deadlock of nuclear administration is the issue of candidate sites for high-level radioactive waste disposal facilities discussed in Chap. 3. In the 1980s, the responsible agency, Andra, proposed candidate sites for underground research facilities and tried to select them, but due to the top-down explanation from a scientific and technological perspective, it faced strong opposition and reached a deadlock. The Prime Minister at the time instructed a reevaluation of the policy, and based on the report created by Representative Bataille, the Radioactive Waste Management Act (commonly known as the Bataille Law) was established in 1991, allowing for the study of three disposal options for 15 years, and the selection of candidate sites for underground research facilities was advanced through consultations between the government and local governments. The Regional Information Committee (CLI) was established, local understanding progressed, and the underground research facility was built. In 2005, the results of 15 years of research were presented at a public hearing, and the geological disposal method was adopted. As a result of these, the Radioactive Waste Management Plan Act was established in 2006.

Based on this law, not only high-level radioactive waste but also various types of radioactive waste management and disposal are being promoted. For example, Andra is required by this law to regularly estimate and disclose the amount of radioactive waste generated. Since most radioactive waste is generated by the decommissioning of nuclear facilities, it is necessary to accurately evaluate the amount of radioactive waste generated by decommissioning. The author have heard that the point of France's decommissioning measures is "calculation", but the law requiring this has been created and implemented in this way. The amount of radioactive waste that France reports to the IAEA under international treaties on safety and radioactive waste is determined in this way.

France's Cour des comptes (Audit Court, National Administrative Court, or Accounting Inspection Institute) released a report titled "Cost of the Nuclear Power Sector" in 2012. The costs spent on the first generation of power reactors, 58 light water reactors, and the nuclear fuel cycle were investigated based on the investigative authority of the Accounting Inspection Institute, and the facts were reported plainly. The results and the pros and cons of the policy are not mentioned. Responses from administrative agencies and related organizations to the report are introduced as appendices. For example, even if the current policy of reprocessing spent fuel is changed in future and direct geological disposal of reprocessed spent fuel is decided, the French law (Spent Fuel and Radioactive Waste Management Plan Act) requires Andra, the agency responsible for radioactive waste, to be able to handle it, as stated in a joint response from French ministries. France's nuclear policy is robust and transparent. The author would recommend that those involved in nuclear power in Japan read this report.

4.3.3 Japan's Nuclear-Related Laws and Organizations

The Japanese legal system is shown in Fig. 4.5 The Constitution of Japan was established in 1946, declaring that sovereignty resides with the people (popular sovereignty), and discusses the Emperor, renunciation of war, rights and obligations of the people, the Diet, the Cabinet, judiciary, finance, and local autonomy. To amend the Constitution, a proposal must be initiated by the Diet with the approval of two-thirds of the members of each House, and a majority of the votes in a national referendum is required. The Constitution is established by the people, and *the Constitution is the supreme law that the people have the state protect. Laws are rules set by the state for the people, established by the Diet*. Cabinet orders, also known as enforcement orders, are orders issued by the Cabinet to implement the Constitution and laws. Unless delegated by law, cabinet orders cannot establish rules that impose penalties, limit the rights of the people, or impose obligations on the people. Prime Minister's Office orders, ministerial orders, and regulations are rules established to implement laws and orders for administrative affairs handled by each minister, and cannot be established unless delegated by laws and other regulations. Notices are public announcements made by each minister about designations, decisions, dispositions, etc., while directives are issued by ministers or directors to their respective agencies and staff, and notifications are used when bureau directors and others make public announcements. Notices, directives, and notifications do not legally bind the people. In addition, there are ordinances established by local governments.

The Japanese nuclear law system is shown in Fig. 4.6. The Atomic Energy Basic Act is a basic law on the use of nuclear power in Japan, enacted in 1955, stating that "research, development, and use of atomic energy shall be limited to peaceful purposes, aimed at ensuring safety, conducted autonomously under democratic management, and the results shall be made public". "Democracy, autonomy, and disclosure" are referred to as the three principles of Japan's use of nuclear power.

The Atomic Energy Basic Act establishes the Nuclear Regulation Authority, the Nuclear Disaster Prevention Council, and the Atomic Energy Commission, and

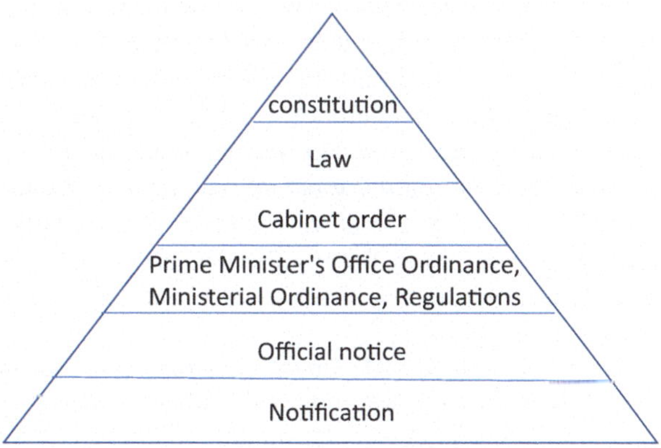

Fig. 4.5 Japanese legal system

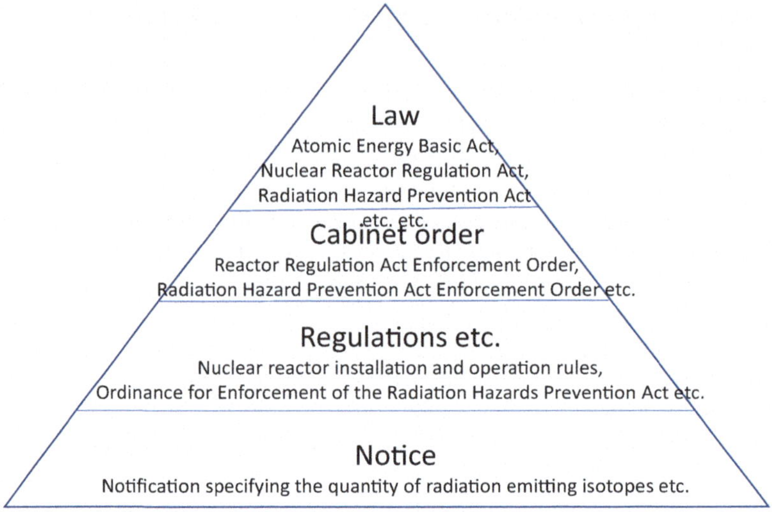

Fig. 4.6 Japanese atomic energy law

requires compliance with government regulations set by law on the management of nuclear fuel materials and reactors, and the prevention of radiation damage. In accordance with this, the Act on the Regulation of Nuclear Source Material, Nuclear Fuel Material and Reactors (Act on the Regulation of Nuclear Reactors, etc.) and the Act on the Regulation of Radioisotopes and Radiation Generating Devices (RI Regulation Act, formerly known as the Radiation Injury Prevention Act until its amendment in 2019) have been established and are in operation. Before the TEPCO Fukushima accident, the Nuclear and Industrial Safety Agency was established as an external bureau of the Ministry of Economy, Trade and Industry, but after the accident, the Atomic Energy Basic Act was amended from the perspective of regulatory independence, and the Nuclear Regulation Authority was established. Along with this, the Nuclear Safety Commission has been abolished.

The Reactor Regulation Law is a law for the safety regulation of nuclear energy use, stipulating the refinement, enrichment and processing of nuclear fuel materials, the installation and operation of power generation and experimental research reactors, the storage and reprocessing of spent fuel, the treatment and disposal of radioactive waste, security measures and nuclear material protection. Organizations (nuclear operators) conducting these operations need to obtain regulatory approval for their plans. For example, for power reactors, it is stipulated that they must apply for and obtain permission for installation, approval of construction plans, in-service inspections, regular inspections, etc. Internationally, Japan ratified the Nuclear Safety Convention in 1995.

The Radioisotope (RI) Regulation Law is a law aimed at preventing radiation damage and ensuring safety by regulating the use of radioactive isotopes and radiation generating devices, and things contaminated by them. Radioactive isotopes and radiation generating devices (accelerators) require application and approval for use. Dose limits for radiation workers and the public are also set. Both the Reactor Regulation Law and the RI Regulation Law are under the jurisdiction of the Nuclear Regulation Authority (the Secretariat is the Nuclear Regulation Agency).

In addition to these, nuclear-related laws and regulations include those related to the transportation of nuclear fuel, nuclear disaster prevention and emergency response, and the import and export of nuclear fuel materials and technology.

Based on the Reactor Regulation Law and the RI Law, their enforcement orders and enforcement regulations are stipulated by cabinet orders and ministerial ordinances.

The rules related to the installation and operation of reactors include those for commercial power reactors, those for power reactors at the research and development stage, and those for reactors for experimental research. There are also rules regarding the use of nuclear fuel materials, reprocessing rules, etc. There are various notifications, for example, the notification that defines the quantity of isotopes emitting radiation, i.e., the lower limit quantity (strength of radiation) and concentration that needs to be treated as a radioactive isotope, is stated in the enforcement order of the RI Law.

In the law concerning the use of nuclear power generation, there is the "Electricity Business and Security of Electrical Works Law", also known as the Electricity Business Law, which is closely related to the business of power companies. Power companies are required to conduct their business in accordance with this law in terms of power generation and transmission. The supervising ministry of this law is the Agency for Natural Resources and Energy of the Ministry of Economy, Trade and Industry. Technical standards for electrical products are established based on the Electricity Business Law, contributing to the security of electrical facilities and power generation facilities. The secretariat is the Japan Electric Association, and private standards such as the Electrical Technical Regulations (JEAC) and Electrical Technical Guidelines (JEAG) have been established. In addition to these, there are standards for nuclear power generation and related facilities, including those for machinery and civil engineering, so cooperation is being sought with the Japan Society of Mechanical Engineers, the Civil Engineering Society, the Japan Atomic Energy Society, and other academic association standards.

Regarding the use of nuclear power, if we talk about the institutional framework of the responsible ministries and agencies, the Cabinet Office has the Nuclear Disaster Prevention Council (the Nuclear Disaster Response Headquarters in emergencies), chaired by the Prime Minister. The Nuclear Regulation Authority is an external bureau of the Ministry of the Environment, responsible for safety regulations for nuclear reactors and the use of radiation, emergency response to nuclear power, and nuclear security. The Atomic Energy Commission also has its affairs handled by the Cabinet Office.

The Ministry of Economy, Trade and Industry (METI) is responsible for the stable and efficient supply of energy and the use and technological development of nuclear energy, promoting the use of nuclear energy such as nuclear power generation, processing and enrichment of nuclear fuel, reprocessing and storage of spent fuel, and disposal of radioactive waste. The Agency for Natural Resources and Energy (ANRE) is an external bureau (an independent administrative organization that performs specific duties) of the Ministry of Economy, Trade and Industry, responsible for stable energy supply policies and energy-saving/new energy policies for oil, electricity, gas, etc. Its role includes stable and efficient energy supply, promotion of energy use, and ensuring industrial safety. Within the Electricity and Gas Business Department, there are the Nuclear Energy Policy Division, the Nuclear Power Site and Nuclear Fuel Cycle Division, and the Radioactive Waste Countermeasures Division. The Nuclear Energy Policy Division is responsible for nuclear energy policy and nuclear energy technology development. The Agency for Natural Resources and Energy has the Comprehensive Resource Energy Research Council as a deliberative council, and there are several subcommittees such as the Electricity and Gas Business Subcommittee. Within the Electricity and Gas Business Subcommittee, there is a Nuclear Subcommittee, where members express their opinions.

The Basic Energy Plan indicates the basic direction for energy demand and use, and is determined by the government based on the Basic Energy Policy Law enacted in 2002. The responsible parties are the Ministry of Economy, Trade and Industry and the Agency for Natural Resources and Energy. It is revised at least every three years, and the 6th Basic Energy Plan was decided by the Cabinet in October 2021. The use of nuclear energy is also discussed in this plan. The Ministry of Land, Infrastructure, Transport and Tourism (MLIT) is responsible for the administration of radioactive material transportation. Land transportation is regulated by the Reactor Regulation Act and the RI Regulation Act, and is carried out in cooperation with the Nuclear Regulation Authority. Maritime transportation is governed by the Ship Safety Act, and air transportation is governed by the Civil Aviation Act.

The Ministry of Education, Culture, Sports, Science and Technology (MEXT) is responsible for basic nuclear research, development of fast breeder reactors and high-temperature gas-cooled reactors, and research and development of radioactive waste treatment and disposal and decommissioning of nuclear facilities.

The Disarmament, Non-Proliferation and Science Department of the Ministry of Foreign Affairs (MOFA) has a Non-Proliferation and Science Nuclear Division, which carries out tasks related to the non-proliferation of weapons of mass destruction and related materials, and diplomatic policies related to the peaceful use of nuclear power.

Advisory committees and councils include the Atomic Energy Commission of the Cabinet Office, the Reactor Safety Expert Review Committee of the Nuclear Regulation Commission, the Nuclear Fuel Safety Review Committee, and the Radiation Council. The Atomic Energy Commission is responsible for planning, deliberating, and deciding on important matters related to nuclear power utilization policy, coordination of inter-agency affairs related to nuclear power utilization, collection and investigation of materials related to nuclear power utilization. The Atomic Energy Commission has been given the authority to recommend to the heads of related organizations through the Prime Minister when it deems necessary for its jurisdiction.

Discussing the main organizations related to nuclear power utilization, first, the public research and development institutions include the Japan Atomic Energy

Agency (JAEA) and the Quantum Science and Technology Research Development Organization (QST), both of which are national research and development corporations. The Japan Atomic Energy Agency conducts basic and applied research on nuclear power, development of nuclear fuel cycle technology, treatment and disposal of radioactive waste, provision to users of the organization's facilities (joint use), training of researchers and technicians, and organization and provision of information on nuclear power. The supervising ministry is the Ministry of Education, Culture, Sports, Science and Technology, and the development of nuclear fuel cycle technology is jointly managed with the Ministry of Economy, Trade and Industry. In terms of nuclear safety, it serves as a technical support organization (TSO) for the Nuclear Regulatory Agency (NRA).

The Quantum Science and Technology Research Development Organization conducts basic research and foundational research and development on quantum science and technology, research and development on the human impact of radiation and its prevention, diagnosis and treatment, and medical use of radiation. A quantum is a very small unit of matter or energy that has both particle and wave properties, and a quantum beam is a broad term for radiation.

The main semi-public nuclear-related organizations include the Nuclear Damage Compensation and Decommissioning Facilitation Corporation (NDF), a licensed corporation based on special laws, and Nuclear Waste Management Organization of Japan (NUMO), a public interest corporation. The NDF provides financial support to Tokyo Electric Power Company for compensation due to the TEPCO Fukushima accident, and conducts research and development and provides advice for the decommissioning measures of the Fukushima Daiichi Nuclear Power Station. NUMO is a business entity that carries out geological disposal projects for high-level radioactive waste. The Ministry of Economy, Trade and Industry is involved as a government agency (NEA 2017).

Japan's power companies were reorganized into nine regional power companies, including Tokyo Electric Power and Kansai Electric Power, in 1951. All are private companies. The nine power companies and Japan Atomic Power Company (JAPC), which was established by the power companies, are engaged in nuclear power generation. The Electric Power Development Corporation (EPDC, now J-POWER), established in 1952 as a special company of the country, is a power generation and transmission operator that has been engaged in hydroelectric power, thermal power, or overseas business. It was privatized in 2003. J-POWER is currently constructing a nuclear power plant in Aomori Prefecture.

The liberalization of electricity in Japan began with the liberalization of wholesale electricity in 1995, and the liberalization of retail electricity was carried out for special high-voltage electricity for large factories and large consumer companies in 2000, for high-voltage electricity for small and medium-sized buildings and factories in 2004, and for low-voltage electricity for households and shops in 2016.

Among the large private research institutions related to nuclear power generation, there is the Central Research Institute of Electric Power Industry (CRIEPI), a general foundation, which conducts research and development on science and technology, economics, and policies related to the electric power business. CRIEPI was established in 1951 and is operated by continuous contributions from nine power companies.

Regarding the government's budget for nuclear power utilization and safety assurance, there are general accounts and special accounts for energy measures related to nuclear power energy utilization. It is desirable for the national accounting to manage as a whole with a single accounting (general accounting) so that the national policy can be comprehensively viewed. This is called the principle of budget unity (single accounting principle). However, in Japan, special accounts that manage separately from general accounting with specific income and specific expenditures have been established under the provisions of the 1947 Finance Law. In fiscal 2019, there were 13 special accounts, among which those closely related to nuclear power generation are the Special Account for Energy Measures and the Special Account for Reconstruction from the Great East Japan Earthquake. The former involves the Cabinet Office, the Ministry of Education, Culture, Sports, Science and Technology, the Ministry of Economy, Trade and Industry, and the Ministry of the Environment, and the latter involves many ministries and agencies.

The mechanism of the Special Account for Energy Measures is shown in Fig. 4.7. As shown, the Special Account for Energy Measures consists of the Energy Supply and Demand Account, which uses the petroleum and coal tax as a source of funds, the Power Source Development Promotion Account, which uses the power source development promotion tax as a source of funds, and the Nuclear Damage Compensation Support Account, which uses the general account as a source of funds. The petroleum and coal tax is levied on crude oil, petroleum products, gaseous hydrocarbons, and coal. The power source development promotion tax is levied on electricity sold by general power transmission and distribution businesses. Both taxes are entirely accounted for in the general account, and the necessary amount is transferred to the special account.

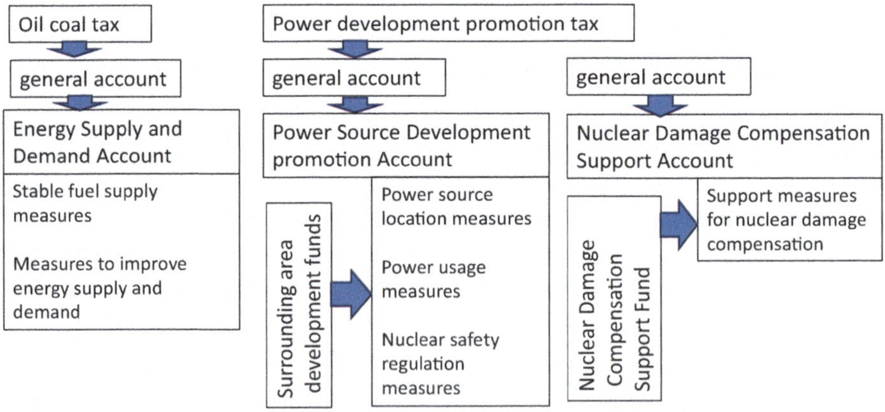

Fig. 4.7 Special accounts for energy measures in Japan. *Source* Special account guidebook, Ministry of Finance (in Japanese). https://www.mof.go.jp/policy/budget/topics/special_account/fy2019/2019-kakuron-6.pdf

The Energy Supply and Demand Account is used for oil stockpiling, promotion of the development and use of non-fossil energy, and operating expenses of the National Research and Development Corporation for New Energy and Industrial Technology Development Organization (NEDO). For example, it is used for energy conservation measures and promotion of the introduction of renewable energy facilities. The Power Source Development Promotion Account is granted to local governments as power source location measures expenses for the purpose of improving public facilities and the welfare and convenience of local residents in the prefectures and municipalities where power sources are located, in order to facilitate the installation and operation of power generation facilities. It is also used for the operating expenses of the Japan Atomic Energy Agency, efforts for safety regulations of nuclear power generation facilities, etc. It is also used to provide information on the effects of radiation and prevention of harmful rumors. The Nuclear Damage Compensation Support Account is allocated to the interest on borrowed money related to the repayment of government bonds granted to Tokyo Electric Power Company for accident compensation support. The Cabinet Office, Ministry of Economy, Trade and Industry (METI), Ministry of Education, Culture, Sports, Science and Technology (MEXT), Ministry of the Environment (MOE), and Nuclear Regulation Authority (NRA) are using the Special Account for Energy Measures. In addition, a subsidy system called the fixed price purchase system, in which electricity generated by renewable energy is purchased by power companies at a fixed price for a certain period of time to introduce renewable energy generation facilities, was established by a special measures law created for this purpose in 2011, and its cost is added to the electricity charges and collected from consumers (Ministry of Finance 2019).

Japan also has a Board of Audit, and audits are conducted. Audits are also conducted in the field of nuclear power, but no report worth mentioning in this book was found. Audits are often discussed in relation to fiscal reform. In Japan, improvements in audits have been made since the 1990s, referring to the transition of the U.S. GAO's audits since the 1960s. Although it is not directly related to the nuclear power sector, it is necessary to understand the overall picture of the issues, so I will discuss the improvement of audits and administrative reform.

Regarding improvements in inspections in Japan, the results of investigations and research by the U.S. GAO have been published in the "Accounting Inspection Research" issued by the Board of Audit of Japan since 1989. For example, in the second issue of 1990, in addition to traditional accounting inspections such as "compliance" and "economy and efficiency", it was stated that "program evaluation" is necessary, and when conducted in Japan, it is important that "each ministry does not actively carry out evaluation activities that may require them to admit their own faults", "information necessary for policy decisions is monopolized by the coalition of industry, government agencies, and politicians, and the general public does not have information to judge the validity of policies", and "in program evaluation, it is necessary to consider efficiency seen by the entire nation, rather than the efficiency in the narrow sense", (Kanemoto 1990). In the ninth issue of 1994, the deployment of GAO's audit standards was introduced, and it was stated that "in our country, there is often discussion about how to objectively create achievement indicators, but the GAO hardly uses such indicators, and reports conclusions and analyses objectively and logically derived from the investigation results, aiming to provide useful information to policy makers" (Kitani 1994).

Inspections by the Board of Audit have been conducted since 1997 from the perspectives of accuracy (whether the financial statements accurately reflect the status of budget execution), compliance (whether accounting processing and budgets are properly processed in accordance with laws and regulations), economy and efficiency (whether operations and projects are economical and efficient), and effectiveness (whether the entire project is achieving its intended purpose and is effective). Policy evaluation was introduced with the reform of central government agencies in 2001. Ministries have an obligation to evaluate their own policies, and the performance and achievement of the policy are regularly measured, evaluated, and published. In 2015, a report on "The Status of Policy Evaluation in Each Ministry" was reported to the Diet and the Cabinet by the Board of Audit, and the report was made public. Furthermore, the Administrative Evaluation Bureau of the Ministry of Internal Affairs and Communications evaluates and monitors the implementation status of each administrative agency's operations. However, it has been pointed out that "since 1990, administrative reforms have been implemented in Japan based on the concept of New Public Management (NPM), and it is now being questioned whether we can move toward post-NPM in response to the complex socio-economic situation caused by declining birthrate and aging population, but the mission of the Board of Audit in our public management is still stuck in Old Public Management" (Kobayashi 2021).

The history of administrative reform in Japan can be summarized as follows.

In the 1960s, while private companies led high economic growth, the number of national civil servants increased, and there was a need to overcome problems unique to the bureaucracy, such as bloating and rigidity. In 1968, a law was enacted to control the number of national civil servants. From 1981, the second Temporary Administrative Investigation Committee, chaired by Toshio Doko (Doko Temporary Investigation), issued five reports under the policy of "fiscal reconstruction without tax increase", promoting expenditure cuts through administrative organization review, privatization of the three public corporations of Japan National Railways, Postal Services, and Telegraph and Telephone, rationalization of licenses, subsidies, and special corporations, establishment of the Ministry of General Affairs, abolition of land transport-related local officials, and integration of regional block agencies such as the Maritime and Land Transport Bureaus. Subsequently, the Temporary Administrative Reform Promotion Council (Administrative Reform Council) was established three times, and in the first term from 1983, it followed up on the Doko Temporary Investigation, in the second term from 1987, it responded to the intensification of Japan-US economic friction and the demand for domestic demand expansion, the emergence of the bubble economy and soaring land prices, and made proposals for deregulation, division of roles between the national and local governments, and transition to a domestic demand-led economy. In the third term from 1990, issues of internationalization and emphasis on people's lives were raised, and a proposal was made for the establishment of a fair and transparent administrative procedure law, which was enacted in 1993. The Administrative Reform Committee, which was established in 1994, issued an opinion on the information disclosure system, which should clarify the status of government activities and fulfill its accountability to the people, and the Information Disclosure Law was enacted in 1999. In 1995, the Local Decentralization Promotion Committee was established, and the Local Decentralization Omnibus Law was enacted in 1999. After 50 years since the end of the war, with the accumulation of cross-ministerial issues such as globalization, declining birthrate and aging population, and advanced information technology, the vertical administration system (a system that does not allow other ministries to interfere in their own jurisdiction, resulting in a failure of overall coordination) was pointed out, and the Administrative Reform Council was established in 1997, and reforms of central government agencies were carried out. From 2001, the strengthening of cabinet functions, the reorganization of ministries from 22 to 12, the establishment of independent administrative corporations for reducing and streamlining administration, the corporatization of postal services, the incorporation of national hospitals, the incorporation of national universities, and the reform of special corporations were carried out, and the policy evaluation system was introduced to ensure the transparency of administration. The consumption tax was raised in 1998. In 2000, due to the large-scale economic measures, the fiscal situation rapidly deteriorated, and after the last decade, a sense of stagnation spread throughout society, leading to the establishment of the Administrative Reform Promotion Headquarters, and in 2006, the Law on the Promotion of Administrative Reform to Realize a Simple and Efficient Government (Administrative Reform Promotion Law) was enacted. The privatization of the Japan Highway Public Corporation, the privatization of postal services, the integration of government-affiliated financial institutions, the reform of special accounts, and the net reduction of the number of staff in national administrative agencies were implemented. The number of national civil servants, which was about 900,000 in 1967, decreased to 890,000 in 1984, 860,000 in 1997, 840,000 in 2000, 310,000 in 2009, and 300,000 in 2013. The number of national university corporations, which are no longer national civil servants, is 130,000, independent administrative corporations are 140,000, and special corporations are 330,000, but if these are added, the number in 2013 is about 900,000, the same as in 1967 (MIC 2012).

In 2007, the revised National Public Service Act was established, and in 2008, the Basic Act on Civil Service Reform was enacted. The civil service reform aimed at "eradicating amakudari (descent from heaven, parachute appointments from government agencies)", "thorough implementation of ability and performance-based principles", "elimination of the negative effects of vertical administration", and "strengthening political leadership" was promoted. The "amakudari problem" was completely banned by each ministry's re-employment assistance, and since 2010, re-employment support has been centralized at the Government and Private Sector Personnel Exchange Center. In the reform of special accounts, consolidation and abolition of special accounts, review of expenditures, and utilization of surplus funds were pursued. In the reduction of national assets, the sale of unused land such as office and dormitory sites, and the sale of government-owned shares of Japan Post and Japan

Policy Investment Bank were promoted. A total review of administrative expenditures aiming at zero waste, which was created by subsidies, was also conducted. In the review of the role distribution between the government and the people, and the nation and the regions, regulatory reforms such as the transfer of authority to municipalities, strengthening the autonomy and freedom of local governments, liberalization of entry into agricultural land, life science fields, private aircraft, and the portability system of mobile phone numbers were implemented. Public service reform through the introduction of competitive bidding was also carried out. In the reform of independent administrative corporations, the number of corporations was reduced, discretionary contracts were reduced, disposal of owned assets, centralization of evaluation agencies, principle of public recruitment of presidents, prohibition of brokerage acts, and introduction of re-employment rules were implemented. In the reform of policy finance, the integration of four financial public corporations such as the National Life Finance Corporation and the Japan International Cooperation Bank into the Japan Policy Finance Corporation was carried out. In the reform of the public interest corporation system, general foundations and general incorporated associations were able to easily establish corporations with only registration, and restrictions on business and supervision by administrative agencies were eliminated. For public interest incorporated foundations and public interest incorporated associations, public interest certification based on clear criteria and preferential tax measures are provided. In efforts to change the foundation and basic rules of public service, the realization of efficient and transparent small government was pursued through the promotion of e-government and e-local government. By effectively implementing policy evaluation, collaboration between the Ministry of Internal Affairs and Communications and the Council on Economic and Fiscal Policy, correspondence between policy evaluation and budget evaluation, and review of unstarted and unfinished public works were carried out. Administrative reform in local public bodies was carried out as reform of the staffing and salary structure of local public servants (Administrative Reform Promotion Office 2009). Currently, the Minister of State is in charge of administrative reform, and the Administrative Reform Promotion Council is active. Administrative project reviews are conducted by each ministry for about 5,000 national projects. Evidence-based policy making is being promoted.

The paper titled "The Limits of Administrative Reform in Our Country" reviews past administrative reforms with a huge fiscal deficit in mind and states the following points (Morita 2012).

"The privatization of the National Railways and the Nippon Telegraph and Telephone Public Corporation, implemented at the second temporary administrative survey meeting in 1983, resulted in improved service and quality by relieving businesses burdened with massive debts and union issues, with the state taking on the debts and facilitating privatization. However, subsequent administrative reforms in our country have aimed at slimming down and improving efficiency by consolidating administrative organizations and reducing the number of public servants, without reducing the quantity of administrative services. The reorganization of ministries and agencies in the 1990s merely consolidated existing ministries and agencies with strong vertical structures, and our ministries and agencies did not change their jurisdictions even after consolidation, maintaining their traditional vertical structures, resulting in the creation of large government agencies that are difficult to manage. It's as if we've simply removed the walls of a condominium to make the rooms larger, but the total area, or the total amount of work, has not necessarily decreased. As for the management of the number of public servants, we have tried to reduce the number of regular national public servants by transferring a certain number of them to organizations outside government agencies. The incorporation of independent administrative agencies was introduced in the UK and other countries as a mechanism to promote efficiency by giving a wide range of discretion to the head of the organization, but in Japan, it was more important to make the organization a separate legal entity. The incorporation of independent agencies in Japan can be said to have aimed at slimming down government organizations on the surface by moving part of the administrative activities outside the government organization. If we truly want to slim down, we need to reform to promote large-scale ITization of operations, as advanced countries are actively doing. Decentralization is slimming for the country as it transfers authority to the regions, but the amount of work in the public sector as a whole does not change much. Our administrative reform has not tackled the slimming of the administrative service itself. In other words, we have not reduced the quantity of administrative services, but have consolidated administrative organizations and reduced the number of public servants".

"The biggest factor in the current expansion of expenditures is the increase in social security costs. Due to aging, the number of recipients of pensions, medical care, and nursing care services is increasing, and the fiscal deficit is growing. It is cruel to pass the bill to future generations by issuing government bonds. Isn't it inevitable that the relatively affluent strata of the same generation will have to support the low-income elderly strata, which are expected to increase even more in future? Because the adjustment of the grant amount according to the degree of necessity is not accurately made, we have been providing benefits to people with a certain level of income or assets, even for things they can do without. The currently planned national ID Under the current system, it would be difficult to adequately capture assets. However, it is believed that developing such a system and realizing a wider redistribution of wealth among the citizens would be the first step toward reducing waste in a substantial sense and steering reform toward fiscal reconstruction. Not only capturing each citizen's income, but also understanding the macro effects, or in other words, the potential "waste", and proposing the content of policies to be implemented with evidence is important. The Board of Audit is expected to collect and disseminate such evidence".

In the United States, administrative accountability (responsibility for results) is ensured through.

1. transparency of information through the creation and publication of reports,
2. improvement activities of responsible ministries, surveillance by the GAO, and
3. review and corrective measures by the federal congress at the time of budget acquisition by administrative agencies, etc.

This ensures robust policies and administration for the citizens are planned, executed, and improved.

In Japan, each ministry conducts its own policy evaluation, and the Administrative Evaluation Bureau of the Ministry of Internal Affairs and Communications conducts policy evaluations as necessary from the perspective of ensuring the unity of the entire government (Law No. 86 of 2001).

The administrative evaluation in Japan is basically a self-evaluation by each ministry, which is different from the United States. Can self-evaluation point out and evaluate matters that essentially need improvement? There are still many companies in Japan that depend on the policies of ministries (the country). Isn't this one of the causes of Japan's long-term stagnation? *Isn't it necessary to consider and build a system that meets the above three requirements in Japan as well?*

In Japanese ministries, meetings and committees composed of external members such as university professors are held to discuss policy directions. The time available for speaking at meetings is extremely limited, and it is rare for reports to be prepared and published in advance on the topics to be discussed from various perspectives. The secretariat of the meetings and committees is served by administrative officials, and the general direction is often already considered. The deliberation results are published and decided as an "interim summary". In Japan, where lifetime employment is still common, there is little mobility of personnel between the government and the private sector, and both have limited experience. A vast amount of knowledge and experience in both fields is necessary to consider a system that meets the three requirements. In the United States, personnel who gain experience by moving between the government, the private sector, and consulting firms contribute to robust policies and their improvement.

As a prerequisite for the above item 1, it is necessary that knowledge and experience in the field are compiled and made public. Japan is weak in this regard. Comparing Japan and the United States in the field of nuclear power, important policies of the U.S. Nuclear Regulatory Commission are based on reports, explanations, and training materials that are created and made public with the cooperation of consultants and researchers at national research institutes. For example, NUREG/CR-6042, NUREG/BR-0308, and NUREG/CR-6349, etc. In Japan's nuclear power field, such reports and explanations are not found. In the field of nuclear power in Japan, the Japan Atomic Energy Agency plays a role equivalent to that of the U.S. national research institutes. However, even if you look at the Japan Atomic Energy Agency Act or its medium- to long-term goals, it only says "conduct research", and it is not explicitly written to compile and systematize the knowledge obtained from the research and make it usable. Improvement is necessary. The reports of the research and development results of the Japan Atomic Energy Agency are reports of their own research and annual reports. Research papers are mostly the results of research in specialized fields. It is not good to conduct research vertically (within the confines of a narrow field), but it is better to do it horizontally (across fields). There is not necessarily a need to redo the same research as in Europe and the United States. However, there is no introduction of reports and explanations that constitute the knowledge base for the use of nuclear power in Europe and the United States.

There is a lack of attitude and results to make them usable as the knowledge base for the use of nuclear power in Japan. During my tenure on Japan Atomic Energy Commission (JAEC), I pointed out the need to build a knowledge base for the use of nuclear power in Japan in the Basic Thoughts on the Use of Nuclear Power (JAEC 2017), and in the response to the medium- to long-term goals of the Japan Atomic Energy Agency (JAEC 2019).

The items 2 and 3 are matters related to accountability, as I have already mentioned. Having a mechanism for improvement is essential for robust policy and administration, regardless of the use of nuclear power. Closely related to this, Japan's fiscal reform is significantly delayed compared to other countries, despite years of effort and slogans. By the way, South Korea carried out administrative reform during the economic crisis of the 1990s and has transformed into a slim competitive society. I hear that Korean companies make decisions quickly. In Japanese companies, universities, and corporations, aren't they wasting a lot of time in meetings and making responsibilities vague? These points should also be improved as the country's administrative reform progresses. The Japan-centric perspective should also be corrected.

The Structure of "Captivity"

In the field of nuclear power in Japan, the committee investigating the accident at the Tokyo Electric Power Company's Fukushima Daiichi Nuclear Power Plant, which occurred in March 2011, has been significantly questioned for its responsibility, and a report has been compiled following an investigation. In addition to the parliamentary and government accident investigation committees, several other accident investigation committees have been organized and have compiled reports. The contents of these will not be reintroduced in this book. The chairman of the parliamentary accident investigation committee, professor Kiyoshi Kurokawa, has stated in his book that the nuclear power operators had made the regulatory administration their "captive". Reflecting on the accident, an independent Nuclear Regulation Authority was established as an external bureau of the Ministry of the Environment. Chairman Kurokawa, in his message from the chairman in the English version of the parliamentary accident investigation report, states, "What must be acknowledged is that this was a 'Made in Japan' disaster. The fundamental cause lies in customs rooted in Japanese culture. Our reflexive obedience, our reluctance to question authority, our 'stick-to-the-plan' attitude, our groupism, and our insularity. Even if other Japanese people were in a position to take responsibility for this accident, the outcome might have been the same" (Kurokawa 2012, p. 9). This is an extremely serious point.

Furthermore, when searching for the English version of the parliamentary accident investigation report, there are instances where this message is not included. If the administrative officer in charge of compiling the report omitted this part, considering it as the personal view of the chairman, would the reader realize that this in itself is a serious problem unique to Japan? Furthermore, even if this issue is investigated, public servants have a duty of confidentiality, so it is unlikely to be made clear why it was omitted. From this, readers should realize that there is a problem in Japan where there is no institution legally granted the power to investigate administrative outcomes. While the above is not a crime, could the police investigate a crime without investigative authority? Furthermore, if there were an institution with the authority to investigate administrative outcomes, this problem should be preventable even without an investigation.

In addition, the Fukushima Daiichi accident has many lessons to be shared with the world, such as the increase in disaster-related deaths due to evacuation, so it would be a mistake to dismiss the Fukushima Daiichi accident as unique to Japan. However, it is expected that those involved in nuclear power will not misunderstand Chairman Kurokawa's message, but will think about it and take it to heart.

There are other examples of being "captive". Executives of Kansai Electric Power were given expensive gifts by a powerful local figure (now deceased) related to their nuclear power plants, in order to secure construction orders for this influential person's related companies. This is an example of a local power figure "capturing" a power company. A report by a third-party committee was published in March 2020. In another field, the former chairman of a large private university in Japan was accused by the Tokyo Tax Bureau in December 2021 of tax evasion for not declaring a large amount of money received from contractors who frequented the university. This is an example of a contractor "capturing" the former chairman. The president of the Association of Private Universities in Japan expressed concern about a proposal from the supervisory agency to transfer the authority of the university's board of directors to the board of councilors to prevent such incidents, stating that even if the authority is transferred, there will always be people who will infiltrate the board of councilors. In Japan, it is socially acceptable to give gifts as a token of gratitude during mid-year and year-end gift-giving seasons, or when someone has taken care of you. However, isn't the problem of Kansai Electric and the former chairman of the large private university connected to this extension? Of course, tax evasion is a crime, but this cultural characteristic of Japan is a weakness. Influenced by Confucianism, Japan has a strong sense of dependence on the country among its citizens, as does South Korea. In order to leave a better Japan for future generations, isn't it necessary to recognize the cultural characteristics and weaknesses of the Japanese, carry out administrative reforms and audits, improve transparency, and realize the three requirements for robust policies?

4.3.4 Structure and Concept of Ensuring Nuclear Safety

There are two main types of legal systems: "continental law" and "Anglo-American law". "Continental law" is adopted in countries like Germany and France, with ancient Roman law as a common origin, and the French Napoleonic Code and the German Civil Code as its foundation. Continental law is also called "Civil Law" because Roman law was enacted for citizens. After the Meiji Restoration, Japan referred to continental law, and the Constitution of the Empire of Japan was heavily influenced by the Prussian Constitution of Germany. The Civil Code was influenced by French law. "Anglo-American law" is a legal system that developed as case law in the king's courts since the establishment of the Norman dynasty in Britain around the eleventh century, and was subsequently inherited by countries like the Unites States. Countries that were once British colonies have adopted this system. Anglo-American law developed based on the "Common Law", which was established as a medieval customary law, hence it is also referred to as "Common Law". It is also referred to as "statutory law". In civil law, statutory laws made by parliaments or governments are considered the primary source of law, and the principle of "positivism", which respects all legal domains, is adopted. Positivism is also referred to as "statutory law". In common law, case law is considered the primary source of law, and precedential binding force is recognized in case law created by courts. This is called "case law principle". Civil law adopts a system centered on public law, while common law adopts a system centered on private law. Civil law adopts a theoretical system centered on substantive law that starts from individual will, while common law adopts a litigation-centered system. Civil law is said to be passive in constitutional review, while common law is said to be active. However, "civil law" and "common law" are merging over time. Japan is said to have a hybrid legal system of civil law and common law because the constitution created after World War II was influenced by the United States.

The structure of nuclear safety regulation is shown in Fig. 4.8. The constitution is at the top. The constitution is a legal norm that establishes basic principles related to the establishment of a nation, and also describes the structure and organization of the political unity of the nation. The constitution discusses the rights of citizens, such as basic human rights, and the constitution can be considered a "law that citizens have the state protect", and the law can be considered a "law that the state has citizens protect". In other words, it can be considered that citizens are limiting state power by the constitution. A majority agreement of the people is required to amend the constitution. Laws are created by the decision of the Diet. Based on the law, regulatory rules are established by ministries and agencies. Further down, there are guidelines that explain regulatory rules and technical standards for the use of nuclear power, which do not have legal binding force.

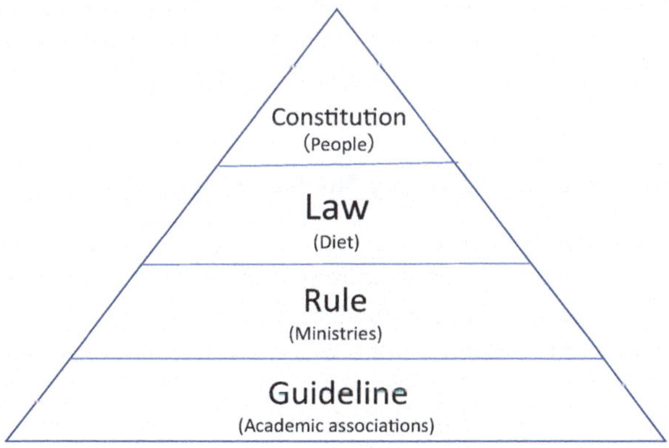

Fig. 4.8 Structure of law

Regardless of "civil law" and "common law", there are several common ideas and principles in the laws and regulations of nuclear regulation. First, *there is the concept of risk and benefit.*

The Concept of Risk (Cost) and Benefit

Nuclear power generation can provide cheap and stable electricity, and the use of radiation and accelerators is carried out because there are various benefits such as diagnosis, treatment, and analysis. Not limited to the use of nuclear power, *technology is used because it has benefits. On the* other hand, the use of technology involves risks to individuals and the environment. When considering laws, it is necessary to consider both risks and benefits. If it brings no benefit, it would be sufficient to prohibit its use, so regulation would become unnecessary. The perspective of considering both risk and benefit is fundamentally important in nuclear legislation and its operation.

A specific example of how risk and benefit analysis is used in nuclear regulation is in the United States, where the cost of safety improvement measures is quantitatively compared with the risk reduced by them to decide whether to adopt those safety measures. Specifically, the cost required to reduce exposure by 10 millisieverts per person is set at 5100 US dollars (in 2014 dollars), which is used as a criterion for judgment. This figure is revised every few years. Furthermore, in the Protective Action Guidelines (PAG Manual) for evacuation during nuclear and radiation accidents in the United States, it is a basic principle that the risk associated with evacuation must be smaller than the risk avoided by evacuation.

The concept of balancing risk and benefit is stated at the beginning of the International Atomic Energy Agency's handbook on nuclear law (IAEA 2003, p. 3). However, there is no concept of considering risk and benefit in Japan's safety, regulation, and disaster prevention. Not limited to nuclear power, when regulating a technology by law, it is required as a basic principle to consider the balance between risk and the benefit of use. *It is a major flaw that consideration of risk and benefit is not required in Japan's safety, regulation, and disaster prevention for nuclear power use.* Ambiguity creates waste and increases the burden on the public through electricity charges, and also causes harm to the public, such as an increase in disaster-related deaths, in evacuation from nuclear disasters. A response is required.

The consideration of risk and benefit is a basic concept in the use of nuclear power and its laws, but there are also several principles in the nuclear law as mentioned next.

Principles for Ensuring Nuclear Safety

Principle of Safety

Laws, rules, and explanations related to nuclear power state that ensuring safety is an essential requirement in the use of nuclear power. If we further classify the principle of safety, there are "principles of prevention", "principles of protection", and "principles of precaution". The principle of prevention is stated as a basic goal of nuclear law to prevent damage associated with use. The principle of protection is the idea of protecting public health, security, and the environment when the risk associated with use is considered to be much greater. The principle of precaution is used when it is difficult to balance risk and benefit, for example, when it is difficult to foresee risk. Strict and detailed rules and standards are applied to activities that may have significant radiation effects, while only limited restrictions are imposed on activities that are only expected to have minor effects.

Principle of Security

Technologies and nuclear materials that lead to the production of nuclear weapons are dangerous if they fall into the hands of terrorist groups or terrorist states. They can cause chaos by scattering radioactive materials. Whether intentional or not, it is necessary to prevent the proliferation of these technologies and materials. The Atomic Energy Act ensures security in terms of nuclear material protection, safeguard measures, import and export restrictions, transportation of radioactive materials, and emergency response.

Principle of Responsibility

Many organizations are involved in the use of nuclear power. It is necessary to clarify who is primarily responsible for ensuring safety. While all parties involved have some degree of responsibility depending on their involvement, the primary responsibility for ensuring safety lies with the reactor operators, nuclear business operators, and handlers of radioactive materials.

Principle of Permission

Not limited to nuclear regulations, things not regulated by laws can be used without obtaining government permission. This is called the principle of permission. The Atomic Energy Act requires prior permission for handling nuclear fuel materials and radioactive materials, but those items approved by the regulatory authority do not need permission. What is approved is made clear in the public documents and notifications of the regulatory authority.

Principle of Continuous Monitoring

Even after the operator has obtained permission, the regulatory authority continuously monitors the operator's activities in accordance with the permission conditions. For this, it is necessary to give the regulatory authority's inspectors the authority to freely enter the business premises.

Principle of Compensation

The use of nuclear power carries a risk of causing damage to individuals, property, and the environment. Therefore, the Atomic Energy Act requires the state to prepare appropriate compensation measures.

Principle of Regulatory Independence

Decisions by the regulatory side regarding safety assurance must not be obstructed or distorted by the side conducting nuclear power business or using nuclear power and radiation. For this, it is necessary for the safety regulatory authority to be independent from the promoting and using sides.

Principle of Transparency

Explaining and disclosing as much as possible about the risks and benefits of nuclear power use enables robust nuclear power use and regulation. Organizations involved in nuclear regulation are required to create and disclose not only regulatory rules but also regulatory philosophies and policies, and to disclose information related to safety such as accidents and malfunctions.

Safety Culture

Safety culture is not a principle defined by the Atomic Energy Act, but it plays an important role in ensuring the safety of nuclear power operators. Safety culture is "a state where members of an organization share values and behaviors that prioritize safety, or the way an organization realizes it". Safety culture is a perception that members of an organization should share in order to prevent accidents that can have serious impacts on organizations and society.

In Japan, operators (power companies) that carry out nuclear power generation have established the Japan Nuclear Safety Institute (JANSI) after the TEPCO Fukushima accident. JANSI has created the following seven principles of safety culture:

(1) Value of safety first: The value of safety first is recognized by the organization and individuals.
(2) Leadership at the top: The top demonstrates a strong commitment to safety through leadership.
(3) Safety assurance mechanism: A mechanism for ensuring safety is incorporated into operations and activities.
(4) Smooth communication: Maintain smooth communication with the internal organization, related organizations, and the general public.
(5) Questioning and learning attitude: The organization and its members have an attitude of questioning, learning, and taking responsibility for corrections.
(6) Risk recognition: The organization and its members recognize the potential risks of operations and facilities.
(7) Vibrant workplace environment: A workplace environment with freedom of speech, vitality, and creativity.

This discusses important activities for ensuring safety outside of laws and regulations. *In the United States, the Institute of Nuclear Power Operations (INPO) plays a major role in ensuring the safety of nuclear power generation.* As mentioned in Chap. 1, the frequency of accidents and failures has significantly decreased since 1990, and the amount of power generated (operating rate) has greatly improved. These achievements have also become the basis for improvements in regulations in the United States. The key point of their activities is, as the phrase "*Hostages of Each Other*" suggests, that the president of a nuclear power company with many accidents and failures is hung up at an INPO meeting by the presidents of another companies to encourage improvement. It is a mandatory improvement activity among nuclear power operators. INPO is not a "good friends club". An accident at a nuclear power plant affects their own company even if it is an accident at another company. Encouraging power companies with poor safety performance to improve is beneficial to all power companies that operate nuclear power generation.

Behind safety issues are organizational management issues. First and foremost, safety issues should be addressed in management to prevent accidents. Accidents and malfunctions above a certain level are made public by the regulatory side, and legal measures are taken. Many of the seeds of accidents and malfunctions can be corrected by the management of the organization in question. Management is a company's exclusive decision, and there is no need to disclose this part. In the United States, there has been a dispute in federal court over whether or not to disclose the information handled by INPO, but the request for disclosure has been rejected because the benefits of non-disclosure are greater. If forced to disclose, there is a risk that the information necessary for improvement will not be presented to INPO, and improvement will not proceed. This is a negative for the American people. In Japan, we often hear opinions that everything should be disclosed, and while the Japanese are strongly group-oriented, it is necessary to understand this point. U.S. nuclear power operators can be considered to have fulfilled their responsibility for the Three Mile Island nuclear accident in 1979, which caused a public uproar, by reducing the accident rate and increasing the amount of electricity generated by improving the operation rate, etc., since 1990, through the establishment and operation of INPO. Japanese nuclear power operators will be required to produce visible results like INPO in future. The author's experience of being involved in the management and operation of the University of Tokyo's research reactor facility for many years also shows that organizational management is important for ensuring safety. We look forward to seeing what results JANSI will show to the public.

Improving Transparency and Passing on Experience

Not limited to nuclear policy, for the robustness of policy, as already mentioned, it is first necessary to have information about decisions and actions taken by responsible organizations. This information includes not only information directly related to the administration of ministries and agencies, but also reports related to the overall picture and experience of large projects implemented with tax money, and accident reports. By creating and disclosing these, we can pass on the experience of failures and successes and promote robust use of nuclear power. In the United States, not only GAO reports, but also reports on large projects are created and disclosed. For example, reports from before the digitization of information, like the Clinch River Fast Breeder Project report, are collected and disclosed by the Nuclear Regulatory Commission.

More than 50 years have passed since the start of nuclear power utilization in Japan. There are many projects that have completed decommissioning measures or are in the process of decommissioning. In Japan too, it is necessary to compile and publish such reports in order to advance a robust nuclear policy in future. Examples of projects that require reports include the fast breeder reactor "Monju", the advanced thermal reactor "Fugen", nuclear fuel cycle projects such as Tokai reprocessing and uranium enrichment conducted by the former Power Reactor and Nuclear Fuel Development Corporation, and the materials testing reactor and related projects conducted by the former Japan Atomic Energy Research Institute. In relation to the Agency for Natural Resources and Energy, there are main test researches at the Nuclear Power Engineering Corporation (NUPEC) and the Nuclear Safety Research Association (JNES). Many of Japan's projects, such as fast breeder reactors, often have precedents in the United States, so reports on similar projects in the United States and France can be useful for promoting robust use in Japan. In relation to accidents, the JCO accident and the TEPCO Fukushima accident can be mentioned. These accident reports have been made, but if you carefully compare the reports of the criticality accidents and the Three Mile Island nuclear accident in the United States with the reports of the JCO accident and the TEPCO Fukushima accident, including the background of the accidents, the responses and measures in both countries afterward, there should be differences and points that need to be improved. Through these, I hope that robust nuclear power utilization will progress in Japan.

4.4 US Nuclear Regulatory Commission

The US Nuclear Regulatory Commission (NRC) and its activities not only serve as a reference for Japan's nuclear regulation, but also as an example of understanding the accountability of US administrative agencies. The US Nuclear Regulatory Commission has the Information Digest (NRC 2021a), Citizen's Guide (NRC 2003), and Strategic Plan (NRC 2022) as basic documents introducing its overview and activities.

The Information Digest (NRC 2021a) has been published annually since 1989, describing the activities of the Commission, the subjects of licensing, and the responsibilities of the Commission. It also provides information from the International Atomic Energy Agency, the US Energy Information Administration (EIA), and the US Department of Energy. The Information Digest, which used to be only a document, is now also provided as a book with audio and PowerPoint diagrams. Industry and overseas information is updated every two years and posted, but the NRC states that it has not reviewed the validity of the content.

The Information Digest is published to arouse the interest of NRC staff in the efforts the NRC is making to meet the expectations of the public in line with its goals and fulfill its role. In other words, the Information Digest plays a role in explaining NRC activities to the public and making them known to NRC staff.

The NRC is responsible for creating rules and guidelines, policy-making, licensing and decommissioning of nuclear power plants, certification, research, monitoring and enforcement, accident and failure response, and emergency planning and response.

The NRC is composed of five commissioners (nuclear regulatory commissioners) and their staff, and is headquartered in Maryland, near Washington DC. There are four regional offices in Pennsylvania, Georgia, Illinois, and Texas. At each nuclear power plant, there are at least two resident NRC inspectors. The total number of staff in 2021 is 2868. The budget is about $880 million, and the licensing revenue obtained from applicants is $720 million, so about 82% of the budget is obtained from licensing revenue. The research budget is $77 million.

The NRC, with its licenses, operates a total of 93 nuclear power plants at 55 locations in 28 states across the United States, supplying 19% of the country's electricity. The NRC is responsible for extending operating periods, site permits for new reactors, construction and operation permits, and design certification of reactors. It also oversees research reactors, test reactors, and medical isotope facilities. Facilities handling radioactive materials for medical, academic research, and industrial use are regulated by the NRC at about 2,200 facilities and by the states at 16,000 facilities. In terms of nuclear fuel materials, it licenses nine facilities that process, convert, and enrich uranium. It also tracks radioactive sources such as cobalt-60. The NRC is responsible for safeguard measures within the United States for nuclear non-proliferation. In terms of radioactive waste, there are 10 facilities for reducing low-level radioactive waste and four state disposal sites. In high-level radioactive waste management, it is responsible for permitting 81 used fuel storage facilities in 35 states, inspecting the transport of used fuel, and decommissioning facilities such as retired nuclear power plants. In security and emergency response, the NRC inspects the emergency response training conducted by each nuclear power plant every two years, and the Federal Emergency Management Agency evaluates it.

Following these descriptions in the Information Digest, the highlights of the year's achievements are reported in sections on coronavirus response, power reactors, non-power reactors, radioactive materials and nuclear materials and waste, activities across the NRC, international activities, operational management, public hearings, news and information.

Detailed reports are made for each item, and through these, the American people and Congress can understand, evaluate, and monitor the activities of the NRC. This is an example of the accountability of U.S. administrative agencies.

Figure 4.9 shows how the NRC regulates.

1. Create rules and guides for applicants and implementers.
2. Licensing to use nuclear materials, operate nuclear facilities, and decommission facilities.
3. Supervision: Monitor whether the power plant operation and facility management of the implementers (Licensee: power companies operating nuclear power plants with permission) comply with NRC rules, investigate for any errors, and follow-up to ensure appropriate corrections are made when necessary.
4. Evaluate the operating experience of facilities and activities that have received approval.
5. Conduct research, hold public hearings, and obtain independent evaluations to aid in regulatory decisions.

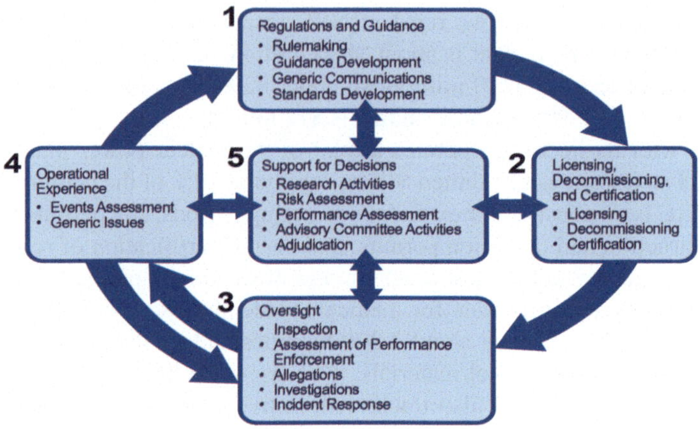

1. Develop regulations and guidance for applicants and licensees.
2. License or certify applicants to use nuclear materials, operate nuclear facilities, and decommission facilities.
3. Inspect and assess licensee operations and facilities to ensure licensees comply with NRC requirements, respond to incidents, investigate allegations of wrongdoing, and take appropriate follow-up or enforcement actions when necessary.
4. Evaluate operational experience of licensed facilities and activities.
5. Conduct research, hold hearings, and obtain independent reviews to support regulatory decisions.

Fig. 4.9 How US NRC regulates. *Source* US NRC

The NRC has set forth the following five principles for good regulation.

- Independence: Regulate with the highest level of ethics and professionalism.
- Openness: Nuclear regulation is a public task and must be handled openly and honestly with the public.
- Efficiency: The highest technical and organizational capabilities are required. These should always be the goals of the Regulatory Commission.
- Transparency: Regulations must be consistent, logical, and practical. The position of the Regulatory Commission must be immediately understandable and easily applicable.
- Reliability: Regulations must be based on the best knowledge obtained from research and operating experience.

The NRC shares the responsibility for public safety with the implementers and states that have entered into agreements (states with organizations that regulate nuclear materials), but ultimately states that the responsibility for safety lies with the implementers (for example, companies operating nuclear power plants).

The NRC is composed of five members, each serving a term of five years, with the terms of each member staggered by one year, so one member changes each year. One of the members is the chairman, who oversees the NRC and speaks on behalf of the commission. Members must declare their support for either the Republican or Democratic party before being appointed, and no more than three members can be from the same party. There is a secretariat for the operation of the commission, and the secretary-general implements the decisions of the commission. The secretariat has rooms for each responsibility. The main rooms are the Reactor Regulation Room, Nuclear Regulation Research Room, Nuclear Material Safety and Safeguard Measures Room, Nuclear Security and Incident Response Room, and Regional Room. In addition, there are two advisory committees on nuclear safety and medical isotopes. Advisory members provide independent advice. Advisory committee meetings are open to the public, and citizens can express their opinions orally if they wish.

The NRC's rule-making process is as follows.

1. Proposal: Proposals are made by congressional or administrative orders, instructions from regulatory commissioners or the NRC Executive Director for Operations (EDO), needs identified by NRC staff, or applications for rule-making.
2. Review and approval of the proposal by the Regulatory Commission.
3. Public Involvement & Stakeholder Engagement: Advance notice of rulemaking, justification for regulation, and drafting of proposed rules.
4. Hearing opinions from stakeholders.
5. Review by the Regulatory Commission and approval of the proposed rule draft: The Regulatory Commission issues a memo detailing requests to the office staff, who then work to resolve the comments.
6. Publicizing the proposed rule to receive comments: Drafting of the environmental assessment, analysis of the rule from a cost-benefit perspective, creation of guideline drafts, and holding public hearings.
7. Analysis of public comments by NRC staff.
8. Approval of the final rule draft by the Regulatory Commission: The Regulatory Commission releases a memo regarding staff requests. The staff resolves the comments from the Regulatory Commission.
9. Publication of the rule: Announcement of the final environmental assessment, final regulatory analysis, and final guidelines.
10. Establishment of the rule: The rule comes into effect. Note that there is a deadline for rulemaking.

A document justifying the regulation is created. This document describes the technical, legal, and policy reasons for changing the regulation, why the rule is being changed, how the problem will be resolved by changing the rule, and the results of considering whether the problem can be addressed in other ways. It also includes considerations about the costs and benefits of each option, retrospective application of the rule (backfitting), and future applications. Once the justification document is approved, it is made public to receive comments from the public. All comments on this document are considered during the rule-making process.

The Information Digest describes changes in the NRC's budget and staff numbers. The budget and staff numbers were $1.054 billion and 3,992 people in 2011, but by 2021 they had decreased to $879 million and 2868 people. As mentioned earlier, it is legally determined that the NRC covers most of its expenses through licensing fees. These fees are paid to the Treasury by September 30 each year. The amount in 2021 was \$724.1 million. The breakdown of the budget and staff by field is also shown. The fact that the budget is not fixed each year and is mainly covered by licensing fees is rational and indicates an inherent mechanism for efficiency. The decrease in the 2021 budget and staff compared to 2011 is thought to be due to the fact that the 2000s were a period known as the nuclear renaissance, during which there were many licensing operations such as type approval of new light water reactors in the United States. The high job turnover in the U.S. employment environment makes this possible. Note that the highest number of operable nuclear power plants in the United States was 112 in 1990, and this number has decreased to 93 in 2021.

Information Digest, following this information, explains the world's nuclear energy, U.S. reactors, nuclear materials, radioactive waste, security, and emergency response. As an appendix, it includes data on operating power reactors, power reactors under construction, power reactors under licensing, power reactors under decommissioning and permanently shut down, power reactors that were under construction or licensed but had their plans canceled, parent companies of commercial power plants, year of operation permits issued for commercial nuclear power plants, expiration dates of operation permits for operating power plants, and data on operating and decommissioning test research reactors regulated by the NRC. Furthermore, it includes diagrams and tables on radiation doses and regulatory values, licensing of commercial nuclear power plants, number of nuclear material licenses by state, major nuclear fuel cycle facilities in the U.S., NRC type approval of dry storage containers for spent fuel, licensed dry storage facilities, volume reduction facilities for low-level radioactive waste, complex material sites under decommissioning, nuclear power plants in various countries, types of power reactors in the world, Native American reservations within 50 miles of operating nuclear power plants, universities that received competitive funding from the NRC in 2020, significant improvement recommendations made by the NRC in 2020, U.S. laws related to the NRC, treaties on nuclear safety, security, and international safeguards, international organizations and their committees in which the NRC participates, countries with which the NRC has cooperation agreements, export/import licenses that came into effect in 2020, main radioactive isotopes used in the U.S. and their uses, terms and their explanations, and finally, web links. With these, U.S. citizens can understand the activities of the NRC.

The Citizen's Guide to U.S. Nuclear Regulatory Commission Information (NRC 2003) is a document that explains information and programs about the NRC to its stakeholders, and is designed to help them find the information they want to know. It is available not only in English but also in Spanish. It first discusses the "Establishment of the NRC" and explains its role and work. Then, under the title "Opportunities for Stakeholder Participation", it states that the NRC's stakeholders are the public (citizens), Congress, NRC licensees, Indian tribes, local governments, managers and workers in the nuclear industry, and the international community, and that the NRC places emphasis on informing these people about its activities. It states that stakeholder participation in NRC activities is essential for appropriate regulation, and tells where on the NRC's website related NRC documents can be found. It then provides information on the NRC's regulatory programs, including reactors, nuclear materials, radioactive waste, decommissioning, corrective actions, security and safeguard measures, nuclear research, international activities and nuclear non-proliferation, and the Office of Inspection General, and further discusses the location of regulatory rules and related information, and national standards and criteria.

While the Information Digest introduces the content of the activities, the Citizen's The Guide explains the items of information for public involvement and their summaries. First, it explains how the regulations carried out by the NRC are made and about the stakeholders of the NRC. It introduces the information and information systems operated by the NRC, and the electronic library. Finally, it explains the whereabouts of information sources inside and outside the NRC, by type of information. From this document, it is understood that it is necessary for the NRC to do a good job to explain to stakeholders where the information is.

The Strategic Plan (NRC 2022) is a document that describes detailed planning, execution, monitoring, and measurement to achieve the NRC's goals, and is issued every four years. The creation and submission of this document is mandated by law. This document reveals the NRC's operating policies regarding performance goals, priorities, personnel deployment, and budget planning. Whether the goals have been achieved is determined by performance indicators listed in the documents when the federal congress decides the NRC's budget. The 2022–2026 Strategic Plan lists three points as goals: safe and worry-free use of radioactive materials, being a sound organization (NRC), and eliciting trust in the NRC from stakeholders. These three goals are each developed into several more specific goals, which are shown on the homepage.

The three basic NRC documents introduced here are examples of the accountability of U.S. administrative agencies. Accountability is the explanation of administrative policies by administrative agencies (in this case, the NRC) and responsibility for the results. Based on this information, evaluations of administrative results and achievements in Congress and budget committees, and reflections in budgets and laws are made. If necessary, investigations are conducted by the GAO (Government Accountability Office), and reflected in administrative improvements. The Information Digest is a document that explains the work done by the NRC, the Citizen's Guide explains the whereabouts of information on NRC activities and their rationale. The Strategic Plan shows the goals and methods of NRC administration. These documents ensure the NRC's accountability to the public and stakeholders.

In other federal government agencies in the United States, the creation of documents equivalent to the Strategic Plan is also mandatory. For example, the Office of Nuclear Energy in the Department of Energy publishes a Strategic vision, which sets goals to enable the continued operation of existing nuclear power plants, enable the development of new types of reactors, develop advanced fuel cycles, maintain U.S. leadership in nuclear technology, and become a high-performance organization, and shows performance evaluation indicators for each goal.

The U.S. Nuclear Regulatory Commission (NRC) has many points that can serve as references for Japan's nuclear regulation. After the TEPCO Fukushima accident, I contacted acquaintances who had worked at the NRC to understand the overview, asked for literature, and met with these acquaintances to ask questions when I visited the American Nuclear Society, and summarized the results. The results of the Q&A are posted below for reference. This information is from around 2012, but there should not have been any major changes since then.

Q&A Related to the Citizen's Guide

Q: What types of meetings and committees does the USNRC have?

A: There are three types: ①Commissioner's meetings, ②ACRS (Advisory Committee) meetings, and ③public hearings. ① and ③ are open to the public, while ② is often closed. Public hearings are held locally, while ① and ② are held at the capital (NRC headquarters). Open meetings can be attended by anyone without reservation. When a large number of people are expected, there may be admission restrictions due to lack of space, but they try to make it viewable in another room via television.

Q: Who answers questions from the public?

A: NRC staff answer. The applicant (operator) does not answer as a rule.

Q: Doesn't openness cause confusion?

A: Those who disrupt the proceedings are expelled. There are also security guards. There have been no particular problems.

Q: Is there someone in charge of audits?

A: Internal audits are handled by an audit department that is responsible for accounting and ethics. There is no audit of technical content. Another is the GAO (Government Accountability Office, a congressional agency) which audits and publishes reports.

Q: Do new applications ever lead to lawsuits?

A: The NRC is sometimes sued, but it rarely loses. It does serve as a mechanism for delay.

(Reference: The U.S. has pushed to increase public participation in hearings for the construction of new nuclear power plants. Independent judges from the ASLB (Atomic Safety Licensing Board Panel) participate and listen to and address the concerns of local residents and groups directly affected. Various questions are raised at the hearings. NRC public hearings (concerns about the impact of severe accidents) are broadcast, and this can also lead to national debates about the need for energy and the role of nuclear power.)

Q: Who is the most influential stakeholder in the operation of the NRC?

A: The Congress. They decide the budget. They also set the upper limit on the number of staff. The budget and personnel can increase or decrease significantly (30–40%) over several years. The NRC's budget is about 100 billion yen, with 80% going to personnel costs and travel expenses. Most of the budget (90%) is recovered from the application fees of the applicants. The NRC is considered one of the efficient and good government organizations.

Q: What about local government participation in licensing?

A: Local governments share responsibilities related to the environment and emergency evacuation. Even without the consent of local governments, the NRC can issue permits. If the power is subject to state regulation, the decisions of the state power commission may affect the licensing from another perspective.

Q&A Related to the Information Digest

Q: What does it mean to conduct a review but not a verification?

A: A review is an evaluation of applications from applicants, which is done by the NRC. The NRC does not perform verification, which involves independently calculating and confirming the contents of the applicant's application.

Q: What is the role of the NRC commissioner in the operation of the NRC?

A: One of the commissioners is the chairman and is responsible for regulatory administration, but there is a separate secretary for operations, and daily operations, budgeting, etc., are left to the secretary and below. The commissioners give the final approval for the budget.

Q: What is the NRC's role in the qualification (certification) of nuclear reactor operators?

A: They only conduct the tests. Actual training is done by external organizations.

Q: How is the operation and management of nuclear power plants monitored?

A: It is done by resident NRC staff. They have free access to anywhere in the power plant. They mainly give verbal warnings. The power plant staff know more about the power plant, so NRC staff do not correct errors or perform operations themselves. It is prohibited for them to do so.

Q: What is the role of the ACRS (Advisory Committee on Reactor Safeguards)?

A: The role of the ACRS is different from that of NRC staff and commissioners. It reviews what the NRC and licensees are doing. This is determined by law. All 15 members are part-time. There are 20 NRC staff who assist with the work of the ACRS. The ACRS can also hire consultants. Many of the consultants are former ACRS members. ACRS members are university professors specializing in safety, researchers at national research institutes, and retired corporate employees.

Q: What is the licensing process like?

A: Everything is done in writing. It is possible to exchange information about the manufacturer and its reactor before the application is submitted.

Q: Can NRC staff become corporate employees after retirement?

A: Commissioners cannot be employees of a company for life. The prohibition period varies depending on the status of the staff. Normally, it is not allowed for 2–3 years. In the opposite case, they are not allowed to handle cases from their former company.

Q: What is the purpose of the research conducted by the NRC?

A: To create a technical basis for regulation. About 50% of the research is on severe reactor accidents.

Q: What is the composition of the NRC staff?

A: About 50% of the staff are technical, and the number of people with a doctorate is... 20–30%, this ratio is very high even among government agencies.

Q: How is the NRC's activities made known to the public?

A: Mainly carried out by the public affairs department.

Q: Do you publicize when you find trouble or malfunctions at the power plant?

A: No, we just file the documents. The media needs to search for the filed documents themselves.

Q: What about press releases?

A: Every 2–3 weeks.

Q: What about the disclosure of the plant's condition?

A: Quarterly.

Q: Can former company employees who have become NRC staff talk to their former colleagues?

A: Possible, but they must not leak information such as regulations。It is the same as personal information management.

Next, discussing the participation of citizens in the public meetings of the U.S. Nuclear Regulatory Commission (NRC 2021b).

Depending on the level of participation, there are three types of meetings: ① Observation meeting, ② Information Meeting with a Question and Answer Session, ③ Comment-Gathering Meeting. Participation in the meeting requires following procedures. If a participant disrupts the progress of the meeting, they will be expelled.

The Observation meeting is where regulatory and technical matters are discussed, and operators who are applying for licenses and their associates can also participate. Topics of this meeting include license renewal, amendment/exclusion rules, review of applicant's technical reports, combined licenses for design/construction, early site permits, design certification, discussions on plant performance as part of the reactor oversight process, and final reports on accident/failure investigations.

The Information Meeting with a Question and Answer Session is a meeting to inform participants about NRC regulations and corrective measures and to conduct a question and answer session. The purpose of this meeting is for NRC staff to directly discuss regulations and technology with participants. Participants can also ask questions and express opinions if they wish.

The Comment-Gathering Meeting is held to gather opinions that are useful for regulatory issues and activities. NRC staff present regulatory issues and responses, and the opinions at the meeting are reflected in the licensing and regulatory rules. The participants are those who express opinions, views, or suggestions for improvement.

These meetings are notified at least 10 days in advance. Documents and participants are made public. Remote meetings are also held if technically possible and necessary. NRC staff answer questions raised at the meeting, and follow-up on questions that could not be answered during the meeting is done later by phone or email. Concerns can also be communicated to NRC staff by phone, in writing, or by email. Comments are made public, and how they were handled can be found on the NRC's public website. Exceptions that are not handled in this way are also stipulated.

4.5 Safety Review of Nuclear Power Plants in Japan

The safety review of nuclear power plants in Japan is conducted in accordance with the Act on the Regulation of Nuclear Source Material, Nuclear Fuel Material and Reactors (Law No. 166 of 1957). Article 43-3-29 states, "For the improvement of the safety of nuclear power reactor facilities, the installers of nuclear power reactors must evaluate the safety of their nuclear power reactor facilities at intervals specified by the Nuclear Regulation Authority rules". Power companies, as the installers of nuclear power reactors, create and submit to the Nuclear Regulation Authority a document (a self-assessment of safety) for a reactor installation change permit application in order to undergo a safety review. Based on this document, the Nuclear Regulation Authority conducts a review of compliance with the new regulatory standards established after the TEPCO accident.

The Nuclear Regulation Authority was established in 2012 as an external bureau of the Ministry of the Environment. The establishment of the Nuclear Regulation Agency as its secretariat, and its designation as an administrative commission called a "3-article commission" as defined by the National Government Organization Act, are stipulated by the Nuclear Regulation Authority Establishment Act. The Nuclear Regulation Authority is composed of five members, one of whom is the chairman. The term of office for members is five years. The Nuclear Regulation Agency consists of the Director-General's Office, the Nuclear Regulation Department, and local nuclear regulation offices and branches. There is also a Nuclear Safety Personnel Training Center (NRA 2024a). The number of staff at the Nuclear Regulation Agency is about 1,000.

The Nuclear Regulation Department has a review group and an inspection group. The review group is divided into practical reactor review, research reactor review, nuclear fuel facility review, and earthquake and tsunami review departments. The Director-General's Office has administrative organizations such as the General Affairs Division, as well as a Technical Infrastructure Group and a Radiation Group, the former being responsible for technical infrastructure and research departments, and the latter for radiation regulation, nuclear security, and safeguard measures.

In addition to these, there are independent advisory councils, such as the Nuclear Reactor Safety Expert Review Council, the Nuclear Fuel Safety Expert Review Council, the National Development Corporation Review Council, and the Radiation Review Council. The Nuclear Reactor Safety Expert Review Council and the Nuclear Fuel Safety Expert Review Council are established by the Nuclear Regulation Authority Establishment Act, and their role is to investigate and deliberate on the safety of nuclear reactors when instructed by the Nuclear Regulation Authority. Review members are appointed from knowledgeable and experienced persons such as university professors and are part-time. The number of members is about 20 to 30. Although they are called review councils, their role is not to conduct safety reviews of nuclear power plants, but to express opinions on safety. The Nuclear Reactor Safety Expert Review Council and the Nuclear Fuel Safety Expert Review Council have three subcommittees: the Nuclear Reactor Safety Basic Subcommittee, the Nuclear Fuel Safety Basic Subcommittee, the Earthquake and Tsunami Subcommittee, and the Volcano Subcommittee.

The safety review of nuclear power plants is the responsibility of the review group of the Nuclear Regulation Division of the Nuclear Regulation Authority.

At the review meetings, one commissioner of the Nuclear Regulation Authority in charge, several officials from the Nuclear Regulatory Agency, and employees of the power company attend. The proceedings are conducted under the chairmanship of the commissioner. The review meeting does not end in one session, but the power company explains the responses to the comments made at the previous review meeting, and a question and answer session is held. Even for one application, the review is divided into the Practical Reactor Review Department, the Nuclear Fuel Facility Review Department, and the Earthquake and Tsunami Review Department, depending on the theme.

Once the power company has passed the safety review for the contents of the application for permission to change the installation of the reactor, it submits the design and construction plan approval/notification and carries out the construction. Once the construction is completed, it undergoes a pre-use operator inspection, and if it passes, it can operate the facility. For this, in addition to ensuring the integrity of these facilities (hard regulation), it is necessary to receive approval for safety regulations for ensuring the validity of safety activities (soft regulation) and approval for nuclear material protection regulations as nuclear material protection regulations before starting operation. Even at the operation stage, there is an obligation to maintain technical standards, and it is necessary to undergo regular operator inspections. There is also an obligation to comply with nuclear material protection regulations. By receiving these regulations, the power company can operate its nuclear power plant (NRA 2024b).

The establishment of the Nuclear Regulation Authority and the history of the formulation of new regulatory standards are summarized in the reference materials of the "Recommendation: Issues and Desired State of Nuclear Safety Regulation" by the Japan Science Council (Science Council of Japan 2020).

Comparing nuclear regulation in Japan and the United States, in the United States, a large number of underlying documents are created, collected, and disclosed. Japan's nuclear regulation is centered on oral exchanges at review meetings, and while there are application documents from power companies for review items, there are few underlying documents. The time for each review member to express their opinion orally is limited. In order to conduct robust safety reviews, it is necessary to enrich the information on the basis, as in the United States.

Not only in nuclear regulation, but also in Japan, it is necessary to enhance the explanation of policies by the administration and the investigation and audit of the results. Without this, robust nuclear regulation cannot be conducted.

4.6 Summary and Discussion

The author discussed the legal framework for the use of nuclear power. Nuclear disarmament and non-proliferation were introduced through the Treaty on the Non-Proliferation of Nuclear Weapons and the Treaty on the Prohibition of Nuclear Weapons, and I discussed the legal framework for nuclear non-proliferation. The author introduced the international framework for nuclear damage compensation and the nuclear damage compensation systems in the United States and Japan. The author introduced the laws and organizations related to the use of nuclear power and safety regulations in the United States, France, and Japan, and then explained the structure and concept of nuclear law. As an example of the accountability of administrative agencies, the author introduced the basic documents of the U.S. Nuclear Regulatory Commission. The author discussed the safety review of nuclear power plants in Japan.

The constitution is the supreme law that the people have the state protect. Laws are rules that the state has the people protect. Parliamentary democracy, where the three powers of the legislature (parliament), the executive (government), and the judiciary (courts) are separated, forms the basis of the operation of countries around the world. Administrative agencies (ministries) are organizations that work for the people based on laws and government (cabinet) policies. Therefore, administrative officials are public servants. In the United States, a system to question the accountability of administrative agencies is functioning. Accountability refers to the explanation of administration and policies to the people and the responsibility of the responsible administrative agency for the results of policies. However, it is not correct to translate accountability as "explanation responsibility" because responsibility cannot be fulfilled by explanation alone. It is better to translate it as "responsibility for results" or to use "accountability" as it is. The term "explanation responsibility" is often heard in Japan, but there is a risk that it will be understood as ending with a verbal explanation, which may be a cause of Japan's stagnation.

Accountability Is Ensured by Meeting the Following Three Conditions

1. There is information about the decisions and actions taken by the responsible person or organization,
2. There are people or organizations who receive or discover this information and examine, investigate, and report on it,
3. There are actions to correct defects based on this information.

   The item 2 is the role of the GAO (Government Accountability Office, Audit Office, Administrative Inspection Office) in the United States, and the item3 is the role of the Federal Congress and its Budget Committee.

Organizations and policies that do not incorporate criticism will deteriorate. The GAO does not comment on policies or their outcomes, but rather, based on its investigative authority, it plays a role in revealing facts. In the judiciary, the police and prosecutors investigate legal violations and play a role in revealing facts. There are many things that cannot be clarified without investigation, not limited to crimes. Once the facts are clarified, they can be considered and contemplated. It is the role of the GAO to investigate and clarify facts based on the authority given by law regarding administration and policy. In the U.S. Congress and its budget committees, hearings are also held, and the results of GAO investigations are referred to, and reflected in policies and budgets. This is not limited to the field of nuclear power use and safety. In the UK, there is the NAO (National Audit Office), and in France, there is the Cour des Comptes (French Court of Audit), both of which perform similar functions.

Japan also has a Board of Audit. The "Accounting Audit Research" published by the Board of Audit contains research results and papers for improvement. As early as the 1990s, issues were raised in comparison with the GAO. Is Japan's Board of Audit contributing to the improvement of administrative accountability like the GAO in the United States? However, this issue is not appropriate to consider only within the framework of the Board of Audit's work, but is related to the consciousness of the people and the issue of administrative reform in Japan. Regarding administrative reform, the author introduced the lead article by a legal scholar in the 2012 "Accounting Audit Research" titled "The Limits of Administrative Reform in Our Country".

Looking at the Board of Audit Act, the role of Japan's Board of Audit is not to investigate the results and policies of the administration, but to conduct accounting audits in a narrow sense. In contrast, the U.S. GAO changed its name from the General Accounting Office to the Government Accountability Office in 2004, and its role is to clarify the accountability of government agencies. The Boards of Audit in Japan and the United States have the same name, but their roles are different.

In Japan, each ministry conducts its own policy evaluation, and the Administrative Evaluation Bureau of the Ministry of Internal Affairs and Communications conducts policy evaluations as necessary from the perspective of ensuring the unity of the entire government. The basic principle of administrative evaluation and policy evaluation in Japan is self-evaluation by each ministry, which is different from the United States. Shouldn't we consider and build a system that meets the above three requirements in Japan as well? Isn't the reason why we can't build such a system because many Japanese companies have been dependent on ministries and have been working with a domestic perspective? Isn't this the cause of Japan's long-term stagnation? I repeat, countries and organizations without a mechanism to incorporate criticism will deteriorate.

First and foremost, administrative reform will not progress unless the workload of the administration is reduced. For instance, the use of information technology is a means to this end, and it has already been implemented in many countries. However, in Japan, as the challenges revealed in the response to the coronavirus have shown, the My Number card, which connects citizens to the administration, has not been distributed to the entire population, preventing the efficiency of the administration from being improved through the use of information technology. This also caused problems in government spending for coronavirus measures.

Administrative reform is easier said than done. However, for the future of Japan, fiscal reform is unavoidable. If we are avoiding painful reforms, isn't that the responsibility of politics and the citizens who have a strong sense of dependence on the country? I am not an expert in this field, but I hope that improvements will be made using the expertise and practical experience of administration and politics.

In the use of nuclear power in Japan, it is necessary to clearly define the responsibilities and roles of the government, power companies, reactor manufacturers, and nuclear research and development institutions, isn't this the first step to make the improvement mechanism work?

This is not unique to the nuclear power sector, but applies to all sectors. As a country, it might be good to make the Board of Audit or the Administrative Evaluation Bureau of the Ministry of Internal Affairs and Communications independent, give them the function of investigating and auditing administration and policies, and make the PDCA (Plan, Do, Check, Action) mechanism work. Crimes are investigated by the police who have the authority to do so, but *facts resulting from policies cannot be revealed unless they are investigated by an independent institution with the authority to do so*. Policy evaluation in Japan is a mechanism for internal improvement of the administration. As we have a parliamentary democracy, isn't it the role of the government and the parliament, like in the United States, to make the mechanism for improving the administration work? I am not saying that Japanese civil servants are negligent or not excellent. They are excellent, diligent, and have a high sense of responsibility and professional consciousness.

There are various other mechanisms for improvement, such as auditors, external directors, and external evaluation committees. These mechanisms will not function unless people are selected for them and their opinions are incorporated for improvement. Are these functioning in organizations where Japanese tax money is invested? The continuous safety improvement activities of the Institute of Nuclear Power Operations (INPO) in the United States are a mechanism for improvement by industry groups. The construction plan for the Clinch River Fast Breeder Reactor in the United States was canceled because the soaring construction costs were not borne by the power companies. This is an example of a check mechanism involving private companies working.

The chairman of the Fukushima Accident Investigation Committee of the National Diet, in his book, states that nuclear power operators had "enslaved" regulatory administration. Examples of this "enslavement" include the issue of receiving money and goods from influential locals by executives of Kansai Electric Power Company, and the tax evasion issue of the chairman of a large private university who received a gratuity. In Japan, it is socially accepted to send gifts as a token of gratitude when someone has taken care of you. It seems that there is a structure of "enslavement" on top of this cultural characteristic. Caution is needed.

Technology, not limited to nuclear power generation technology, is used because it is beneficial. The International Atomic Energy Agency's Handbook on Nuclear Law states the concept of risk (cost) and benefit as the first of several principles incorporated into the law. Regardless of nuclear power, when regulating a certain technology by law, it is required as a basic principle to consider the balance between risk and the convenience of use. As introduced in Chap. 2, in the regulation of nuclear power plants in the United States, a cost-benefit evaluation of safety improvement measures required by regulation is quantitatively required. This is also stated in the Strategic Plan, one of the basic documents of the NRC introduced in this chapter.

A major drawback in Japan's use of nuclear power, in terms of safety, regulation, and disaster prevention, is that risk and benefits are not quantitatively considered. Behind this, I feel that the promoters of nuclear power generation in Japan, namely power companies and responsible ministries, have avoided talking about costs (electricity charges), or in other words, the goal of providing cheap electricity. Whether this can be corrected is primarily the responsibility of the people, and then of politics. In terms of safety improvements, costs are not considered, and it is unclear whether risks have been reasonably reduced. In Japan, nuclear power plants are shut down and new regulatory standard compliance reviews are being conducted, causing economic losses to the public through electricity charges. The return period of a large tsunami is long, and it would have been better to respond to the lessons of the accident while operating. Other countries responded while operating. Even in the Chernobyl accident, the nuclear reactor adjacent to the accident reactor was operating.

As an example of the accountability of U.S. administrative agencies, I explained about the Information Digest, Citizen's Guide, and Strategic Plan, which are basic documents of the U.S. Nuclear Regulatory Commission (NRC). The Information Digest describes the role and activities of the NRC. The Citizen's Guide is designed to help find the information you want to know. The Strategic Plan is a document that describes the NRC's goals and indicators to achieve them, and is published every four years. Whether the goals have been achieved is determined by performance goals stated in the documents when the Congress decides the NRC's budget. Through these, the NRC's accountability to the public and stakeholders is ensured. In other federal government departments as well, the Strategic vision is created and published by the policy makers. I hope to improve the accountability of Japanese ministries and agencies. It should be noted that about 80% of the NRC's budget is covered by fees paid by applicants for licenses and other approvals. It is a system that increases or decreases each year depending on the amount of work, and a mechanism for improving administrative efficiency is built in.

The UK government has been promoting digital government since 2010, and each department of the UK government creates and publishes a digital government strategy every year. All departmental online services are provided as www.gov.uk (Gov.UK 2022; Myojin 2015). The UK government's information is provided centrally under this URL. There is a sense of crisis that not only the improvement of public services, but also the prosperity and competitiveness of the UK depends on the use of digital technology. This information was very useful in writing this book, for example, on radioactive waste and decommissioning measures. I will not introduce Gov.UK here, but have listed it as an exercise problem.

Exercise

1. Summarize the overview and logic of the Nuclear Non-Proliferation Treaty and the Treaty on the Prohibition of Nuclear Weapons, and discuss the Treaty on the Prohibition of Nuclear Weapons from the perspective of protecting your country's independence.
2. Summarize the history and current status of North Korea's nuclear development, the transition of UN Security Council resolutions, the overview of nuclear tests, and the responses of various countries.
3. Summarize the history of Iran's nuclear development suspicions, the transition of UN Security Council resolutions, sanctions against Iran by the United States and the EU, the Comprehensive Joint Action Plan (ICPOA), and the subsequent history.
4. Investigate the interference of Western countries in Iran's oil resources after World War II, and the changes in Iran's political system up to the present, and consider their relationship with the suspicion of Iran's nuclear development.
5. Investigate and summarize the efforts of the Japanese government for nuclear disarmament and non-proliferation.
6. Investigate and summarize the compensation paid in connection with the accident at Unit 2 of the Three Mile Island Nuclear Power Plant in the United States in 1979.
7. Investigate and summarize the compensation for the JCO accident that occurred in Tokai Village, Ibaraki Prefecture in 1999.
8. Investigate and summarize the compensation paid in connection with the accident at the TEPCO Fukushima Nuclear Power Plant in 2011.
9. Describe the outline of the institutional framework for the use of nuclear power in the United States (organization of the government, etc.).
10. Investigate and summarize the main reports on nuclear-related matters by the U.S. Government Accountability Office (GAO) since the 1970s. Investigate and summarize the results of hearings and committee deliberations in the U.S. Congress related to them.
11. Describe the institutional framework (government organizations, etc.) for the use of nuclear power in France.
12. Analyze and compare the reports on the Three Mile Island nuclear accident in the United States, the Challenger explosion accident, and the reports of the National

Diet and Government Accident Investigation of the TEPCO Fukushima accident in Japan, including the background of the accidents and recommendations. Summarize the subsequent responses and measures of the U.S. government and compare them with the response of the Japanese government to the TEPCO Fukushima accident.
13. Compare and discuss the organization and operation of the Nuclear Regulation Authority in your country and the Nuclear Regulatory Commission in the United States.
14. Investigate the homepages of the Nuclear Regulation Authority in your country and the Nuclear Regulatory Commission in the United States, and discuss the types of information provided, the content (including technical depth), and searchability, etc., by comparing them.
15. Investigate and discuss the rules and their operation regarding the regulation of nuclear power plants by the Nuclear Regulation Authority in your country and the Nuclear Regulatory Commission in the United States, using the safety review of nuclear power plants as an example.
16. Investigate and discuss the public meetings and public comments of the Nuclear Regulation Authority in your country and the Nuclear Regulatory Commission in the United States by comparing them.
17. Summarize the overview of the UK government's digital service, Gov.UK. Compile examples in the field of nuclear power.
18. Investigate and discuss the methods and contents of administrative evaluation in your country (evaluation of their own policies by each ministry and evaluation by the board of audit) and the administrative evaluation by the GAO in the United States, including the basis of the laws. Summarize the differences in administrative evaluation between your country and the United States.

References

Administrative Reform Promotion Office (2009) Administrative reform—past efforts. Cabinet Office Administrative Reform Promotion Office, May (in Japanese). https://www.gyoukaku.go.jp/siryou/souron/pdf/090610_panfuretto.pdf. Accessed 1 July 2024

ASN (2003) Joint convention on the safety of spent fuel management and on the safety of radioactive waste management. First national report on the implementation by France of the obligations of the Convention, March. ASN. https://www.french-nuclear-safety.fr/international-relations/international-conventions. Accessed 1 July 2024

Breeder Reactor Cooperation (1985) Final report. The clinch river breeder reactor plant project, produced by breeder reactor corporation, January. https://www.nrc.gov/docs/ML1806/ML18064A893.pdf. Accessed 1 July 2024

GAO (2001) Department of Energy, Fundamental reassessment needed to address major mission, structure, and accountability problems, GAO-02-51, December. Government Accountability Office. https://www.gao.gov/products/gao-02-51. Accessed 1 July 2024

Gov.UK (2022) Policy paper. UK digital strategy, Updated 4 October. https://www.gov.uk/government/publications/uks-digital-strategy/uk-digital-strategy. Accessed 1 July 2024

IAEA (2003) C. Stoiber, A. Baer, N. Pelzer and W. Tonhauser, Handbook on nuclear law, July. IAEA. https://www-pub.iaea.org/MTCD/Publications/PDF/Pub1160_web.pdf. Accessed 1 July 2024

ISCN (2024) Safeguards, integrated support center for nuclear non-proliferation and nuclear security (ISCN). Japan Atomic Energy Agency (in Japanese). https://www.jaea.go.jp/04/iscn/archive/sg_is/index.html. https://www.jaea.go.jp/04/iscn/org/intro_en.html (in English)

JAEC (2017) Basic thoughts on the use of nuclear power (draft), 25th Atomic Energy Commission Document No. 1-1. Japan Atomic Energy Commission (in Japanese). https://www.aec.go.jp/kaigi/teirei/2017/siryo15/1-1_haifu.pdf. Accessed 1 July 2024

JAEC (2018) About the review of the nuclear damage compensation system, nuclear damage compensation system subcommittee October 30. Japan Atomic Energy Commission (in Japanese). https://www.aec.go.jp/kaigi/teirei/2018/siryo38/1-3_haifu.pdf. Accessed 1 July 2024

JAEC (2019) About the change of the goals (medium and long-term goals) to be achieved by Japan atomic energy agency in business operation (report), 7th Atomic Energy Commission Document No. 2. Atomic Energy Commission (in Japanese). https://www.aec.go.jp/kaigi/teirei/2019/siryo07/2_haifu.pdf. Accessed 1 July 2024

JAERO (2023) Compensation for nuclear damage. Japan Atomic Energy Relations Organization (in Japanese). https://www.jaero.or.jp/sogo/detail/cat-07-02.html. Accessed 1 July 2024

JAIF (2024) Research and study on nuclear damage compensation (in Japanese). http://www.jaif.or.jp/ja/seisaku/genbai/genbaihou_activity.html. Accessed 1 July 2024

Kanemoto Y (1990) Program evaluation by the board of audit—what to learn from the American GAO, audit research, No. 2, July. Board of Audit (in Japanese). https://www.jbaudit.go.jp/koryu/study/mag/2-2.html. Accessed 1 July 2024

Kitani S (1994) The development and factors of GAO's audit standards, audit research, No. 9, March. Board of Audit (in Japanese). https://www.jbaudit.go.jp/koryu/study/mag/9-4.html. Accessed 1 July 2024

Kobayashi M (2021) The origin of the challenges of public enterprises in our country—The importance of the board of audit's mission, Audit inspection research, No. 63 lead article, March/ Board of Audit (in Japanese). https://www.jbaudit.go.jp/koryu/study/mag/pdf/j63d01.pdf. Accessed 1 July 2024

Kurokawa K (2012) Message from the Chairman, The official report of The Fukushima nuclear accident independent investigation commission, executive summary. The National Diet of Japan. https://reliefweb.int/report/japan/official-report-fukushima-nuclear-accident-independent-investigation-commission. Accessed 1 July 2024

Law No. 86 of 2001 (2001) Law concerning the evaluation of policies implemented by administrative agencies (in Japanese). https://elaws.e-gov.go.jp/document?lawid=413AC0000000086. Accessed 1 July 2024

MEXT (2024) Nuclear damage compensation system. Ministry of Education, Culture, Sports, Science and Technology (in Japanese). https://www.mext.go.jp/a_menu/genshi_baisho/. Accessed 1 July 2024

MIC (2012) The history and philosophy of administrative reform, materials distributed at the first reform council. Administrative Management Bureau, Ministry of Internal Affairs and Communications (in Japanese). https://www.cao.go.jp/sasshin/kondan/meeting/2012/0704/pdf/ss1.pdf. Accessed 1 July 2024

Ministry of Finance (2019). 2019 edition of the special accounts guidebook, Part 1: general theory of special accounts, Part 2: Chapter 6 energy measures special account. Ministry of Finance (in Japanese). https://www.mof.go.jp/policy/budget/topics/special_account/fy2019/2019-zentaiban.pdf. Accessed 1 July 2024

MOFA (2021) Overview of the treaty on the non-proliferation of nuclear weapons (NPT), June 11. Ministry of Foreign Affairs (in Japanese). https://www.mofa.go.jp/mofaj/gaiko/kaku/npt/gaiyo.html. Accessed 1 July 2024

MOFA (2023a) Nuclear disarmament and non-proliferation, June 22. Ministry of Foreign Affairs (in Japanese). https://www.mofa.go.jp/mofaj/gaiko/hosho_02.html. Accessed 1 July 2024

MOFA (2023b). Overview of the nuclear suppliers group, July 28, Ministry of Foreign Affairs (in Japanese). https://www.mofa.go.jp/mofaj/gaiko/kaku/nsg/index.html. Accessed 1 July 2024
Morita A (2012) The limitations of administrative reform in our country, accounting audit research, No. 46. Foreword, Board of Audit, September (in Japanese). https://www.jbaudit.go.jp/koryu/study/mag/pdf/j46d01.pdf. Accessed 1 July 2024
Myojin A (2015) Digital service reform in the UK (GOV.UK)—promotion of government digital strategy in unity with residents and private sector, July 21. Aegis Research Institute Inc. (in Japanese). https://www.ogis-ri.co.jp/rad/webmaga/1234958_6728.html. Accessed 1 July 2024
NEA (2011) Nuclear legislation in OECD and NEA countries: France. OECD/NEA. https://www.oecd-nea.org/law/legislation/france.pdf. Accessed 1 July 2024
NEA (2016) Nuclear legislation in OECD and NEA countries: United States (2016 update). OECD/NEA. https://www.oecd-nea.org/jcms/pl_24255/nuclear-legislation-in-oecd-and-nea-countries-united-states-2016-update. Accessed 1 July 2024
NEA (2017) Nuclear legislation in OECD and NEA countries: Japan. OECD/NEA. https://www.oecd-nea.org/jcms/pl_24201/nuclear-legislation-in-oecd-and-nea-countries-japan-2017-update?details=true. Accessed 1 July 2024
NRA (2024a) Organizational chart, nuclear regulation authority (in Japanese and in English). https://www.nra.go.jp/nra/gaiyou/nra_chart.html https://www.nra.go.jp/english/e_nra/outline/index.html. Accessed 1 July 2024
NRA (2024b) Overview of nuclear safety regulation. Nuclear Regulation Authority (in Japanese). https://www.nra.go.jp/NuclearRegulation/jitsuro_anzen_index.html. Accessed 1 July 2024
NRC (2003) Citizen's guide to U.S. nuclear regulatory commission information (NUREG/BR-0010, Rev. 4) U.S. NRC. https://www.nrc.gov/reading-rm/doc-collections/nuregs/brochures/br0010/index.html. Accessed 1 July 2024
NRC (2021a) Information digest, 2021–2022 (NUREG-1350, Volume 33). U.S. NRC. https://www.nrc.gov/docs/ML2130/ML21300A280.pdf
NRC (2021b) Enhancing participation in NRC public meetings, nuclear regulatory commission [NRC–2016–0178]. Federal Register/Vol. 86, No. 52/Friday, March 19/Notices. https://www.federalregister.gov/documents/2021/03/19/2021-05787/enhancing-participation-in-nrc-public-meetings. Accessed 1 July 2024
NRC (2022) Strategic plan (NUREG-1614) U.S. NRC. https://www.nrc.gov/reading-rm/doc-collections/nuregs/staff/sr1614/index.html. Accessed 1 July 2024
Nuclear White Paper (2019) 2019 edition of the atomic energy white paper, Chapter 4: peaceful use and nuclear non-proliferation & nuclear security assurance. Atomic Energy Commission (in Japanese). http://www.aec.go.jp/jicst/NC/about/hakusho/hakusho2020/index_pdf01.htm. Accessed 1 July 2024
Sano T (2022a) Can the treaty on the prohibition of nuclear weapons protect Japan. J Atomic Energy Soc Japan 64(4):50–51 (in Japanese)
Sano T (2022b) Can the treaty on the prohibition of nuclear weapons protect Japan. Shinzansha (in Japanese)
Science Council of Japan (2020) Reference material 2 the establishment of the nuclear regulation authority and the process of formulating new regulatory standards, issues and desired state of nuclear safety regulation, June 30. Science Council of Japan, Committee on Engineering, Subcommittee on Nuclear Safety (in Japanese). https://www.scj.go.jp/ja/info/kohyo/pdf/kohyo-24-t290-6.pdf. Accessed 1 July 2024
WNA (2021) Liability for nuclear damage. World Nuclear Association, March. https://world-nuclear.org/information-library/safety-and-security/safety-of-plants/liability-for-nuclear-damage.aspx. Accessed 1 July 2024

Chapter 5
Conclusions

As a conclusion, the author will express his opinion on nuclear power generation in Japan.

Nuclear power generation has three characteristics: safety, self-sufficiency, and affordability/stability. Efforts are needed to ensure that these characteristics are demonstrated in Japan's nuclear power generation as well as in the world.

Safety was discussed in Chap. 2. According to the report of the United Nations Scientific Committee on the Effects of Atomic Radiation, the number of public deaths due to radiation from nuclear power plant accidents is zero for the Fukushima accident by TEPCO, and eight children died of thyroid cancer at the Chernobyl accident. Nuclear power has supplied a vast amount of electricity worldwide so far. Based on statistics, when comparing the death rate of the public per unit of power generation with other power generation methods, nuclear power has the lowest. *Contrary to public perception, nuclear power is the safest source of energy.*

In the TEPCO Fukushima accident, in addition to psychological instability and damage from rumors, the delay in lifting the evacuation increased disaster-related deaths and posed a challenge to community recovery. An analysis of the disaster-related death data released by the government showed that the risk associated with evacuation was 2.8 times greater than the radiation risk avoided by evacuation. Considering that the radiation risk for the elderly is one-fifth of the average age, it is 14 times greater. Considering that the 2 times difference in risk conversion factor between prompt exposure by atomic bombing and low-dose rate exposure of the accident, the result is 28 times greater. Long-term evacuation was carried out in accordance with the principle of keeping radiation exposure as low as reasonably achievable (ALARA), but this result does not justify the application of ALARA to the evacuation in low-dose area.

Unnecessary abortions were performed in Greece and Italy immediately after the Chernobyl accident, and the birth rate has declined. *At low doses, the side effects of fearing radiation are far greater than the actual risk.* In the event of a reactor accident, *it is better not to apply the ALARA principle to the public in low-dose areas (e.g.,*

Y. Oka, *Nuclear Power and Society*, Nuclear Science and Technology,
https://doi.org/10.1007/978-981-97-8951-1_5

below 100 millisieverts per year). Nuclear power plants, during normal operation, expose the public to radiation far less than natural radiation, so this policy can significantly improve the situation during an accident. *It is the role of the International Commission on Radiological Protection (ICRP) to "not apply ALARA to the public in low-dose areas during nuclear and radiation accidents". The Japanese government should not follow the recommendation of ALARA at nuclear and radiological accidents in low-dose area.*

The limit of radiation exposure is given by the average age dose rate, which does not reflect the difference in radiation risk by age. The radiation risk for the elderly is less than the average age. Allowing the elderly to return voluntarily and engage in decontamination and agricultural resumption activities can prevent both disaster-related deaths due to disuse syndrome (inactive lifestyle diseases) and community collapse. *The countries such as France, South Korea, and China where nuclear power plants are located in agricultural areas should learn from this lesson.* Improving Japan's nuclear disaster prevention guidelines is the role of the Nuclear Regulation Authority.

Not limited to nuclear power generation, but *technology is used because it is beneficial.* The risks associated with the use of technology need to be managed so that they do not outweigh the benefits. This is referred to as the risk-benefit or cost-benefit approach. This approach is mentioned at the beginning of the handbook on nuclear law by the International Atomic Energy Agency. In U.S. nuclear regulation, metrics have been established to quantitatively compare the cost of safety improvements and the amount of radiation exposure that can be avoided in order to consider risk-benefit. In the U.S. Protective Action Guide (PAG), traffic accidents during evacuation are considered as a risk associated with evacuation. Disaster-related deaths were not considered because there was no quantitative data. *Japan should quantitatively incorporate the risk-benefit approach into nuclear safety, regulation, and disaster prevention.*

In Japan, disaster-related deaths have been recognized since the Great Hanshin-Awaji Earthquake in 1995. Data on disaster-related deaths were also released in the Great East Japan Earthquake and the TEPCO accident. Because of this data, *it was possible to quantify the risks associated with evacuation.* In the U.S., evacuations sometimes occur due to hurricanes and wildfires, but the number of disaster-related deaths is not always counted. Most disaster-related deaths are among the elderly.

Japan has a system for certifying the need for care and support. If the relationship between the disaster-related deaths from the TEPCO accident and the level of care needed at the time of the accident is clarified, quantitative data to improve the problem of disaster-related deaths should be obtained. Just as the results of studies on the survivors of the atomic bombings in Hiroshima and Nagasaki became the basis for subsequent health effects of radiation, *research on disaster-related deaths and their level of care due to the TEPCO accident should be useful globally in the future to prevent evacuation from disasters and disaster-related deaths.* The results will be useful not only for nuclear accidents but also for evacuation measures from disasters.

Self-sufficiency and cheap and stable supply were discussed in Chap. 1. Japan's energy self-sufficiency rate is extremely low among developed countries, at about

11% (FY 2020). Japan depends heavily on oil and natural gas for most of its energy. These resources are concentrated in countries with unstable political situations, and prices fluctuate greatly depending on the balance of supply and demand. Since they are imported, they are also affected by exchange rate fluctuations. When prices soar and the yen weakens, import costs increase, and prices rise through energy prices. *Since Japan relies almost entirely on imports for its fossil fuels, it is losing money that should be going to domestic economic activity and employment to foreign countries.*

Since the beginning of the twenty-first century, the prices of natural gas and crude oil have soared worldwide due to increased global demand, such as China's economic development. *Russia's invasion of Ukraine not only caused a surge in prices, but also made us realize once again that Japan's energy security is vulnerable.* After 2022, in Japan, energy prices such as electricity charges have risen significantly due to the effects of a weak yen. *The vulnerability of Japan's power supply, which heavily depends on fossil fuels, has been exposed.* Even more than ten years after the TEPCO accident, Japan still has many nuclear power plants that have not been restarted because safety inspections have not been completed. If electricity charges rise, it will have a significant impact not only on prices but also on social services. There is a risk of a decline in international competitiveness and job losses.

Nuclear power generation is a quasi-domestic energy source that can be self-sufficient in Japan. It can generate power for several years even if the import of nuclear fuel materials is interrupted. Nuclear fuel is small in volume and easy to stockpile. Furthermore, uranium resources are abundant in developed countries such as Australia and Canada, so there is little concern about stable supply. In France, 70% of the electricity is generated by nuclear power. Electricity charges in France are cheap. The cost of nuclear power generation (power production cost) in the United States is cheap, second only to hydroelectric power. Natural gas prices fluctuate greatly. Even in 2022, prices soared, but in Japan, the electricity charges of power companies with operating nuclear power plants are cheaper than those of power companies without operating nuclear plants.

A nuclear power plant is like a hydroelectric power plant. It has no lifespan. Hydroelectric power plants require a large amount of money for construction, but the operating and maintenance cost is small. Nuclear power plants also have high construction costs, but the proportion of fuel costs in power generation costs is small. After the construction cost is amortized, power can be generated cheaply. This is demonstrated by the fact that the cost of nuclear power generation (power production cost) in the United States is cheap. Both hydroelectric and nuclear power plants have no lifespan because they replace aging equipment such as power turbines after long-term operation. In the United States, some nuclear power plants have obtained operating permits for 80 years from the start of operation by applying for operating permit renewal every 20 years.

The power generation cost (levelized cost of electricity), which is calculated by setting the payback period for power plant construction investment and the interest rate on borrowed money, does not represent the actual power generation cost for hydroelectric and nuclear power because the payback period for construction investment and the actual usage period are different. *Nuclear power plants, like hydroelectric power plants, have the advantage of being more beneficial the longer they are used.* Moreover, the power output of a single unit is enormous, for example, the output of Egypt's Aswan High Dam (2.1 million kilowatts) is smaller than that of two large nuclear power plants (about 1.35 million kilowatts per unit). Nuclear power plants have small sites and do not submerge vast upstream areas. In terms of site efficiency, nuclear power plants are more environmentally friendly than solar or wind power plants. Many of Japan's nuclear power plants are located in agricultural areas, creating local employment. This includes not only direct employment but also construction companies, building companies, and facilities such as accommodation for workers and shops. Farmers can work these jobs part-time, eliminating the need to go out of their town in winter for earning money for living. *Nuclear power plants contribute to environmental protection and regional revitalization.*

In the US and UK, economic liberalization and power liberalization have progressed since the 1980s. In a liberalized environment, power plants with low investment risk, i.e., those with small investment amounts and short investment recovery periods, are preferred. *In the United States, new construction of nuclear power plants did not progress after the 1980s. The reason was not concerns about safety, but the avoidance of the huge investment required to build nuclear power plants.* Power plants such as natural gas, solar, and wind power have much smaller investment amounts per unit and shorter periods required for investment recovery than hydroelectric or nuclear power plants. In other words, the investment risk is small. Therefore, a large number of natural gas power plants were built in the United States from the 1990s. Investment in renewable energy power plants progressed in the 2000s. However, as shown in Chap. 1 with the example of the United States, the power production cost of natural gas power plants is higher than that of hydroelectric or nuclear power. *In a liberalized economic environment, power generation methods with high costs but low investment risks were preferred. Under economic liberalization, the perspective of providing cheap electricity to the public stably and for a long period of time is lost.* Electricity rates in deregulated states in the United States are higher than in non-deregulated states. In the United States, due to power deregulation, investment in transmission lines was not made. As a result, large-scale power outages occurred in the United States, and the progress of power deregulation stopped. The investment environment for large-scale power sources varies by country. Power deregulation is also progressing in Japan, but *Japan should adopt an energy policy that does not lose sight of providing cheap electricity stably.*

In Japan, the liberalization of electricity is progressing slower than in the United States. Furthermore, since the Tokyo Electric Power Company (TEPCO) accident, the introduction of renewable energy has been promoted at the expense of the public, with costs added to electricity bills through a fixed price purchase system. Along with the suppression of this burden, the stable supply of electricity (prevention of power outages) is a major issue that needs to be addressed with the mass introduction of renewable energy. In a liberalized electricity environment, investments in large power plants and transmission lines will cease, so *in countries like Japan where the mass introduction of renewable energy and electricity liberalization are progressing in parallel, the stable supply of electricity becomes a more serious issue, requiring careful policy.* Electricity is also a public good, so its supply cannot be left solely to the free market. *The goal of providing cheap and stable electricity is important.* This does not mean that Japanese power companies, protected by electricity business regulations, do not need to make management efforts to reduce generation costs.

On the other hand, from the perspective of power companies, if a private company is running the power business, the predictability of the business, that is, reducing the risk associated with the business, is extremely important for its management. Currently, for Japanese power companies such as TEPCO, Kansai Electric Power Company, etc., there are *three types of business risks related to nuclear power generation: regulatory risk, judicial risk, and electricity liberalization risk.* Regulatory risk is a risk related to the safety regulations of power generation reactors, such as the long time required for safety inspections of nuclear power plants due to new regulatory standards after the TEPCO Fukushima accident. Judicial risk is a business risk related to lawsuits to stop operation. If a nuclear power plant that has been built with a huge investment cannot operate, it becomes a major business risk. The management risk associated with electricity liberalization is a risk related to the predictability of electricity business regulations, such as whether investments in power generation facilities and transmission facilities can be stably recovered over a long period. In Japan today, all of these risks are significant. In Japan, in order to use nuclear power stably, it is required to reduce these business risks.

On the other hand, in Japan, because power companies had regional monopolies (each power company's power supply area was determined), competition did not work, and moreover, electricity rates were determined by the cost-plus method, which accumulates costs, there was no mechanism for power companies to aim for cheap electricity rates. As a result, Japan's electricity rates are about twice as high as those in South Korea. Therefore, in addition to reducing business risks, a system design that also meets the perspective of providing cheap electricity stably is required by the responsible ministries, and efforts to establish and improve cost awareness are required by power companies.

Electricity cannot be stored in large quantities. Stabilization of the power system is necessary to prevent power outages. This requires four points: securing power supply capacity, securing power quantity, securing frequency adjustment capability, and securing inertial force and synchronization force. Securing power supply capacity means having a power source that exceeds the maximum power demand throughout the year. Securing power quantity refers to securing generation capacity that can respond to short-term fluctuations in power demand. Securing frequency adjustment capability refers to the ability to respond to rapid changes in the amount of power flowing through the power system. Securing inertial force and synchronization force refers to suppressing fluctuations in AC frequency, a role fulfilled by turbine generators used in thermal power generation, nuclear power generation, and hydroelectric power generation. Renewable energy is left to nature for power generation and cannot generate power according to power demand, nor does it have frequency adjustment capability or inertial force and synchronization force. Nuclear power generation is a generation method that has all of these characteristics. Renewable energy is being introduced in Europe, but European countries are connected by power grids and can share power. In Germany and Italy, they were heavily dependent on imports of natural gas from Russia and dealt with the output fluctuation problem of renewable energy through natural gas thermal power generation. Japan cannot import power from overseas. The transmission capacity between Hokkaido and Kyushu, which have many suitable locations for renewable energy, and Honshu is also small. Storage by secondary batteries has not yet become widespread in terms of capacity or cost. Even after widespread use, facility renewal will be necessary due to aging. *Nuclear power generation is an indispensable power source for Japan in terms of long-term stable power supply and national security.*

In Japan, a one-time 20-year extension of the operating period for nuclear power plants has been decided by law after the TEPCO Fukushima accident. Because the cost of construction to comply with the new regulatory standards reflecting the lessons of the TEPCO Fukushima accident cannot be recovered within the 20-year extended operating period, including nuclear power plants that power companies have abolished, *there are about 20 nuclear power plants that have transitioned to abolition measures after the TEPCO Fukushima accident*. If they had been used for a long time, they could have supplied cheap power, which is a *national loss. Japan should reconsider its one-time 20-year operating permit renewal system.* This is not for the power companies, but to prevent the loss of the people. How many people want to pay high electricity bills? There are also people who cannot pay. Changing the law is the role of the Nuclear Regulation Authority, which is in charge of this law, and the Diet.

More than ten years have passed since the TEPCO Fukushima accident, but there are still few nuclear power plants that have resumed operation after completing safety inspections under the new regulatory standards and renovation work. After the accident, in other countries, stress tests (examinations by regulatory authorities to see if power plants can withstand the lessons learned from the accident) were conducted, and necessary measures were taken in sequence. However, no country has stopped operating nuclear power plants like Japan and carried out response work for the TEPCO Fukushima accident. Even in the Chernobyl accident, the reactors adjacent to the accident reactor were in operation. No major tsunamis occurred in the 12 years after the TEPCO accident. *Even if the new regulatory standards were met while in operation, the risk would not have changed.* In Japan, all nuclear power plants were shut down and the power supply to replace them was provided by thermal power generation. This resulted in the loss of foreign currency due to fuel imports, worsening domestic employment and the economy. This could have been prevented if the plants had not been shut down. In the future, *Japan should first prioritize restarting the power plants that are currently shut down.* Furthermore, in the future, to improve energy self-sufficiency, reduce electricity rates, and ensure stable power supply, *the proportion of nuclear power generation should be increased*. Reducing the proportion of nuclear power generation is a suicidal act for Japan.

In recent years, there has been a worldwide heatwave. Japan also experienced extreme heat. No one seems to point it out, but if a large-scale power outage had occurred in the metropolitan Tokyo area and air conditioning could not be used, many elderly people would have died. Energy issues are related to the survival of the nation and the lives of its citizens. It is a major issue that will determine Japan's future, and this has not changed from the past. Japan should not take lightly the fact that it is a country with few energy resources. It affects not only national security but also industry, employment, life, and health. In order for Japan to remain a prosperous country, it is necessary not to make mistakes in dealing with energy and environmental issues.

Nuclear power generation is inexpensive, stable, and safe, contributes to local employment, requires a small site for power plants, and is an environmentally friendly power source. Japan's primary energy self-sufficiency rate is about 11%, the lowest in the world. There is a twofold difference from South Korea, which is the next lowest. France supplies 75% of its electricity through nuclear power generation, and its primary energy self-sufficiency rate is 55%. Not only stable, inexpensive, and safe power supply, but also the need to address global environmental issues, so in the long term, *Japan should aim to supply 75% of its electricity through nuclear power generation, like France.* We hope that renewable energy will do its best and that prices will come down. However, batteries are necessary for their stable supply.

I would also like to mention the radioactive waste, as there are misunderstandings. *In Japan, radioactive waste seems to be considered a major unresolved issue, but this is not the case worldwide.* As mentioned in Chap. 3, treatment and disposal are progressing in the West. For low-level radioactive waste, measures are being taken, including the relationship with general waste and reuse, and treatment and disposal are progressing. *Radioactive waste is a type of industrial waste. Its treatment and disposal are activities for a circular society.*

Even for the geological disposal of high-level radioactive waste, the world is making progress, albeit with difficulty. It's not just Finland. France is also making steady progress. France does not call it final disposal. It is a disposal that might become final. That's because the plan takes 100 years from the start of disposal to the closure of the disposal site. The current generation alone cannot decide what is final. In Sweden, a geological disposal site plan was approved in 2022. The United States is struggling because the government decided on the disposal site from the top down, facing opposition from the local community. Building trust with the local community takes time. It is necessary to gain a careful understanding. There is no need to directly seek the consensus of the entire nation on the disposal site for radioactive waste. *The general rule, regardless of nuclear power issues, is to proceed within a parliamentary democracy, seeking the understanding of the local community.*

The siting of general waste disposal sites, as the term "Suginami Garbage War" suggests, used to be a social issue in Japan, but is now established in society. Did they seek the consensus of all residents for the siting of a waste incineration plant? There are many respectable heads of local governments, and there should be potential sites where the suitability of geological disposal sites can be examined. Geological disposal sites require digging holes and examining the geology underground to narrow down potential sites.

The following is mainly a message for those involved in nuclear power in Japan. This is because these are primarily the responsibilities of nuclear-related organizations and their members.

There are issues at the interface between nuclear power use in Japan and society. Several issues can be raised. First of all, *it is the clarification and execution of the responsibilities of each nuclear-related organization.* In other words, in the use of nuclear power in Japan, it would be better for each organization, such as government agencies, research and development institutions, local governments, power companies, and nuclear reactor manufacturers, to be aware of their roles and responsibilities and fulfill their respective roles. The responsibility may be vague due to collectivism.

Collectivism is the act of unconsciously following the opinions of loud groups without verifying the basis for oneself. Industry magazine articles reflect the intentions of the industry sponsors. News in academic journals often collects the intentions of the source who wants to secure budgets or orders. An academic society is a group of researchers, and researchers want to secure research funding. In Japan, there is a few investigation or presentation of the history and past facts of nuclear power use in the West. There are not necessarily many experts who search in English, or search and select search results, and have a broad cultivation that can judge. Nuclear engineers and researchers do not make it their job to investigate and think about the problems at the interface between nuclear power generation and society. The author wonders if they are not swallowing what was written based on hearsay and what they heard.

The use of nuclear power in Japan is ambiguous in terms of individual and organizational responsibility, and audits are not functioning, so collectivism does not notice and correct what is wrong. Even if there are rare nuclear power experts who notice, they cannot correct it, so they give up and leave the nuclear power industry. The author knows an excellent nuclear expert who disappeared from the nuclear power industry around 2000, saying, "Nuclear power is crazy". There are experts who are trying to understand and improve, but the wall of Japanese collectivism is thick. The United States and the UK clearly define individual and organizational responsibilities and are not collectivist, so such a situation is not left unattended. Japan's nuclear power generation was far ahead of Korea around 1990. Now, including exports, it has been overtaken. China's nuclear power generation is also steadily advancing. The TEPCO accident caused trouble to the world. Those involved in nuclear power in Japan should reflect not only on the accident itself but also on such a situation. One of the reasons that the author decided to write this book is that even if the author expressed his opinion to the parties concerned as the chairman of Japan Atomic Energy Commission, there are limits to oral explanations and it was not understood. It is difficult to understand objectively without understanding the matters and knowledge behind the opinions. This book lists references and describes them.

Although it is a story about Japan as a whole, in Japan, committees of ministries and agencies gather university professors and others to examine the validity of policies, but the members of the committee do not necessarily survey and understand the policies to be examined in an overview. Since science and engineering university professors do not make policy examination their job, they usually do not investigate and think deeply about it. As a result, it is easy to discuss whether they agree or disagree with the industry's opinion.

For improvement, accountability is necessary for organizations operated with public funds in Japan. Accountability is the explanation of the role of each organization and the work in writing and disclosure, and the responsibility for the results of the work. It is a mistake to translate accountability as an explanation responsibility. It's a problem if it ends when you explain it in words.

To improve accountability, it is first necessary for Japanese administrative agencies to clearly document and disclose to the public the issues, directions, and results of their policies. Organizations responsible for nuclear power utilization with public funds, such as research and development institutions and foundations, should document and disclose their achievements and work results. Furthermore, it involves enabling audits of each administrative agency.

As an example of the accountability of U.S. administrative agencies, the author introduced the accountability of the Nuclear Regulatory Commission (NRC) in Chap. 4. The NRC explains its policy and content in its Strategic Plan. It explains various information and thoughts in the Information Digest. Furthermore, it has created a Citizen's Guide to help citizens find the information they need. The Strategic Plan is also used to evaluate NRC activities. The accountability of the NRC ensures the trust of the American people in the NRC. This has earned the trust of the American people in the safety of nuclear power use. These documents also play a role in making NRC activities known to NRC staff. It is mentioned that building trust takes a long time, but losing trust can happen in an instant.

Not limited to the nuclear power sector, Japan's audits are not functioning. No matter how excellent a group or organization is, it will deteriorate if audits do not function. In the West, administrative accountability (explanation of work in documents and responsibility for the results), audits by the Board of Audit, and monitoring and correction by the parliament are carried out. Audits of administrative agencies are the responsibility of the Board of Audit, and corrective measures are the responsibility of the Federal Parliament. The facts of a crime are revealed by police investigation. The Board of Audit does not comment on good or bad, but reveals facts based on its investigative authority and points out problems. Facts and results are not always self-evident, and are revealed by investigation by the Board of Audit, which has investigative authority. These should be evaluated and reflected by the parliament and the government. These need to be documented and implemented. The Japanese Board of Audit, looking at the Board of Audit Act, is not set up to play a role like the U.S. GAO or the UK's NAO. The evaluation of the results of the work of Japanese ministries and agencies is self-evaluation by the ministries and agencies themselves. This may not be expected to bring about essential improvements.

For a robust nuclear policy, as introduced in Chap. 4, what the U.S. government is doing, not limited to the nuclear field, can be a reference. That is, in the United States,

1. There is information about decisions and actions taken by responsible individuals or organizations (role of the agency).

2. There are people or organizations who receive or discover that information and examine, investigate, and report on it (role of the Board of Audit: not to comment on the good or bad of the policy. The role of collecting, compiling, and disclosing information based on investigative authority).
3. There are actions to correct defects based on such information (role of the Federal Parliament, Budget Committee).

These are being implemented through robust policies. These three points are the very issues of Japan's fiscal reform, but they are also what the author hopes to achieve in the use of nuclear power.

There are various other mechanisms for a general audit of an organization. These include auditors, external directors, and external evaluation committees. If the side being evaluated decides on these committee members and evaluation indicators, the audit will not function. Whether or not to allocate a budget is also part of the audit mechanism. It should be avoided to follow only the reports and wishes compiled by the side receiving the budget.

The nuclear power industry should not depend too much on the country or group, and should not make responsibilities vague through meetings. It is required that each organization fulfills its responsibility with a common understanding that nuclear power is for the people. "Nuclear power for the people", that is, each nuclear-related organization plays its role with a common goal of supplying cheap, stable, and safe electricity from nuclear power generation, and new developments can be made. This goal should aggregate the goals of each nuclear-related organization. "For the people" is not unique to Japan, but is a common goal in any country worldwide.

The use of nuclear power in Japan needs to break away from the precarious state of hearsay-centered collectivism. There is a term called "nuclear village". The definition of a nuclear village may vary, but here it is defined as a "group of nuclear-related people who are hearsay-centered and collectivist". Hearsay includes things that have been heard and written. Collectivism is the uncritical following and dependence on what organizations or people who are thought to have authority, such as countries and industries, have said. The reason for defining it this way is that when writing this book and searching on the internet, referring to reports from government agencies in Europe and America, the author noticed that there are many things that are different from the information circulating in Japan's nuclear power industry, and many things that are not known. The basis is not clear in hearsay. There is biased information circulating in Japan's nuclear power industry. There are things that have not been transmitted in the precedents of Europe and America.

Therefore, *it is good to search and think in English. There are many precedents in Western countries for the use of nuclear power in Japan. We can learn from their failures and experiences.* Based on these historical facts and evidence, it is good to create strategies. We live in an era where anything can be searched on the internet. Western governments and international organizations publish policy documents and their results. The Board of Audit also publishes many reports. It feels like a different era from the past when you had to go to the library to get information. You can search for information from around the world even at home. What you choose from what is displayed when you search on the internet requires a sense of problem, ability, and experience, just like research, but generally, policy documents and reports from Western agencies, reports from the Board of Audit, and reports submitted by each government to international organizations according to treaties can be referenced. Established papers can also be helpful. These are not created by hearsay.

Based on these pieces of information, in light of Japan's situation, it is good to think logically and inductively based on historical facts and evidence. The problem of contact (interface) with society is that there is no norm of mother nature, so it is impossible to think deductively based on natural laws (equations of physics, etc.). If you can properly recognize the past failures and lessons at home and abroad, and the surrounding environment that caused them, you should be able to prevent the same failures. In the United States, reports summarizing project experiences are created and published. The U.S. Nuclear Regulatory Commission also plays a role in collecting and publishing important documents related to nuclear power use.

Sometimes, reports summarizing the hopes of a group of researchers are created by scientific academies or societies. If the administration budgets based on these hopes, the results need to be clarified. If it is unclear, the Board of Audit, which has the power to investigate, will investigate and create a public report stating the facts revealed from the results of the investigation, and the parliament needs to evaluate the policy and reflect improvements in the budget and policy.

There is no cost-benefit (risk-benefit) approach to the safety, regulation, and disaster prevention of nuclear power in Japan. This is an issue that needs improvement. The concept of cost-benefit is the idea that a balance must be considered between the cost required for measures and the benefits brought about by those measures. It is absurd to enhance safety while ignoring costs. Not only the burden on the public, but also the achievement of the goal of providing cheap electricity to the public through nuclear power generation requires a cost-benefit approach. In Japan, where collectivism is strong, some people may think that the cost should be borne by the government or power companies, but that burden becomes a burden on the public in the form of taxes and electricity charges. In the rules of the U.S. Nuclear Regulatory Commission, the cost (in dollars) required for regulatory improvement measures is used as a judgment indicator, defined in relation to the amount of radiation reduction improved by those measures. If you want to use probabilistic risk assessment for regulation and safety improvement, you need an indicator to judge cost-benefit.

The U.S. protective action guides states that the risk associated with evacuation must not exceed the risk that can be avoided by evacuation. In this case, cost does not necessarily mean the amount of money. The health effects (benefits) avoided by the dose due to evacuation must not exceed the health effects (costs) caused by evacuation, the increase in disaster-related deaths due to evacuation, and the psychological and mental health effects. The reason why nuclear power plants are being shut down and new regulatory standards are being implemented in Japan is also due to the lack of a cost-benefit approach. The Japanese nuclear industry has always seemed to avoid talking about costs, but as a result, power companies are being forced to bear a heavy burden due to the shutdown of nuclear power plants due to compliance with new regulatory standards. This burden is reflected in the electricity bill, which is a big minus for the public.

Improving the understanding of Japanese nuclear-related personnel about risk communication, risk perception and psychology is also an issue at the interface between nuclear use and society. Despite the fact that nuclear power generation is the safest method of power generation in statistical data, why have many people come to think of it as the most dangerous method of power generation? The information about safety was supposed to be on the side of the nuclear-related personnel, but isn't the lack of understanding of nuclear engineers about risk perception and communication the cause? Perhaps the nuclear engineers thought that if the public understood the safety they understood, and as a result of their efforts of talking about safety, nuclear power came to be considered dangerous. The understanding of this problem by nuclear-related personnel is insufficient, and the approach may have been wrong.

First of all, is the public's risk perception rational? Research results on risk perception show that the public's risk perception is irrational, but can it be changed by risk communication with the public just because it is irrational? The risk communication mentioned here is not a stakeholder dialogue conducted with a small number of people in local areas, but communication (dissemination of safety information, or in other words, risk information) to large groups such as the entire nation or prefecture.

Risk communication is not scientific education for the public. In the UK, the Royal Society has been promoting science communication since the 1980s, but a 2006 report expressed concern that many scientists are participating in science communication activities with the intention of educating the public with their knowledge, not for genuine dialog. Nuclear experts forget how many years it took them to understand their expertise, such as the health effects of radiation. Not just experts, human understanding progresses little by little through dialogue. This is why dialog at seminars are necessary in university education.

Risk communication destinies to fail if it is not two-way. This is stated in the conclusion of the famous paper by a renowned risk psychologist, Professor Paul Slovic. Dialogue cannot be effectively conducted unless it is with a small number of people. Having a dialogue with the entire nation or prefecture would require an extremely large number of meetings, which is practically impossible. Furthermore, *the purpose of communication is not merely understanding risk, but building trust.* Psychologically, talking about safety is talking about danger, so the more you talk about safety, the more the image of danger is impressed. There is an introductory book on risk psychology that first features the phrase "I'm aware of it, so I care about it" along with an illustration.

Negative information is impressed three times stronger. Information about safety is psychologically negative information. The reason why nuclear power came to be considered the most dangerous might be because nuclear power specialists tried to make the public understand the safety of nuclear power and radiation without understanding these points. The media may have spread the information, but shouldn't we first consider that the root of the problem lies with nuclear power specialists?

The creation and disclosure of safety information and other risk communication is the role of regulatory agencies. Risk communication at the prefectural and national levels is also the role of regulatory agencies. There is no need for the nuclear power promotion side to particularly talk about safety to the public. *The U.S. nuclear industry association, Nuclear Energy Institute does not communicate to the public about the safety and risks of nuclear power.* The U.S. Nuclear Regulatory Commission also does not conduct risk communication activities for the public (oral explanations of safety to the entire public). After all, ensuring safety is the responsibility of the nuclear power plant operators, so it is logically contradictory for the Nuclear Regulatory Commission to say on their behalf, "Nuclear power generation is safe". Staff of the U.S. Nuclear Regulatory Commission should be explaining risks at public hearings and local meetings. This is a dialogue with a small group, and a manual for risk communication when they interact with participants has been created. It is NUREG/BR-0308, entitled "Effective Risk Communication". The audience for the nuclear industry to explain safety and risks is not the public, but the Nuclear Regulatory Commission.

Communication through dialogue does not have an effect unless it is with a small group. A facilitator is necessary to promote dialogue. In the UK, a group of environmental consultants are cooperating in dialogue meetings on geological disposal. This should also be a reference in Japan. In Fukushima Prefecture, after the TEPCO Fukushima accident, various dialogue activities were conducted, including voluntary meetings and meetings held by public institutions, on the health effects of radiation. This is a good example of stakeholder dialogue.

In order to improve the public's understanding of nuclear power, it is necessary to enhance documented information, not limited to information on risks and safety. Information should be created and made public, and it should be made available for the public to find. After the TEPCO Fukushima accident, information on radiation exposure risks has been enhanced. The Ministry of the Environment has created and published unified explanatory materials on the health effects of radiation. Fukushima Prefecture has created a portal site to aid in the understanding of the prefectural, national and international citizens. The Agency for Natural Resources and Energy is providing energy-related information for the public as "Special contents". These are also good examples. From the perspective of improving transparency to the public, we hope that not only policy information but also various basis information will be created and provided. The United States is already in this state.

The safety that nuclear technology experts talk about is a technology for preventing accidents and reducing the release of radioactive substances. This is *"safety technology", not "safety" that directly relates to the public.* In technical terms, there is a concept of "defense in depth" for ensuring reactor safety, but up to the fourth layer, it is a concept for preventing accidents and suppressing the release of radioactive substances, which is about "safety technology". The public is affected after radioactive substances are released into the environment. This is the "safety" of the fifth layer of defense in depth.

"Safety technology" and "safety" are different. "Safety technology" is something that experts should strive for within the industry, not something to explain to the public. Is there any other industry that explains its safety technology to the public on a large scale? For example, in the field of civil engineering, do they communicate the

safety of bridges to the public? Isn't "safety technology" the responsibility within the industry? Nuclear power experts who think that "safety technology" should be explained to the public may not understand this point. The fifth layer is not the job of nuclear technicians. For the public, disaster prevention and emergency response are important. The responsibility to make the fifth layer function substantially lies with nuclear-related organizations, disaster prevention organizations, and emergency response organizations. Progress of accident and amount of radioactivity release need to be informed from the fourth player people of the electric power company to the fifth layer people of the government. It did not work at TEPCO Fukushima accident.

Chapter 2 presented a diagram of the public's risk perception factors and the distribution of various risk perceptions. The public's risk perception is characterized by "unknown factors" and "fear factors". As this suggests, the public's risk perception is subjective. The safety perceived by the public can be called "subjective safety". In contrast, the safety perceived by engineers is "objective safety". Objective safety is safety based on theories and analytical results of natural science and engineering derived from natural laws.

Both "objective safety" and "subjective safety" are correct. What the author wants to point out here is that *the two logically never intersect*. However, research and activities to somehow connect the two have been conducted in the nuclear field in Japan. The author believes that these activities will not be able to achieve the goal of getting the public to understand safety. Rather, it is spreading the image of danger to the public. This may be the reason why nuclear power generation is considered extremely dangerous. It is a waste of effort to try to intersect things that do not logically intersect. Japan may be the only country that is still conducting oral risk communication with the general public without realizing this. The nuclear industry in the West does not communicate safety information orally to the public. EDF Energy in UK stopped safety TV commercials in 2015. Local stakeholder dialogue is different from oral risk communication with the public. *Technology, not limited to nuclear power, is used based on objective safety*; this is the responsibility of engineers.

The words "safety" and "peace of mind" are sometimes used in Japan. Safety corresponds to objective safety, and peace of mind corresponds to subjective safety. Peace of mind is a matter of individual hearts and is not something given by the country or others. Technology, not limited to nuclear power generation, is used based on objective safety.

Regarding research and development, the author touched on it a little in Chap. 1. In the sister book "Nuclear Power Reactor Development", the author explains them in detail. *Western precedents for research and development of new reactors, etc., are being ignored.* Ignoring past experiences and lessons will lead to repeating the same mistakes. It is necessary to first recognize the lessons and issues obtained and think about measures. This is not to say that research is unnecessary.

Japan's use of nuclear power has a weak knowledge base. A knowledge base refers to the collected and systematized knowledge, computational codes, and the personnel who have acquired them, which can respond to various situations. Its accumulation is primarily the role of nuclear research and development institutions.

However, in Japan, there are hardly any systematically compiled reports. Safety-related computational codes that collect and utilize acquired knowledge are also not systematically maintained. Research papers are being produced on individual themes in narrow fields, but very few works are being done to compile the results obtained organizationally. *If the acquired knowledge is not collected and compiled, it will be lost. It is necessary to systematize knowledge so that it can be used in various cases in the future. The foundation of the information age is various knowledge. Without this, Japan will lag behind in international competition.*

The confusion in the TEPCO accident was also due to the fact that in Japan, knowledge of core meltdown accidents was not compiled and shared. There may have been some knowledge about severe accident countermeasure technology, but at the time of the accident, knowledge and experience of nuclear disaster prevention and emergency response are necessary. Knowledge is not rules or procedures, but the knowledge and experience necessary to imagine various situations and think about countermeasures. It is not the job of government officials to create these, but the job of research and development institutions. Not only in the case of core meltdown accidents, but Japan has not systematically compiled knowledge for the use of nuclear power. There are very few people who have it. If you think you can rely on someone in collectivism, you will repeat failures. Failure is not just about things breaking.

To repeat, those involved in nuclear power in Japan need to *break away from the hearsay-centered collectivist nuclear village, clarify the responsibilities of each organization, and strive to achieve their roles.* This should also lead to the resolution of issues between the use of nuclear power and society in Japan. We need to be aware of the captive structure and collectivism. To do this, it might be good to *be aware of the origin of providing cheap and stable electricity to the people and strive for it.*

To repeat, this is not unique to the nuclear field, but a story about Japan as a whole, but Japan does not have a mechanism to clarify the accountability of ministries. The reason why this is not the case may require recognition that the dependence of Japanese companies and people on the country is behind it. When trying to promote a certain policy in Japan, it tends to be a story that a budget is needed, but isn't it first necessary to arrange a system for that, including an audit mechanism? Administrative officials are excellent people with a strong sense of responsibility. However, not only ministries but also organizations will deteriorate if they do not operate an audit mechanism.

The lack of a system for publishing investigation reports on the results of the administration by the Board of Audit may be the cause of Japan's long-term stagnation. The Board of Audit of Japan investigates the appropriateness of budget execution in terms of accounting, but it is not given the authority or role to clarify the results of the administration. This function exists in the audit institutions of Western governments.

From the end of 2023 to 2024, the management of political funds by political groups of political parties became an issue in Japan. This is an example of the result of the lack of administrative accountability mentioned above. The background to the industry providing political funds in the first place is the expectation of support from the country in terms of policy and budget for the industry. The fact that many of Japan's industries have been dependent on national policies and budgets since the high-growth period may be the cause of Japan's long-term stagnation. Isn't this structure that makes this institutionally possible still present in the operation of Japan's legal system? The executive branch of national policies and budgets is the administrative agency. The industry's wishes are conveyed to the administrative agency through politicians or through industry groups. Not all wishes are realized, but responses are made in terms of budget and policy. The result is that the public has no way of knowing unless a government agency with investigative authority (the Board of Audit) publishes a report. This function does not exist in Japan's legal system. Unless it is established as a system, the administrative agency will not implement it.

The results of Japan's administration are self-evaluations by the administrative agencies. Furthermore, there is an Administrative Evaluation Bureau in the Ministry of Internal Affairs and Communications, which conducts topical evaluations. *Western audit institutions do not evaluate, but play a role in clarifying the facts.* It is the role of the parliament to reflect evaluations in the budget and personnel. Evaluation requires values. Since diverse values are robust and good, they may not be evaluated. It is first necessary to investigate and disclose the facts.

In Japan's administrative agencies, deliberations are held in councils, etc., gathering experts and stakeholders. The members express their opinions orally, and these are summarized by the secretariat and the chairman of the council. It is rare for there to be a document that is the basis for each opinion in the committee. If it is not a document, the basis is not clear. In the West, there are many disclosed documents and reports. In Western ministries, each policy handled by the head of the administrative agency is explained and disclosed in a document.

In Japan, the term "responsibility of explanation" is often used, but accountability refers to "explaining policies by administrative documents and publishing investigation reports on the results of the administration by the Board of Audit", not "responsibility of explanation". In Japan, the practice of documenting and disclosing policies handled by the administrative agency itself is not carried out. If you just explain and finish, you can't implement robust policies.

There is a term "White Paper" in Japan, but in the West, a "White Paper" is a paper refers to a document explaining policies. However, in Japan, a white paper is not a document explaining policies, but an archive of the activities of each ministry and agency for the year. When the author was the chairman of the Atomic Energy Commission, I was asked several times to explain the Atomic Energy White Paper at meetings with the West, but the author had to comment that the Japanese white paper is not a policy document, but an archive of the year. It was embarrassing.

As seen in the examples of "the Board of Audit not revealing the results of administration", "oral aggregation of opinions", and "responsibility of explanation", *in Japan, the structure that allows the industrial world to depend on the country has been preserved.* Japanese companies that have achieved international expansion may not need to depend so much on Japanese administrative agencies, but correcting defects in the legal system is outside the scope of these companies, so they are not corrected. The author sometimes feels that the dependence of the Japanese people on the country has allowed these to be preserved, but are we really *entrusting the future of Japan with a future to the younger generation?* First of all, the author would like them to notice the above points. The author is not sure whether legal scholars and economists have pointed out this problem in the past and whether efforts have been made to respond and improve.

The Ministry of Internal Affairs and Communications' homepage for central government reform contains a collection of materials on central government reform (reports, laws, etc.). Central government reform was considered and discussed in Japan around the year 2000. The materials from that time are stored in the National Diet Library. This central government reform was carried out with the aim of improving the efficiency of ministry and agency operations (reorganization of central government agencies), and accountability may not have been recognized as the main issue. Subsequent reports include an investigation report by an auditing firm (Ernst & Young ShinNihon LLC 2017). The references of this chapter include a book by an expert from the Board of Audit (Higashi 2011). The U.S. GAO changed its name from the Government Accounting Office to the Government Accountability Office in 2004, shifting its focus from auditing accounting fraud to auditing administration. If we investigate the reports that summarized the discussions and discussions in the United States at that time, we may get useful suggestions for Japan.

Accountability will not be implemented unless it is established as a legal system. Isn't there a fundamental flaw in Japan's parliamentary democracy in this regard?

References

Ernst & Young ShinNihon LLC (2017) Research on check institutions for government activities in the United States and the United Kingdom, March (in Japanese). https://www.soumu.go.jp/main_content/000485250.pdf

Higashi N (2011) Audit system of the board of audit. Chuo Keizai Sha (in Japanese)

Afterword

This book can be used as a reference for historical and factual verification, but it also includes parts written based on the author's experiences and conclusions. It is not just a book that collects facts and history about the intersection of nuclear power generation and society, nor is it a book that only presents arguments and opinions, but it describes both. The author believes that such books are not common, but the author thought this format was good as a way of describing problems in an academic and inductive way.

In this book, the author tried to write in an inductive and logical way. The method of consideration is the same as the method of summarizing papers in scientific and engineering research. The author intended to gather facts recorded in historical facts, reports of government agencies in various countries, audit bureaus, international organizations, etc., consider them inductively, and summarize them logically. There is many hearsay-based information, secondary information, and claims about social issues. The author did not adopt these pieces of information as a basis.

In the field of science and engineering, there are natural laws that can be referred to as norms, and it is possible to consider deductively based on them, but when considering problems at the intersection with society, this is impossible, so the author did not consider deductively from a certain proposition. The author was careful not to follow a deductive way of thinking. The use of nuclear power is a premise, but this book is not a description for justifying it. In this book, referring to the West, the author pointed out many issues about the use of nuclear power in Japan. Many people involved in nuclear power may feel that the description is negative, but isn't it first necessary to recognize the problem for future improvements?

Most of the experts engaged in education, research and development, and industry for the use of nuclear power are experts in nuclear technology, and many of them may not have spent time thinking about problems at the intersection with society as part of their work. The author was the same until he was appointed as the chairman of Japan Atomic Energy Commission. These problems can lead to biased understanding and judgment if you do not take the time to consider them based on evidence. Unlike

Y. Oka, *Nuclear Power and Society*, Nuclear Science and Technology,
https://doi.org/10.1007/978-981-97-8951-1

the West, Japan has weak administrative accountability, and there is no organization to investigate the results of policies and administration and perform audit functions. Therefore, there may be cases where biased understanding and results are not corrected.

Facts are not always clear. This is also the case with the use of nuclear power. Industry magazines have sponsors, so their information may be biased. It is often the case that information from the promotion side is conveyed through industry magazines and newsletters. When a report is created by gathering committee members from the promotion side, even the reports of the science academy may be biased. In the West, audit bureaus independently investigate administrative accountability (the results of administration and responsibility for them) and play a role in clarifying facts. Their reports are useful for fact-checking.

The reports from Western accountability inspection agencies and U.S. Congressional hearings have been very informative. What the author noticed is that Japan's accounting inspection agency is only responsible for investigating accounting fraud, and unlike Western accountability inspection agencies, it is not given the authority and role to investigate administrative results, and the work of ministries and agencies is centered on self-evaluation. This problem is not limited to the nuclear power sector in Japan, but also leads to dependence on the country's systems and budgets by Japanese companies and organizations, which the author feels is the root cause of Japan's long-term stagnation.

The problem of Japan's huge fiscal deficit may also be due to the lack of administrative accountability in Japan. Argentina was the fourth largest economy in the world in the first half of the twentieth century, with exports of beef to Europe, but by the end of the twentieth century, it had defaulted on its debt due to accumulated fiscal deficits, and had experienced currency crises and hyperinflation several times. In Japan, the assets of the people are collateral, so it is often heard that the fiscal deficit is not a problem as long as the Bank of Japan prints more money, but this is a mistake from a domestic perspective. There are many examples of financial attacks from abroad causing difficulties for national finances, such as the Asian currency crisis. Isn't it necessary for Japan to take measures to avoid defaulting on its debt? First, a system needs to be created to question administrative accountability, and the dependence of Japanese companies and various organizations on the country's systems and budgets needs to be changed.

In addition, the analysis of disaster-related deaths due to the evacuation from the TEPCO accident introduced in Chap. 2 was a result of realizing during the writing of this book that the radiation risk for the elderly is one-fifth of the average age, and thinking that the risk of evacuation could be reduced. The author prioritized compiling this part into an English research paper and presenting it, which delayed the completion of this book.

The first half of Chap. 2 is also a report on the TEPCO Fukushima accident, including the 10 years after the accident. The author has researched the history of the development of power reactors and have published a Japanese version, "Nuclear

Power Reactor Development; History, Technology and Lessons". An English version is being prepared for publication from Springer. The author would be grateful if you could use it as a sister book to this one.

Appendix
Why has Nuclear Power Come to be Considered the Most Dangerous?

Nuclear power is widely considered to be extremely dangerous. In the 2011 TEPCO Fukushima accident, the reactor building exploded. Seeing this footage, many people in Japan and around the world may have formed the impression that nuclear power is dangerous. However, the number of deaths from radiation exposure in the TEPCO Fukushima accident is zero. Statistically, the number of deaths per unit of power generated is lowest for nuclear power. So why has nuclear power come to be considered the most dangerous method of power generation? The author will introduce considerations from risk psychology and the situation in the Western nuclear industry, and discuss what should be done.

Number of public deaths due to radiation from nuclear power plant accidents (objective safety)

The United Nations Scientific Committee on the Effects of Atomic Radiation (UNSCEAR) has reported on the health effects of radiation from the Chernobyl accident in 1986and the TEPCO Fukushima accident in 2011. This is shown in Table A.1.

Table A.1 The numbers of public death due to radiation from nuclear power plant accidents

Chernobyl accident: 8 people (pediatric thyroid)[a]
TEPCO Fukushima accident: zero[b]

[a] The majority of children with thyroid cancer, which arose due to the delay in restricting intake of radioactive milk during the Chernobyl disaster, were treated with surgery. In the Chernobyl accident, 49 workers died while trying to bring the accident under control. Increases in cancer and death in the public would be difficult to observe in epidemiological studies because the doses are low

[b] No health damage due to radiation from TEPCO Fukushima accident will be observed

Y. Oka, *Nuclear Power and Society*, Nuclear Science and Technology, https://doi.org/10.1007/978-981-97-8951-1

In the Chernobyl accident, radioactive iodine released into the atmosphere adhered to grass. Due to delayed intake restrictions, children who drank milk from cows that ate the grass developed thyroid cancer. Most were treated with surgery, but eight died. In the TEPCO Fukushima accident, there were no deaths due to radiation exposure. Figure A.1 shows the number of deaths per unit of power generated by various power generation methods. Power generation methods that burn fuels such as coal, oil, natural gas, and biomass have high mortality rates due to air pollution and other factors. Hydropower has caused many deaths downstream when dams have collapsed in developing countries. Solar and wind power have risks such as electric shock and equipment collapse and contact with windmills because they are installed close to the areas where people can access. Based on mortality statistics, nuclear power is an extremely safe source of power. The safety shown by statistics will be referred to as **objective safety**. This is also the safety of equipment that has been made and used based on technical knowledge and experience based on technical standards.

Public risk perception is not rational (subjective safety, psychological safety)

Figure A.2 shows public's risk perception factors and awareness distribution of various risks (**Reference 1**). This figure is by a prominent American risk psychologist and is often cited in introductory books on risk psychology. Various risks can be classified by unknown factors such as "unobservable, unknown, new, delayed onset" and dread risk factors such as "out of control, scary, devastating, deadly, unfair interests", and the larger these are, the greater the risk is perceived. The risk of nuclear reactor accidents and radioactive waste is perceived as high. Public risk perception is not rational, so we will call this subjective safety. It could also be called psychological safety. Although the author is not an expert on global warming, it seems that the risk of warming, which has spread rapidly around the world, can be explained by the large unknown factors and dread risk factors. Things with a high subjective risk may be taken up politically, and this point is common to both.

Why has nuclear power generation come to be considered dangerous?

According to risk psychologists, "risk talk/negative talk leaves a three times stronger impression". Since humans have survived by avoiding risks, this can be considered to originate from human survival instincts. Stories of danger and accidents attract public interest, so the media often covers them. In the case of nuclear power generation, stories of accidents and radiation exposure risks have been repeated. Advocates of nuclear power generation have been talking about safety, thinking that the safety of nuclear power should be understood by the public. The nuclear power industry in Japan has also been focusing on risk communication. In Japan, there is even an abbreviation called "Riskomi". However, safety talk is dangerous talk. Negative talk leaves a three times stronger impression. Could this be the reason why the image of nuclear power generation being dangerous has spread?

Risk communication is not education

Many nuclear engineers seem to think that if the public understands the knowledge of nuclear safety and technology they know, understanding will progress, but they forget

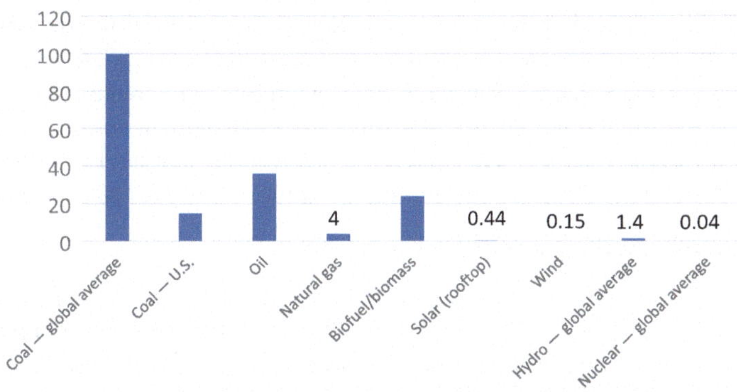

Fig. A.1 Mortality rates for each energy source in deaths per billion kWh produced. *Source* B. W. Brook et al., Why nuclear energy is sustainable and has to be part of the energy mix, Sustainable Materials and Technologies, 1–2 (2014) 8–16, Table 6, Updated (corrected) data from: World Health Organization; CDC; Seth Godin; John Konrad

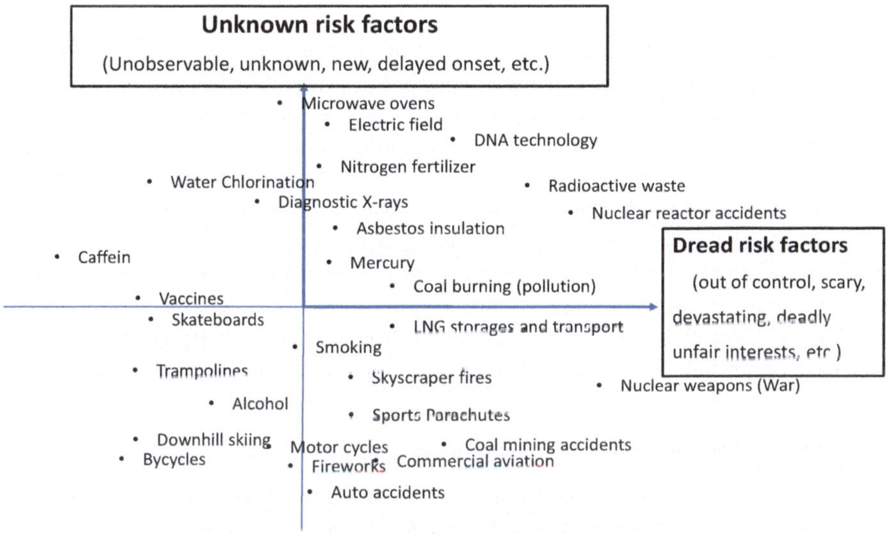

Fig. A.2 Risk perception factors and awareness distribution of various risks. *Source* adapted from Fig. 2 of Paul Slovic and Elke U. Weber, "Perception of risk posed by extreme events", Conference of Risk management strategies in an uncertain world, New York, April 12–13, 2002

how many years it took them to acquire that knowledge. In university, after learning the basics in class, you can acquire specialized knowledge only after experimenting, practicing, and thinking for yourself in seminars and research. It takes 10 years to become an expert in a field. The Royal Society in the UK has been promoting science communication since the 1980s. In a 2006 report, it expressed concern that scientists are participating in dialogue activities thinking that it is about educating their knowledge rather than listening to or dialoguing with the public (**Reference 2**). What nuclear engineers are explaining is "safety technology", not the "safety" that the public is interested in. The safety that the public is interested in is the safety when radioactive materials are released into the environment in an accident, not the technology to prevent accidents and reduce emissions that nuclear engineers are working on. Are civil engineers explaining the safety of bridges to the public?

Unique Remarks in the Nuclear Power Sector

There are comments that saying things like "We must never let such an accident happen again" is unique to the nuclear power industry. In other industries, for example, the oil industry, in the event of a large-scale crude oil spill from a tanker into the ocean, they say "We are making maximum efforts to reduce the impact of the accident. We can respond better from this experience", "In other industries, we do not say that accidents should never happen again, or that safety is the top priority. Oil spills occur frequently, but they do not lead to strict regulatory requirements like nuclear power generation. By saying this in nuclear power generation, it is spreading the image that it is extremely dangerous," a British expert has stated (**References 3, 4**).

This kind of statement has been repeated in nuclear power generation. Nuclear engineers wanted to convey their understanding of safety, but as a result, the public perceived the risk psychologically and thought it was dangerous. This may be the reason why the actual safety of nuclear power generation and the public's perception of it are significantly different.

The nuclear power industry in the West is aware of this problem, and EDF Energy in UK stopped airing safety commercials in 2015. The Nuclear Energy Institute (NEI) in the United States has communication experts in its management (board of directors), who are conducting practical activities with consultants. They have understood this problem for a long time, and neither the nuclear industry nor the Nuclear Regulatory Commission in the United States are conducting risk communication to general public directly.

It's Better Not to Do Risk Communication

Objective safety does not logically intersect with subjective safety (psychological safety). They are two different things. If nuclear engineers who do not understand this try to intersect the two, the image that nuclear power is dangerous will be conveyed, contrary to their intention.

Risk communication (sending safety messages to the general public to understand nuclear safety) should not be done. It is counterproductive. Talking about safety is psychologically a discussion about danger (risk), and negative talk not only leaves a three times stronger impression, but also, since the purpose of communication is to build trust, it is difficult to build trust by talking about danger. Note that the risk communication mentioned here is not about local stakeholder dialogue. These two are different things and need to be clearly distinguished.

Paul Slovic stated at the end of his famous 1987 paper, "Risk communication and risk management efforts are destined to fail unless they are organized as a two-way process" (**Reference 5**). The maximum number of people who can engage in a dialogue is about 40. It is impossible to have face-to-face dialogues with the people of entire nation or prefecture. Dialogue requires facilitation skills to encourage dialogue and a broad knowledge to respond to the interests of the participants, which requires training and experience. It seems that there are many nuclear-related people in Japan's nuclear industry who advocate for risk communication in order to promote the use of nuclear power, but simply disseminating safety information is self-defeating and counterproductive. Stakeholder dialogue is impossible without learning the basics of risk psychology and other fields and practically training experienced people.

What should be done?

It would be good to create various explanations and make them public and link them, improve searchability, so that the public can find them. Good examples after the TEPCO Fukushima accident include BOOKLET to Provide Basic Information Regarding Health Effects of Radiation by Ministry of the Environment, Fukushima Reconstruction Information Portal Site by Fukushima Prefecture, and Special Contents by Agency for Natural Resources and Energy. It would also be good to create, cite, and disclose professional information and explanations that form the basis of the description, so that the public can trace the basis to the point they want to know. In the United States, information from the Nuclear Regulatory Commission (NRC) and others plays this role. The public can know up to the level they want to know. If they search for themselves, their understanding should progress. There is no need to distinguish between professional information and information for the public. The information disclosed by the NRC is like that.

In the past, the author asked a communication expert in the United States, "What should I do if asked about safety?" The answer was, "Just say it's safe." This corresponds to the Japanese government's statement about the release of treated water containing tritium from TEPCO Fukushima site. If you try to explain about safety in detail verbally, you will fall into the trap of subjective safety and psychological safety. The fact that the Ministry of the Environment has created and disclosed BOOKLET to Provide Basic Information Regarding Health Effects of Radiation is helping the public understand the Japanese government's statements about the water.

Lessons in Crisis Management

The government's message at the time of the TEPCO Fukushima accident was superior in the UK. The UK government activated its Scientific Advisory Group for Emergencies (SAGE). The public release of one-voice messages by the scientific advisor and Q&A sessions including experts helped the calm response of British residents in Japan (**Reference 6**). Although it is not written in this paper, the author believes that the UK's estimate of the release amount was reasonable. Western countries that use nuclear power on a large scale overestimated the risk of spent fuel pools and called for widespread evacuation on the assumption that the release amount was large, but it takes more than 20 days for the water in the spent fuel pool to disappear

(**Reference 7**). Spent fuel that has been removed from the reactor and has passed time can be cooled even in the air. In the Tokyo Electric Power Company's Fukushima accident, it was confirmed by the TEPCO employee on a helicopter and in TEPCO meetings that there was water in the spent fuel pool after the detonation, but this was not communicated domestically or internationally. This is a crisis management issue. Germany decided to phase out nuclear power, but it was decided based on incorrect information.

It is a mistake to think that responses in times of crisis should be based on worst-case scenario.

Worst-case scenarios can only be considered in one-dimensional indicators (radiation exposure). Various risks should be considered in response. This paper mentions that the UK was aware of the risks associated with evacuation. Japan only had radiation exposure in mind, so it overlooked the risk of disaster-related deaths due to inactive lifestyle diseases as the evacuation was prolonged (**Reference 8**). If disaster prevention measures are taken in a vertical division of fields, risks other than radiation exposure will be overlooked. Isn't Japan's disaster prevention vertically divided? It is not just compound disasters that require collaboration in the field of disaster prevention. The collaboration mentioned here refers not to the collaboration of procedures that are vertically divided, but to the sharing and collaboration of knowledge and lessons.

The TEPCO Fukushima accident is not only a typical accident at a light water nuclear power plant, but also three reactors have melted down, and it is considered to be one of the largest accidents in terms of the amount of radioactive material released. In the 2000s, there was the Indian Ocean tsunami, and Madras nuclear power plants in India were flooded. The experience and lessons were examined by Japanese nuclear disaster prevention researchers, and international organizations also responded. The nuclear power plant of Japan Atomic Power Co., Ltd. in Tokai Village, Ibaraki Prefecture, had taken measures against flooding based on Ibaraki Prefecture's tsunami evaluation, so it avoided a core meltdown accident due to the large tsunami caused by the Great East Japan Earthquake. However, neither Tokyo Electric Power Company nor the Nuclear and Industrial Safety Agency, which was essentially in charge of nuclear power safety regulations at the time, could take measures. Measures were taken in the subsequent new regulatory standards, but those involved in Japan's nuclear safety, including crisis management, may need to question why they could not respond and use it for improvement. It is an organizational issue.

In the United States, a legal system that clarifies the accountability of the administration is helping to gain public trust in nuclear safety

The U.S. Nuclear Regulatory Commission (NRC) explains its work and plans in the Strategic Plan. It creates an Information Digest and Citizen's Guide so that citizens can find NRC information. The Government Accountability Office (GAO) works for accountability. The Plan is also a reference for investigating and reporting on the results of NRC's work. Based on this, the federal congress makes amendments to the budget and policies and supervises the NRC. The GAO plays a role in revealing facts, but does not comment on the quality of the results. This is a mechanism for ensuring the accountability of U.S. government agencies, not limited to the NRC. Through these, the trust of the American people in the NRC is secured, and as a result, the trust of the people in the safety of nuclear power generation is obtained. Facts such as administrative results cannot be revealed unless an institution with investigative authority investigates. In Japan, self-evaluation of government agencies is central, and the Board of Audit of Japan is not given legal authority like the U.S. Government Accountability Office. This is not a story unique to the nuclear field. The Ministry of Internal Affairs and Communications conducts policy evaluations, but they are basically self-evaluations by administrative agencies. Western accounting audit institutions do not evaluate. Evaluation is the role of the parliament. The recent political fund problem is a result of the absence of this legal system in Japan, not the cause. The author believes this is a major cause of Japan's long-term stagnation, but what do readers think (**Section** 4.6, **Chapter 5 of this book**)? The United States changed the name of the GAO from Accounting to Accountability in 2004. Japan needs a system to generate innovation and competitiveness, not subsidies.

It is dangerous to think that low dose radiation is dangerous and scary

In the TEPCO Fukushima accident, many disaster-related deaths occurred due to prolonged evacuation (**Figure** 2.3 **of this book**, **Table 2 of Reference 8**). Disaster-related deaths in Fukushima are thought to have been mainly residents of nursing care facilities, but this is a misunderstanding. Many residents who evacuated to temporary housing died from inactive lifestyle diseases. The Japanese government responded in accordance with the International Commission on Radiological Protection's principle of "radiation exposure should be reduced as reasonably as achievable (ALARA)". There was no flaw in the government's procedure because it followed ALARA which ICRP recommended, but when we analyzed the risk and benefit of evacuation based on the number of disaster-related deaths and the measured values of environmental radiation doses of the Japanese government agencies, the risk of evacuation was approximately 20 times greater than the radiation risk avoided by evacuation (**Sect.**2.3.3 **of this book, Reference 8**). This is a major lesson from the TEPCO Fukushima accident, which needs to be shared with the world.

The average dose in the first year for 9 out of 11 cities and towns that were forced to evacuate for a long time in Fukushima Prefecture did not exceed 100 millisieverts. Even at the highest dose point, 7 cities and towns did not exceed 100 millisieverts, and the accumulated dose for the entire evacuation period did not exceed 100 millisieverts in 4 cities and towns (**Table 2.3 of this book, Table 5 of Reference 8**). It is unclear whether there is an effect of radiation-induced cancer below 100–200 millisieverts. In the event of a nuclear and radiation accident, it is better to lift the evacuation of residents in low-dose areas as soon as possible and make it possible to decontaminate and restore the town. This can prevent disaster-related deaths due to inactive lifestyle diseases and the collapse of the community. This lesson is important not only in Japan but also in France, South Korea, and China, where nuclear power plants are located in agricultural areas. At the time of the Chernobyl accident, artificial abortions were performed in Greece and Italy, and the birth rate plummeted in January 1987 (**Reference 9**). **It is dangerous to think that radiation is dangerous and scary at**

low doses. The principle of ALARA should not be applied to residents in low-dose areas below 100 millisieverts/year in the event of a nuclear and radiation accident. It is naive to think that measures can be taken with only mental care for evacuees in the event of an accident. Did risk communication for the public work at the time of the TEPCO accident? The reason it didn't work is due to the public's risk perception being subjective and psychological. Risk communication is not education. Dialogue is essential for education and understanding.

Note: This appendix is translation of "Why has nuclear power come to be considered the most dangerous?" in Japanese, authored by Yoshiaki Oka and published from IEEI (International Environment and Economy Institute) on April 24, 2024, https://ieei.or.jp/2024/04/oka_20240423/

References

1. Paul Slovic and Elke U. Weber, "Perception of risk posed by extreme events", Conference of Risk management strategies in an uncertain world, New York, April 12–13, 2002
2. Royal Society. Science Communication, Survey of factors affecting science communication by scientists and engineers, The Royal Society, UK, 2006
3. Grimston M. The Paralysis in Energy Decision Making, Whittles publishing, 2016, pp. 412–416
4. Malcolm Grimston "Public Understanding of Nuclear Energy: It is not (just) just about science" Rebuilding Public Trust in Nuclear Energy, The 47th JAIF Annual Meeting, Tokyo, April 15 2014, https://www.jaif.or.jp/cms_admin/wp-content/uploads/understanding/annual/47th/47-s1_grimston-e.pdf
5. Paul Slovic: "Perception of Risk", Science, Vol. 236 pp 280–285 (1987)
6. Grimes RW, Chamberlain Y, and Oku A, "The UK Response to Fukushima and Anglo-Japanese Relations" Science and Diplomacy, June 2014
7. Gauntt R. et al., Fukushima Daiichi Accident Study (Status as of April 2012), SAND2012-6173, July 2012, Sandia National Laboratories
8. Oka Y. "Risks and benefits of evacuation in TEPCO's Fukushima Daiichi nuclear power station accident", Progress in Nuclear Energy, Volume 148, June 2022, 104222, Elsevier (Open access)
9. BOOKLET to Provide Basic Information Regarding Health Effects of Radiation, Chapter 3, 3.8 Impact on the Heart: Increase in Miscarriages in Europe - Chernobyl Nuclear Accident, Ministry of the Environment, 2022 Edition

SPRINGER NATURE

GPSR Compliance

The European Union's (EU) General Product Safety Regulation (GPSR) is a set of rules that requires consumer products to be safe and our obligations to ensure this.

If you have any concerns about our products, you can contact us on ProductSafety@springernature.com

In case Publisher is established outside the EU, the EU authorized representative is:

Springer Nature Customer Service Center GmbH
Europaplatz 3
69115 Heidelberg, Germany

The manufacturer's authorised representative in the EU is Springer Nature Customer Service Centre GmbH, Europaplatz 3, 69115 Heidelberg, Germany. If you have any concerns regarding our products, please contact ProductSafety@springernature.com

Printed and bound by CPI Group (UK) Ltd, Croydon, CR0 4YY
01/10/2026
02225858-0012